Ionic Conductive Membranes for Fuel Cells

Ionic Conductive Membranes for Fuel Cells

Editor

Riccardo Narducci

MDPI • Basel • Beijing • Wuhan • Barcelona • Belgrade • Manchester • Tokyo • Cluj • Tianjin

Editor
Riccardo Narducci
Industrial Engineering
University of Rome
"Tor Vergata"
Rome
Italy

Editorial Office
MDPI
St. Alban-Anlage 66
4052 Basel, Switzerland

This is a reprint of articles from the Special Issue published online in the open access journal *Membranes* (ISSN 2077-0375) (available at: www.mdpi.com/journal/membranes/special_issues/ionic_conductive_membranes).

For citation purposes, cite each article independently as indicated on the article page online and as indicated below:

LastName, A.A.; LastName, B.B.; LastName, C.C. Article Title. *Journal Name* **Year**, *Volume Number*, Page Range.

ISBN 978-3-0365-2769-7 (Hbk)
ISBN 978-3-0365-2768-0 (PDF)

Contents

About the Editor vii

Riccardo Narducci
Ionic Conductive Membranes for Fuel Cells
Reprinted from: *Membranes* **2021**, *11*, 159, doi:10.3390/membranes11030159 1

Stefano Giancola, Raul Andres Becerra Arciniegas, Armand Fahs, Jean-François Chailan, Maria Luisa Di Vona, Philippe Knauth and Riccardo Narducci
Study of Annealed Aquivion® Ionomers with the INCA Method †
Reprinted from: *Membranes* **2019**, *9*, 134, doi:10.3390/membranes9100134 5

Alia Akrout, Aude Delrue, Marta Zatoń, Fanny Duquet, Francesco Spanu, Mélanie Taillades-Jacquin, Sara Cavaliere, Deborah Jones and Jacques Rozière
Immobilisation and Release of Radical Scavengers on Nanoclays for Chemical Reinforcement of Proton Exchange Membranes
Reprinted from: *Membranes* **2020**, *10*, 208, doi:10.3390/membranes10090208 19

Ahmet H. Avci, Diego A. Messana, Sergio Santoro, Ramato Ashu Tufa, Efrem Curcio, Gianluca Di Profio and Enrica Fontananova
Energy Harvesting from Brines by Reverse Electrodialysis Using Nafion Membranes
Reprinted from: *Membranes* **2020**, *10*, 168, doi:10.3390/membranes10080168 37

Je-Deok Kim, Akihiro Ohira and Hidenobu Nakao
Chemically Crosslinked Sulfonated Polyphenylsulfone (CSPPSU) Membranes for PEM Fuel Cells
Reprinted from: *Membranes* **2020**, *10*, 31, doi:10.3390/membranes10020031 53

Maria-Maddalena Schiavone, David Hermann Lamparelli, Yue Zhao, Fengfeng Zhu, Zsolt Revay and Aurel Radulescu
The Effects of Temperature and Humidity on the Microstructure of Sulfonated Syndiotactic–polystyrene Ionic Membranes
Reprinted from: *Membranes* **2020**, *10*, 187, doi:10.3390/membranes10080187 67

Ayub Shahab Marf, Ranjdar M. Abdullah and Shujahadeen B. Aziz
Structural, Morphological, Electrical and Electrochemical Properties of PVA: CS-Based Proton-Conducting Polymer Blend Electrolytes
Reprinted from: *Membranes* **2020**, *10*, 71, doi:10.3390/membranes10040071 85

Norazlianie Sazali, Wan Norharyati Wan Salleh, Ahmad Shahir Jamaludin and Mohd Nizar Mhd Razali
New Perspectives on Fuel Cell Technology: A Brief Review
Reprinted from: *Membranes* **2020**, *10*, 99, doi:10.3390/membranes10050099 111

Edmund J. F. Dickinson and Graham Smith
Modelling the Proton-Conductive Membrane in Practical Polymer Electrolyte Membrane Fuel Cell (PEMFC) Simulation: A Review
Reprinted from: *Membranes* **2020**, *10*, 310, doi:10.3390/membranes10110310 129

Mohammad Faisal Umar, Syed Zaghum Abbas, Mohamad Nasir Mohamad Ibrahim, Norli Ismail and Mohd Rafatullah
Insights into Advancements and Electrons Transfer Mechanisms of Electrogens in Benthic Microbial Fuel Cells
Reprinted from: *Membranes* **2020**, *10*, 205, doi:10.3390/membranes10090205 . . . . **183**

About the Editor

Riccardo Narducci

Riccardo Narducci received his master's degree in Chemistry at the University of Perugia (2004). He obtained double PhDs in Chemistry (Cum Laude) at University of Rome "Tor Vergata" and Aix-Marseille Université. He was H-Chercheur at Aix-Marseille Université and adjunct Professor at University of Tor Vergata from 2011 to 2014. He won the Vinci Programme (UFI) for 2013 and 2016. He has participated in five European Programs on PEMFCs. He is a Review Editor of *Frontiers in Energy Research* (EPFL), Guest Editor of *Membranes* (Topic Editor), *Polymers* and *Crystals* (MDPI). He is an author and co-author of more than 45 research articles, reviews and book chapters on international journal, one Italian patent, one international patent pending and more than 45 scientific communications (keynote, invitations, etc.) in international congress and meetings. He is currently a member of LIME and is involved in synthesis and characterization of anionic and ampholytic membranes for FCs, inorganic materials and development of INCA method for PFSA membranes. From January 2020, he obtained habilitation as Associate Professor in Chemistry (SC 03/B2, SSD Chim07).

membranes MDPI

Editorial

Ionic Conductive Membranes for Fuel Cells

Riccardo Narducci

Department Industrial Engineering and International Laboratory: Ionomer Materials for Energy, University of Rome Tor Vergata, 00133 Roma, Italy; riccardo.narducci@uniroma2.it

check for updates

Citation: Narducci, R. Ionic Conductive Membranes for Fuel Cells. *Membranes* **2021**, *11*, 159. https://doi.org/10.3390/membranes11030159

Received: 19 February 2021
Accepted: 22 February 2021
Published: 25 February 2021

Publisher's Note: MDPI stays neutral with regard to jurisdictional claims in published maps and institutional affiliations.

The need to reduce pollution and the continuous increase in petrol cost have reinforced the interest in fuel cells (FCs), efficient and clean systems for the conversion of fuel into energy. Polymer electrolyte membrane fuel cells (PEMFCs) exhibit excellent characteristics in their weight, volume, and current density for automotive applications and cogeneration systems. Unfortunately, the high cost of perfluorinated membranes and the low stability of anionic membranes in an alkaline environment still limit their use. During the past years, this kind of membranes has been widely explored in terms of synthesis of new materials, grafting of strong and stable functional groups, mechanical properties, conductivity, FC performance, and so forth. This special issue of Membranes is dedicated to this exciting research field, with some excursions in related fields, focusing on commercial polymers, like Nafion® and Aquivion®, and promising and low-cost sulfonated aromatic polymers (SAPs) and poly(vinyl alcohol) (PVA) and some strategies to enhance stability, like cross-linking (XL), increase crystallinity with the use of high boiling solvents or nanocomposites. All of these properties were studied with classical investigation techniques, like impedance or the new Ionomer n_c Analysis (INCA) method.

To reflect the broad scope of this topic, contributions from leading scientists across the world, whose research addresses ionomeric membranes from different perspectives, sharing a common vision of pollution reduction and the search for sustainable energy sources, have been gathered. Giancola et al. [1] investigated the possibility to increase the working temperature and endurance of short-side-chain perfluorosulfonic acid (SSC-PFSA) Aquivion® membranes with an annealing procedure in the presence of the plasticizing solvent DMSO. It was applied for the first time through the Ionomer n_c Analysis (INCA) method in order to evaluate ionomer thermomechanical properties and to probe the increase of crystallinity during the annealing treatment. Complementary differential scanning calorimetry (DSC) and dynamic mechanical analysis (DMA) measurements confirmed the increase of polymer stiffness over the whole range of temperature.

Akrout et al. [2] focused on the use of bifunctional nanoclay halloysites, grafted with amino groups and embedding with radical scavengers, that is, CeO_2 nanoparticles, to form composite membranes with Aquivion® ionomer. The composite with 4 wt% of CeO_2@HNT-NH_2 showed unchanged tensile properties but presented high proton conductivity and increased stability to radical attack compared with nonmodified Aquivion®.

Avci et al. [3] studied salinity gradient power (SGP) harvesting by reverse electrodialysis (RED) with Nafion 117 and 115 membranes for NaCl and NaCl + $MgCl_2$ solutions in order to measure the gross power density extracted under high salinity gradient and to evaluate the effect of Mg^{2+} (the most abundant divalent cation in natural feeds) on the efficiency in energy conversion. In all tests, Nafion 117 exhibited superior performance when 0.5/4.0 M NaCl was fed through 500 μm thick compartments at a linear velocity of 1.5 $cm \cdot s^{-1}$. However, the gross power density of 1.38 $W \cdot m^{-2}$ detected in the case of pure NaCl solutions decreased to 1.08 $W \cdot m^{-2}$ in the presence of magnesium chloride. Although Nafion membranes exhibited better performance than CMX and Fuji-CEM-80050, their use is limited by high cost, and a significant reduction of membrane price is required for affordable RED applications.

Kim et al. [4] focused on cross-linked sulfonated polyphenylsulfone (CSPPSU) with high ion exchange capacity (IEC) and developed an activation process treatment with alkaline and acidic solutions to remove sulfur dioxide (SO_2), which forms as a byproduct during heat treatment. The membranes obtained using this activation method had high thermal, mechanical, and chemical stabilities. In I-$V_{iR\ free}$ studies for fuel cell evaluation, high performances similar to those using Nafion were obtained, and by using a constant current method, a stability of 4000 h was attained.

Schiavone et al. [5] described the microstructural characterization by small-angle neutron scattering (SANS) method of sulfonated sPS films and sPS–fullerene composite membranes at different temperatures between 20 and 80 °C under a relative humidity (RH) level of 10% to 70%. The water is taken up around the agglomerations of sulfonic groups and gives rise to hydrated domains and grows in size and number by increasing the hydration level, by increasing the temperature at a constant hydration level; due to desorption of some water, these domains shrink, mostly from the bulk amorphous regions. The sulfonated sPS–fullerene composite membranes perform at a high temperature much better than the fullerene-free membranes in terms of proton conductivity in liquid water. Apparently, this may be related to the formation of additional hydrated pathways.

Marf et al. [6] focused on the preparation and characterization of polymer blends based on poly(vinyl alcohol) and chitosan (PVA/CS) incorporated with various quantities of ammonium iodide. Structural analysis from X-ray diffraction (XRD) revealed structural change upon the addition of NH_4I salt. The protruded appearance on the samples' surface was evidently shown at high salt concentrations (~50 wt %) in the field-emission scanning electron microscopy (FE-SEM) images. The system incorporated with 40 wt % of NH_4I salt exhibited a high ion transference number. A potential cutoff of 1.33 V was recorded for the electrolyte system as decomposition voltage.

A review by Sazali et al. [7] examined recent advances and up-to-date modeling in fuel cell technologies, especially towards polymer electrolyte membrane fuel cells (PEMFCs), solid oxide fuel cells (SOFCs), and direct methanol fuel cells (DMFCs).

A review by Dickinson and Smith [8] focused on theoretical models and their parameterization used to describe the proton-conductive membrane in polymer electrolyte membrane fuel cells (PEMFCs), especially for Nafion 1100 materials. Detailed attention was given to methods of coupling proton transport with water uptake and diffusive water transport. Other sections addressed the formulation and parameterization of models incorporating interfacial transport resistances, hydraulic transport of water, swelling, and mechanical properties. Lastly, a section was dedicated to the formulation of models predicting the rate of membrane degradation and its influence on PEMFC behavior.

A review by Umar et al. [9] reported the use of a novel biotechnique called benthic microbial fuel cells (BMFCs), a kind of microbial fuel cells (MFCs) distinguished by the absence of a membrane, for the bioremediation of pollutants and for renewable energy production via different electron pathways.

I am confident that the articles contained in the Special Issue will serve to further stimulate advances in this research area. I thank all our friends and colleagues who contributed papers to the themed issue.

Funding: This research received no external funding.

Institutional Review Board Statement: Not applicable.

Conflicts of Interest: The authors declare no conflict of interest.

References

1. Giancola, S.; Arciniegas, R.A.B.; Fahs, A.; Chailan, J.-F.; Di Vona, M.L.; Knauth, P.; Narducci, R. Study of Annealed Aquivion® Ionomers with the INCA Method †. *Membranes* **2019**, *9*, 134. [CrossRef]
2. Akrout, A.; Delrue, A.; Zatoń, M.; Duquet, F.; Spanu, F.; Taillades-Jacquin, M.; Cavaliere, S.; Jones, D.; Rozière, J. Immobilisation and Release of Radical Scavengers on Nanoclays for Chemical Reinforcement of Proton Exchange Membranes. *Membranes* **2020**, *10*, 208. [CrossRef]

3. Avci, A.H.; Messana, D.A.; Santoro, S.; Tufa, R.A.; Curcio, E.; Di Profio, G.; Fontananova, E. Energy Harvesting from Brines by Reverse Electrodialysis Using Nafion Membranes. *Membranes* **2020**, *10*, 168. [CrossRef]
4. Kim, J.-D.; Ohira, A.; Nakao, H. Chemically Crosslinked Sulfonated Polyphenylsulfone (CSPPSU) Membranes for PEM Fuel Cells. *Membranes* **2020**, *10*, 31. [CrossRef]
5. Schiavone, M.-M.; Lamparelli, D.H.; Zhao, Y.; Zhu, F.; Revay, Z.; Radulescu, A. The Effects of Temperature and Humidity on the Microstructure of Sulfonated Syndiotactic–polystyrene Ionic Membranes. *Membranes* **2020**, *10*, 187. [CrossRef]
6. Shahab Marf, A.; M Abdullah, R.; B Aziz, S. Structural, Morphological, Electrical and Electrochemical Properties of PVA: CS-Based Proton-Conducting Polymer Blend Electrolytes. *Membranes* **2020**, *10*, 71. [CrossRef]
7. Sazali, N.; Wan Salleh, W.N.; Jamaludin, A.S.; Mhd Razali, M.N. New Perspectives on Fuel Cell Technology: A Brief Review. *Membranes* **2020**, *10*, 99. [CrossRef]
8. Dickinson, E.J.F.; Smith, G. Modelling the Proton-Conductive Membrane in Practical Polymer Electrolyte Membrane Fuel Cell (PEMFC) Simulation: A Review. *Membranes* **2020**, *10*, 310. [CrossRef]
9. Umar, M.F.; Abbas, S.Z.; Mohamad Ibrahim, M.N.; Ismail, N.; Rafatullah, M. Insights into Advancements and Electrons Transfer Mechanisms of Electrogens in Benthic Microbial Fuel Cells. *Membranes* **2020**, *10*, 205. [CrossRef]

membranes

MDPI

Article

Study of Annealed Aquivion® Ionomers with the INCA Method †

Stefano Giancola [1], Raul Andres Becerra Arciniegas [2,3], Armand Fahs [4], Jean-François Chailan [4], Maria Luisa Di Vona [2], Philippe Knauth [3] and Riccardo Narducci [2,*]

1 Institut Català d'investigaciò Química (ICIQ), Av. Països Catalans 16, 43007 Tarragona, Spain; sgiancola@iciq.es
2 University of Rome Tor Vergata, Department of Industrial Engineering and International Associated Laboratory: Ionomer Materials for Energy, Via del Politecnico 1, 00133 Roma, Italy; Raul.becerra@students.uniroma2.eu (R.A.B.A.); divona@uniroma2.it (M.L.D.V.)
3 Aix Marseille Université, CNRS, Madirel (UMR 7246), Electrochemistry of Materials Group and International Associated Laboratory: Ionomer Materials for Energy, Campus St Jérôme, 13013 Marseille, France; philippe.knauth@univ-amu.fr
4 Université de Toulon, MAPIEM (EA 4323), CS 60584, 83041 Toulon CEDEX 9, France; Armand.fahs@univ-tln.fr (A.F.); chailan@univ-tln.fr (J.-F.C.)
* Correspondence: riccardo.narducci@uniroma2.it

† In memoriam Prof. Giulio Alberti.

Received: 31 July 2019; Accepted: 14 October 2019; Published: 17 October 2019

Abstract: We investigated the possibility to increase the working temperature and endurance of proton exchange membranes for fuel cells and water electrolyzers by thermal annealing of short side chain perfluorosulfonic acid (SSC-PFSA) Aquivion® membranes. The Ionomer n_c Analysis (INCA method), based on n_c/T plots where n_c is a counter elastic force index, was applied to SSC-PFSA in order to evaluate ionomer thermo-mechanical properties and to probe the increase of crystallinity during the annealing procedure. The enhanced thermal and mechanical stability of extruded Aquivion® 870 (equivalent weight, EW = 870 g·mol^{-1}) was related to an increase of long-range order. Complementary differential scanning calorimetry (DSC) and dynamic mechanical analysis (DMA) measurements confirmed the increase of polymer stiffness by the annealing treatment with an enhancement of the storage modulus over the whole range of temperature. The main thermomechanical relaxation temperature is also enhanced. DSC measurements showed slight base line changes after annealing, attributable to the glass transition and melting of a small amount of crystalline phase. The difference between the glass transition and melting temperatures derived from INCA plots and the ionic-cluster transition temperature derived from DMA measurements is consistent with the different experimental conditions, especially the dry atmosphere in DMA. Finally, the annealing procedure was also successfully applied for the first time to an un-crystallized cast membrane (EW = 830 g·mol^{-1}) resulting in a remarkable mechanical and thermal stabilization.

Keywords: proton exchange membranes; PEMFC; PFSA annealing; hydration; n_c index

1. Introduction

The urgent need for a reduction of pollution and carbon dioxide in the atmosphere together with the price fluctuations of exhaustible fossil fuels, have reinforced the interest in more clean, efficient and sustainable systems for the conversion of energy. Proton exchange membrane fuel cells (PEMFC) and water electrolyzers (PEMWE) [1–3] are expected to play a key role in the near future for sustainable energy production and storage.

Despite their huge potential, the high price of produced electricity and hydrogen and durability issues of device components, including the proton exchange membrane, are jeopardizing their wide development. The implementation of highly conductive and durable membranes would thus mark a turning point in the large-scale commercialization of more efficient and long-term technologies. Currently, one of the challenges that researchers still need to face is to combine low membrane resistance with high mechanical strength.

Perfluorosulfonic acids (PFSAs) are the state-of-the-art membranes for PEM technologies due to their outstanding chemical stability, the high proton conductivity in hydrated conditions, the suitable mechanical strength and the low hydrogen and oxygen permeability, used as such and as composites [4–11]. Nafion® long side chain PFSA (LSC-PFSA) has been the most investigated ionomer for both PEMFC and PEMWE [12–17].

In recent years, ever-increasing attention has been directed towards perfluorinated ionomers with shorter and non-branched pendant side-chain (SSC-PFSA; e.g., Aquivion®) [18–21]. The higher crystallinity of SSC-PFSA (e.g., Aquivion® EW = 870) [22] with respect to LSC-PFSA [23] allows it to shift the balance between proton conductivity and mechanical properties at lower equivalent weight (EW). This ultimately results in better transport properties and higher operating temperature of PFSA membranes. Figure 1 shows the structures of Nafion® EW 1100 and Aquivion® EW 830.

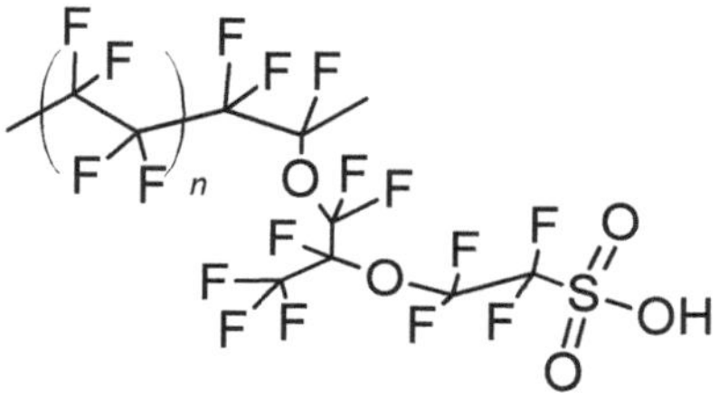

(a)

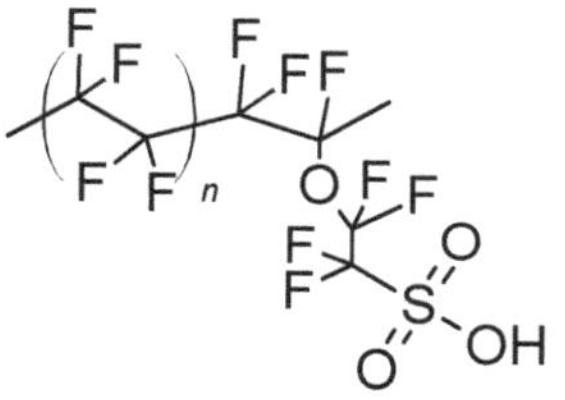

(b)

Figure 1. (**a**) Long side chain perfluorosulfonic acid (LSC-PFSA) Nafion® 1100 (n = 6.6) and (**b**) short side chain PFSA (SSC-PFSA) Aquivion® 830 (n = 5.5).

Despite the tremendous research effort, PFSA properties are not yet fully understood [23]. A deep PFSA comprehension is still needed to exceed the current membrane performance. In this context, we decided to start a systematic research on Nafion® chemical-physical fundamental properties (e.g., memory at room temperature of the water-uptake at higher temperatures, effect of the equivalent weight on the ionomer melting point (T_m), precise relation between equilibrium relative humidity (RH) and water-uptake at the various temperatures of operation, effect of thermal treatments and so on).

We introduced a new methodology, called INCA (Ionomer n_c Analysis), where n_c is an index proportional to the force of the ionomer matrix, which balances the osmotic pressure of the inner proton solution at the different temperatures. The determination of the hydration number and the volume of the samples allow us to find the density, the molality and molarity of the internal solution in addition to the internal osmotic pressure [24–32]. The INCA method is based on the use of n_c/T plots to determine some fundamental properties of an ionomer, such as the degree of crystallinity and the melting temperature. n_c is directly proportional to Young's modulus: one n_c unit is equivalent to an increase of Young's modulus by about 6.5 MPa. n_c depends on both RH and temperature [25]. In particular, it decreases by increasing both T and RH. It is thus a mechanical index characterizing the membrane mechanical properties at fixed hygrothermal conditions. A high n_c means good mechanical properties, related to high performances.

Today's tendency is to maximize the efficiency of PEM devices by reducing both membrane thickness and EW. On the one hand, thin membranes based on low EW ionomers offer advantages in terms of reduced electrical resistance and enhanced hydration especially at low relative humidity (RH). On the other hand, they suffer from poor mechanical strength and high gas crossover. Membrane mechanical properties are thus currently crucial and a consequent part of the effort of researchers is focused on PFSA mechanical stabilization [33].

A further aspect to consider is the morphological stability of PFSA. Alberti et al. reported that when the relative humidity and temperature (RH/T couples) exceeds certain critical values, irreversible processes could take place and provoke Nafion® morphological change leading to a severe reduction of the membrane through-plane conductivity [28,34]. This aspect is crucial especially in PEMWE where the membrane is exposed to both high relative humidity (100% RH) and temperatures (T ≥ 80 °C) and thus subject to very high compressive stress [35].

Thermal annealing is a procedure used to obtain PEM mechanical stabilization [36,37] and reduced gas permeability [38]. This powerful physical reinforcing strategy can be easily applied to both preformed and cast membranes. Moreover, it can be used also in combination with other reinforcing methodologies [18,21,39]. However, annealing has been usually performed in an empirical way through a short treatment at high temperatures (160–200 °C). Some years ago, due to the use of the INCA method, we started a thorough investigation of the annealing of extruded Nafion® 1100 driven by the desire to rationalize it and maximize its efficiency [27,31]. We assumed that the mechanical stabilization was induced by an increased crystallization of a pre-existing semi-crystalline phase due to a macromolecular rearrangement. The annealing temperature (T_{an}) must be chosen between the ionic-cluster transition temperature (T_{α}) and the melting temperature (T_m) of the crystalline phase. Dimethyl sulfoxide (DMSO), which has a high boiling temperature, was added as proton acceptor and plasticizing agent, with the aim to promote the solid-state macromolecular rearrangement. The annealing treatment resulted in a significant enhancement of Nafion® mechanical strength [31,40]. Moreover, no loss of through-plane proton conductivity was observed despite the lower water content of the annealed sample [31]. This might be ascribed to the reduced tortuosity of the material and/or the reduction of hydrophilic domain dimensions translating respectively in improved proton mobility and/or concentration. Based on the INCA analysis, we were able to select and perform a Taylor-made thermal treatment. Very interestingly, annealing allowed to sensibly reduce the formation of low conductive ribbon type morphologies occurring in membranes constrained to swell between two plates at 100% RH and 100 °C. This morphological stabilization prevented the drastic through-plane conductivity drop observed in non-annealed films due to the ribbon formation/reorganization parallel to the membrane surface [28,41]. We would also like to point out that membranes that do not have nanometric or micrometric phases dispersed within them (composite) cost less and can also be more easily reused.

In this paper we apply, for the first time, our INCA analysis to both extruded and cast Aquivion® membranes of different EW with the aim to evaluate their fundamental properties and extent our method also to SSC-PFSA. We are thus able to design a proper annealing treatment to stabilize mechanical properties of both preformed and cast Aquivion® membranes.

2. Experimental

2.1. Chemicals

Aquivion® water dispersions D79-25BS (EW = 790 g·mol^{-1}), D83-24B (EW = 830 g·mol^{-1}), Aquivion® extruded membranes E87-12S (EW = 870 g·mol^{-1}, 120 μm thick), E98-05 (EW = 980 g·mol^{-1}, 50 μm thick) and other reagents were supplied by Sigma-Aldrich (St. Louis, MO, USA).

2.2. Membrane Preparation

2.2.1. Extruded Semi-Crystalline Aquivion® 870 and 980

Before testing, Aquivion® 870 and 980 membranes were first treated for 2 h in 1 M sulfuric acid at room temperature (RT) and then washed in deionized water (DW) several times for 24 h.

2.2.2. Un-Crystallized Aquivion® 790 and 830

Aquivion® dispersions (12 wt %) in a mixture of 1-propanol/water (70/30 wt %) were casted with a doctor blade; the solvent was evaporated first in air for 24 h at room temperature and then for 1 h at 80 °C in a ventilated oven. Before testing, Aquivion® 790 and 830 cast membranes were treated for 2 h in 1 M sulfuric acid at RT and then washed in DW several times for 24 h.

2.2.3. Annealed Semi-Crystalline Aquivion® 870 and 980

Annealed Aquivion® 870 was prepared as follows [27]: a large batch of 1 M solution of DMSO in ethanol was prepared. One piece of anhydrous Aquivion® 870 was cut and weighed corresponding to 0.5 meq (i.e., 0.435 g). This membrane was then placed inside a Teflon bottle. One mL of 1 M ethanol solution of DMSO was added in the vessel to give a calculated value of λ(DMSO) = 2.0 ± 0.2 where λ(DMSO) are moles of DMSO per EW of ionomer. The vessel was closed and the solution was left to equilibrate with the ionomer membrane for about 1 h at room temperature. After evaporation of the ethanol solution under moderate agitation at 80 °C, the vessel was closed again and placed in an oven at 140 °C for the desired time. After cooling, the membrane was treated for 2 h in 1 M sulfuric acid at RT and then washed in DW several times for 24 h. For Aquivion® 980, the same procedure was followed as for Aquivion® 870 except that the oven treatment was performed at 150 °C for 7 days.

2.2.4. Annealed Aquivion® 830

Cast Aquivion® 830 was first treated at 135 °C for 15 h and then left to equilibrate for 1 h at RT in a 1 M solution of DMSO in water λ(DMSO) = 2.0 ± 0.2, we used an aqueous solution because in the ethanol solution the swelling was excessive). After evaporation of the water solution under moderate agitation at 80 °C, the membrane was annealed in a closed vessel at 135 °C for the desired time. After cooling, the membrane was treated for 2 h in 1 M sulfuric acid at room temperature (RT) and then washed in deionized water several times for 24 h at RT.

2.3. Characterization

2.3.1. Water Uptake and n_c Measurements

The previous materials were treated in liquid water for 600 h at different temperatures inside a Teflon container to determine the n_c index.

After this equilibration, the membranes were kept at 25 °C for 24 h in a closed Teflon vessel [24]. The excess of water was carefully wiped off with filter paper and the membrane mass was determined (m_{wet}) using a weighing bottle and an analytical balance; then the samples were dried over P_2O_5 for 3 days and weighed (m_{dry}):

$$WU = \frac{m_{wet} - m_{dry}}{m_{dry}} \times 100 \quad (1)$$

The hydration number was calculated as:

$$\lambda = \frac{n(H_2O)}{n(SO_3H)} = \frac{WU}{IEC \times M(H_2O)} \times 1000 \quad (2)$$

The uncertainty is about 0.5.

The λ values were converted into n_c values by the Equation (3):

$$n_c = \frac{100}{\lambda - 6} \quad (3)$$

This equation is valid for $\lambda \geq 10$ as derived in references [24,25].

2.3.2. Differential Scanning Calorimetry (DSC)

Differential scanning calorimetry (DSC) was performed on a DSC Q100 apparatus (TA Instruments, New Castle, DE, USA). The scans were carried out under a nitrogen purge, and pristine and annealed Aquivion 870 samples (6 mg) were placed in holed aluminum pans. An empty pan was used as the reference. The samples were heated from 20 to 200 °C with a scanning rate of 10 °C/min. To avoid relaxation effects, the glass transition temperature was determined on the second cycle, so that the free water or solvent is evaporated. The midpoint temperature of the heat flow jump was taken as the glass transition temperature (T_g).

2.3.3. Dynamic Mechanical Analysis (DMA)

Dynamic mechanical analysis (DMA) was performed on a DMA Q800 apparatus (TA Instruments, New Castle, Delaware, United States) in extension mode with samples of approximately 12 mm × 7 mm size and 120 µm thickness. The DMA was operated in air at a fixed frequency of 1 Hz with 0.5 N initial static force, force track 125% and an oscillation amplitude of 10 µm. This last value was chosen to keep the linear viscoelastic domain of samples during the experiments. The measurements were conducted with 3 K/min heating rate between 20 and 200 °C [31]. The relaxation temperature (T_α) was considered as the maximum of tan δ.

3. Results and Discussion

3.1. *n_c/T Plots of Un-Annealed and Annealed Semi-Crystalline Extruded Aquivion® 870 and 980*

Extruded PFSA membranes generally had higher mechanical strength and ductility compared to cast ones of same EW. Due to these outstanding properties, extrusion still remained the state-of-the-art fabrication technique for perfluorinated ionomer membranes. For this reason, directly enhancing the mechanical properties of preformed extruded films would be enormously beneficial. We decided thus to investigate the annealing behavior of Aquivion® 870, the commercial membrane with the lowest EW and the heat of the fusion similar to that of Nafion® 1100.

Figure 2 shows the n_c/T plots in liquid water of Aquivion® 870 after annealing at 140 °C in presence of DMSO for respectively 3 and 7 days.

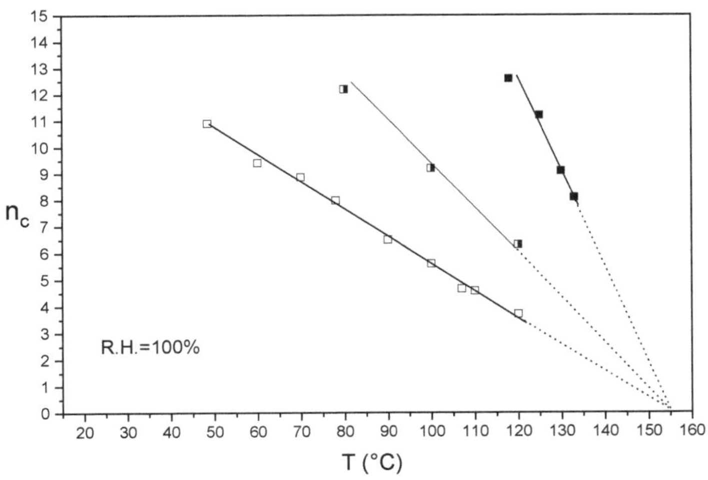

Figure 2. n_c/T plot of semi-crystalline Aquivion® 870 before (**empty squares**) and after annealing at 140 °C in presence of DMSO for 3 (**half empty squares**) and 7 days (**full squares**).

n_c/T plots are characterized by single or multiple straight lines. The extrapolated temperature on the T axis ($n_c = 0$) of each line can be linked to a characteristic ionomer temperature at certain hygrothermal conditions. We remember from the thermodynamic and statistic theory of elastomer deformation that for a semi-crystalline polymer Young's modulus decreased linearly to zero at the melting temperature [42].

The n_c of the un-annealed Aquivion® linearly decreased from 50 to 120 °C and the extrapolation of the plot to the T axis ($n_c = 0$) results to be 155 °C. This temperature was attributed to the melting temperature of an ionomer crystalline phase already present in the as-received membrane [27]. Crystallinity has been indeed detected also in preformed Aquivion® 870 membranes like in other extruded PFSA films [43]. Based on this assumption, we selected 140 °C as the proper annealing temperature, closer to T_m; for Nafion® 1100 the temperature of 130 °C was also used, obtaining for the same time of treatment a lower stabilization [27]. This temperature is indeed slightly lower than the ionomer T_m. Encouraged also by the success already obtained with Nafion® and confirmed by DSC analysis [27], we decided to perform the thermal treatment in presence of DMSO in order to facilitate the macromolecular rearrangement and promote a better crystallization. As shown in Figure 2, the extrapolation of the plots of un-annealed and annealed samples converge at the same temperature (155 °C). Moreover, the annealing treatments remarkably shift the plots towards the right side and increase the absolute slopes.

This indicates a noteworthy enhancement of membrane mechanical strength, also at high temperature, ascribed to an increase of the sample crystallinity. The shift towards right increased by increasing the time of the treatment from 3 to 7 days. It is also possible to quantify the annealing treatment by taking in consideration the change of slope of the plots. In particular, after annealing at 140 °C for 7 days, an increase of slope by 240% was observed.

Using the relation of the osmotic pressure π of a solution with its concentration c, where R is the ideal gas constant and T the absolute temperature:

$$\pi = cRT \tag{4}$$

one can easily derive Equation (5), given the proportionality of the n_c index with the osmotic pressure [29]:

$$\frac{\Delta n_c}{\Delta T} = kcR \tag{5}$$

where k is the proportionality constant between n_c and the osmotic pressure and c is the concentration of the inner proton solution. The absolute slope increases during annealing could thus be ascribed to an increase of the inner proton concentration due to the reduction of the volume of cluster domains. However, we were presently unable to calculate exactly the proton concentration from simple n_c/T plots because of the lack of precise morphological information [29] of PFSA ionomer like Nafion® and Aquivion®.

Alberti et al. for Nafion® 1100 have reported similar annealing behavior. Moreover, both Aquivion® 870 and Nafion® 1100 have similar extrapolated melting temperatures indicating similar crystallite morphology. This can be ascribed to the similar backbone length (polytetrafluoroethylene PTFE repeat unit) of the two ionomers [43].

Figure 3 shows the n_c/T plots in liquid water of thin Aquivion® 980 membranes, before and after annealing at 150 °C in presence of DMSO for 7 days. Similarly to Aquivion® 870, the n_c of the un-annealed membrane decreased linearly in the 50–130 °C temperature range with a plot extrapolation ($n_c = 0$) of 160 °C. Also in this case, this temperature was attributed to the melting temperature of a preformed ionomer crystalline phase [27]. The extrapolation of the plots of both un-annealed and annealed membranes converged at the same temperature (160 °C). The treatments resulted in an enhancement of the membrane mechanical strength proved by a plots shift towards the right side due to the increase of the absolute slopes. The melting temperature was slightly higher than that of Aquivion® 870 due to a higher EW, but the polymer had higher water content throughout the temperature range, evidently due to a lower initial crystallinity.

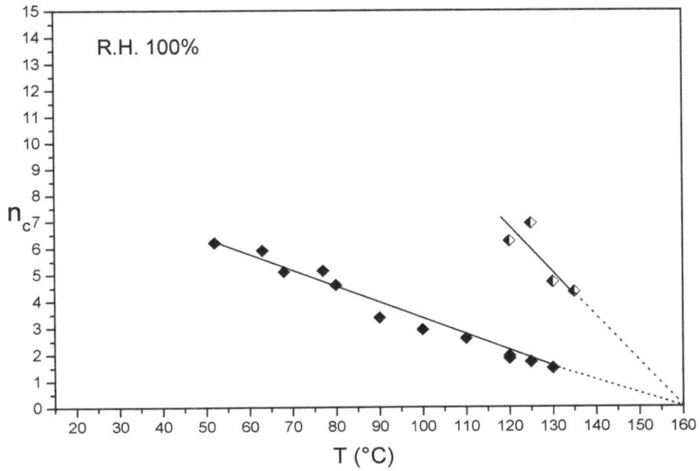

Figure 3. n_c/T plot of semi-crystalline Aquivion® 980 before (**full rhombus**) and after annealing at 150 °C in presence of DMSO for 7 days (**half empty rhombus**).

3.2. *n_c/T Plots of Un-Crystallized Aquivion®*

n_c/T plots in liquid water of un-crystallized Aquivion® 790 and 830 were evaluated and compared with that of un-crystallized Nafion® 1100 already reported [29] (Figure 4).

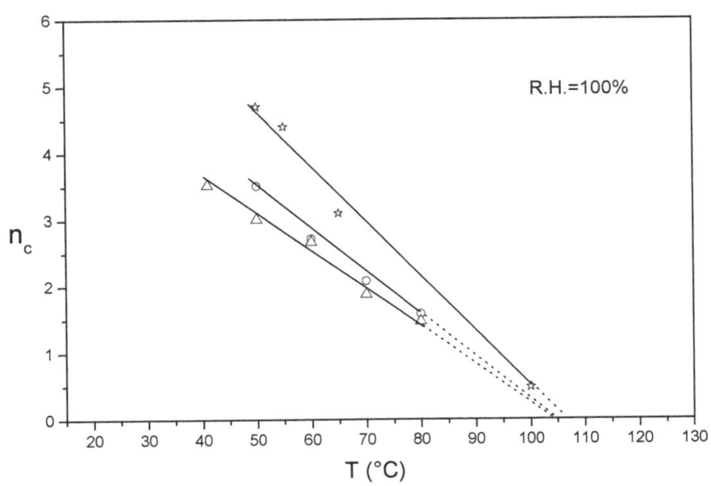

Figure 4. n_c/T plot of un-crystallized Aquivion® 790 (**empty triangles**), Aquivion® 830 (**empty circles**) and Nafion® 1100 (**empty stars**).

For SSC-PFSA, a linear plot was observed by increasing the temperature from 40 to 80 °C with extrapolated temperatures around 105 °C for Aquivion® 790 and 830. This temperature was similar to that of un-crystallized Nafion® 1100 (105–110 °C). Very interestingly, the extrapolated temperatures of the n_c/T plots were similar to the ionic-cluster transition temperature (T_α) obtained by dynamic mechanical analysis (DMA) for low hydrated membranes [31,43]. This temperature has been associated to the onset of long range chain mobility occurring as a result of destruction of electrostatic interactions, including hydrogen bonds and van der Waals interactions present between the chains [23]. Although a direct comparison between INCA and DMA methods is not yet possible due to the different membrane water content and the lack of experimental results on completely amorphous films, we supposed that the extrapolated temperature of the n_c/T plot would coincide with T_α at certain membrane hydration conditions. T_α has been reported to decrease with increasing membrane hydration due to the shielding effect of water acting as plasticizer [23].

The INCA method results thus to be a powerful analytical tool also able to determine ion-cluster transition temperatures (T_α) of both semi-crystalline and un-crystalline PFSA Nafion® and Aquivion® membranes. Moukheiber[43] reported for semi-crystalline Aquivion® a T_α decreasing with decreasing EW. He attributed this behavior to the reduced ionomer crystallinity promoting chain motion.

3.3. n_c/T Plots of Un-Crystallized and Annealed Aquivion® 830

Figure 5 displays the evolution of n_c/T plots of cast un-crystallized Aquivion® 830 after treatment at 135 °C for 15 h and annealing at 135 °C for 7 days in presence of DMSO. We selected for Aquivion® 830 an annealing temperature slightly lower than that used for Aquivion® 870 assuming also a lower melting temperature for the former due to its higher branching degree hindering chain packing and crystallization and due to the lower EW. This behavior is widely common in branched polymers [42]. The dependence of the melting temperature on the EW was verified by Alberti et al. for extruded Nafion® 1100 and 1000 with a decrease of 10 °C for the lowest equivalent weight. The same was checked for Aquivion extruded membranes with different EW: they presented different T_m and as the EW increased the T_m increased accordingly. This knowledge is useful for choosing the annealing temperatures.

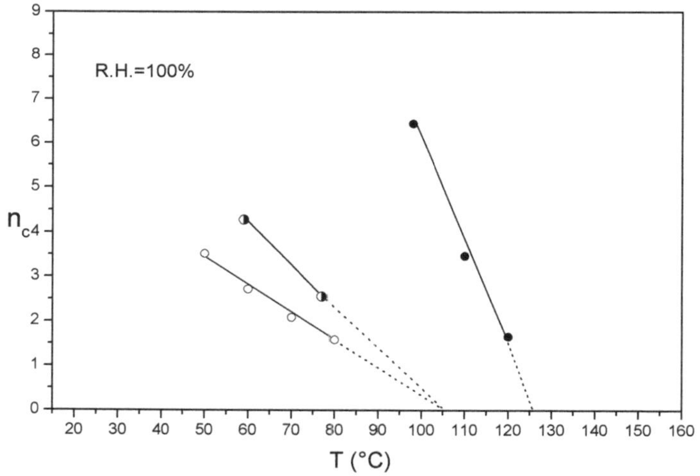

Figure 5. n_c/T plot of un-crystallized Aquivion® 830: as cast (**empty circles**), after treatment at 135 °C for 15 h (**half empty circles**) and treated at 135 °C for 15h and annealed at 135 °C in presence of DMSO for 7 days (**full circles**).

As depicted in Figure 5, the n_c/T plot of the un-crystallized Aquivion® shift towards right after treating at 135 °C for 15 h with no change of the extrapolated ionic-cluster transition temperature. Based on the completely amorphous character of the as-cast film, we ascribed the membrane mechanical stabilization to morphological PFSA changes with no crystallinity onset. It is widely accepted that

PFSA ionomers are in a quasi-equilibrium state with long relaxation time [23]. Thermal treatment could thus promote the change to a more entangled morphological state due to macromolecular motion or/and decrease the internal volume. A remarkable improvement of PFSA mechanical properties was observed after annealing in presence of DMSO resulting in a significant right shift of the plot. Moreover, the extrapolated T_α also increased to 125 °C.

3.4. Dynamic Mechanical Analysis (DMA) of Pristine and Annealed Aquivion® 870

The DMA analysis allows distinguishing the elastic response (storage modulus E′) and the viscous response (loss modulus E″) of the polymers. In the solid state, the elastic part is much higher than the viscous part; the storage modulus is close to the Young modulus obtained from static tensile tests. The ratio between the loss modulus and the storage modulus is the damping (tan δ), a good parameter to find the relaxation phenomena. The largest peak, denominated α, is assigned to the main relaxation process, which is associated to the glass transition in most polymers and concerns the global amorphous phase. In the case of ionomers (Nafion, etc.), this phenomenon is largely impacted by ionic regions, which are sensitive to the water content, the degree of neutralization and/or the ion type. This relaxation is due to main and side chain motions within or near the ion-rich domains [44].

Figure 6 and Table 1 present the DMA results for Aquivion® 870 before and after annealing at 140 °C in the presence of DMSO for 3 days. The main tan δ peak attributed to the α relaxation phenomenon was around 124 °C for pristine Aquivion® 870 and increased to 131 °C for the annealed sample. The annealing evidently enhanced the long-range order in the ionomer, which hindered chain motions, which need more energy to move, leading to a higher relaxation temperature. One can note also that the storage modulus was higher for the annealed sample than for the pristine one over the whole range of temperature from the glassy state to the rubbery state. The slight decrease of the α peak intensity after the thermal treatment was consistent with an increase of the stiffness of the polymer, due to the better long-range order after annealing. Finally, the sharp decrease of the storage modulus at the end of the experiment was attributable to the melting of the polymer at around 185 °C.

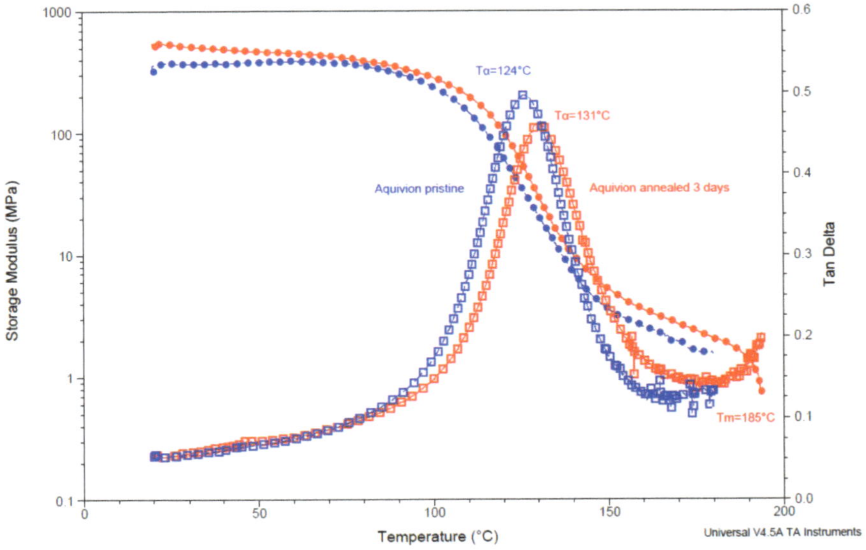

Figure 6. Dynamic mechanical analysis (DMA) curves of Aquivion® 870 before (**blue**) and after annealing at 140 °C in presence of DMSO for 3 days (**red**).

The main limitation of the DMA technique, used essentially for non-ionic polymers, is that the measurements are performed in dry conditions, which levels the effect on the main relaxation process. The higher melting temperature observed in DMA vs. INCA plots might be attributed to the absence of water, which increased the interactions between polymer chains and also to the dynamic conditions in DMA related to the heating rate. The INCA method reflected the mechanical properties in equilibrium in liquid water at working temperature and could be a complementary technique to a DMA analysis in order to study different ionomer samples.

Table 1. Ionomer relaxation temperatures T_α (°C), maximum damping intensity I (tan δ) and storage modulus E′ of Aquivion® 870 before and after annealing at 140 °C in presence of DMSO for 3 days.

Sample	T_α/°C	I (tan δ)	E′/MPa (25 °C)	E′/MPa (50 °C)
Aquivion 870 pristine	124 ± 3	0.49	350 ± 30	360 ± 40
Aquivion 870 3 days annealed	131 ± 1	0.46	540 ± 10	490 ± 30

3.5. Differential Scanning Calorimetry (DSC) of Pristine and Annealed Aquivion® 870

Figure 7 shows the DSC curves obtained for pristine and annealed material. The first run (presented only for the pristine sample) shows the endothermic peak due to the evaporation of water below 140 °C. In the second run, this peak was absent and one could observe slight changes of slope in the annealed sample, at around 120 and 170 °C, which could be attributed, respectively, to the glass transition of the ionomer amorphous phase and the melting temperature of a small amount of crystalline material. These endothermic transitions were related to a higher long range order (consistent with the increase of T_α in DMA) and a slightly increased crystallinity (in accordance with T_m in DMA) after treatment with DMSO, a behavior similar to that seen previously in Nafion 1100 [27]. The crystallinity formed during annealing corresponds to only small portions of the ionomer chains and a large part of amorphous ionomer remains, therefore, linked to the crystalline part. Since crystalline components are joined between them by the amorphous portions, the chains separation becomes only possible after the melting of the crystalline component [27].

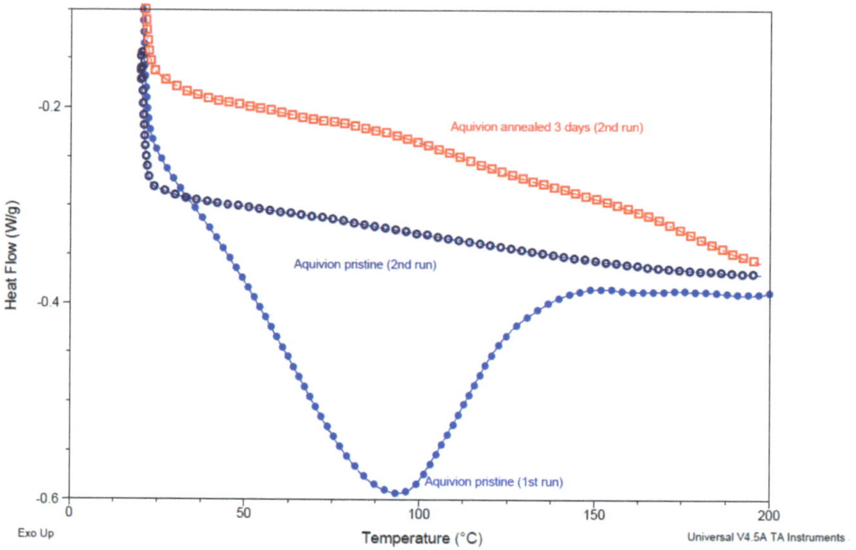

Figure 7. Differential scanning calorimetry (DSC) curves of Aquivion® 870 before (**blue**) and after annealing at 140 °C in presence of DMSO for 3 days (**red**).

4. Conclusions

In this work, the INCA (Ionomer n_c Analysis) method, first developed for Nafion®, was extended to Aquivion® short side chain perfluorosulfonic acid membranes with the aim of understanding their thermo-mechanical properties in specific hygrothermal conditions, and properly improved their mechanical stability by thermal annealing with a plasticizing solvent. Semi-crystalline Aquivion® 870 (EW = 870 g·mol^{-1}) shows a linear n_c/T plot in the examined temperature range (50–120 °C) with an extrapolated temperature of 155 °C (n_c = 0) corresponding to its melting point. After annealing at 140 °C in presence of DMSO as annealing agent, we observed an outstanding thermo-mechanical membrane stabilization represented by a remarkable slope increase of the n_c/T plot. Similar behavior was verified for Aquivion® 980. The INCA analysis of un-crystallized low-temperature casted Aquivion® 830 (EW = 830 g·mol^{-1}) and Aquivion® 790 (EW = 790 g·mol^{-1}) showed linear plots with extrapolated temperature of 105 °C. Based on the completely amorphous membrane structure, we associated this temperature to the ionomer transition temperature T_α, largely impacted by ionic regions. Similar temperatures were observed for un-crystallized Nafion 1100 (≈110 °C). DMA and DSC experiments showed T_α values, which were slightly higher, due to the dry conditions, which enhanced the interactions between chains that need more energy to move. The melting temperature was also enhanced. INCA is thus a powerful analytical tool to evaluate ionic-cluster transition temperatures of amorphous PFSA and T_m for semi-crystalline materials at a certain relative humidity. Finally, an annealing treatment in the presence of DMSO was successfully applied also to cast Aquivion® 830. These results corroborated the reliability and versatility of this method to enhance the mechanical properties of both extruded and cast PFSA membranes of different EW. From the point of view of mechanical properties, the best membrane was the annealed Aquivion® 870, however depending on the temperature of use other materials and treatments may be taken into consideration.

A further improvement of SSC PFSA membranes is possible by optimization of the annealing conditions in presence of an appropriate solvent.

Author Contributions: Conceptualization, R.N., S.G.; Methodology, S.G., M.L.D.V., P.K., R.N., J.-F.C.; Validation, S.G., R.A.B.A., A.F.; Investigation, S.G., M.L.D.V., P.K., R.N., A.F., J.-F.C., R.A.B.A.; Writing—original draft preparation, R.N.; Writing—review and editing, S.G., M.L.D.V., P.K., R.N., J.-F.C.

Funding: This research received no external funding.

Acknowledgments: The authors are indebted to the late G. Alberti (May 1930–October 2017) for having initiated this study.

Conflicts of Interest: The authors declare no conflict of interest.

References

1. Carmo, M.; Fritz, D.L.; Merge, J.; Stolten, D. A comprehensive review on PEM water electrolysis. *Int. J. Hydrogen Energy* **2013**, *38*, 4901–4934. [CrossRef]
2. Barbir, F. PEM electrolysis for production of hydrogen from renewable energy sources. *Sol. Energy* **2005**, *78*, 661–669. [CrossRef]
3. Gotz, M.; Lefebvre, J.; Mors, F.; Koch, A.M.; Graf, F.; Bajohr, S.; Reimert, R.; Kolb, T. Renewable Power-to-Gas: A technological and economic review. *Renew. Energy* **2016**, *85*, 1371–1390. [CrossRef]
4. Kreuer, K.D. On the development of proton conducting polymer membranes for hydrogen and methanol fuel cells. *J. Membr. Sci.* **2001**, *185*, 29–39. [CrossRef]
5. Springer, T.E.; Zawodzinski, T.A.; Gottesfeld, S. Polymer Electrolyte Fuel Cell Model. *J. Electrochem. Soc.* **1991**, *138*, 2334–2342. [CrossRef]
6. Zawodzinski, T.A.; Derouin, C.; Radzinski, S.; Sherman, R.J.; Smith, V.T.; Springer, T.E.; Gottesfeld, S. Water Uptake by and Transport Through Nafion® 117 Membranes. *J. Electrochem. Soc.* **1993**, *140*, 1041–1047. [CrossRef]
7. Borup, R.; Meyers, J.; Pivovar, B.; Kim, Y.S.; Mukundan, R.; Garland, N.; Myers, D.; Wilson, M.; Garzon, F.; Wood, D.; et al. Scientific aspects of polymer electrolyte fuel cell durability and degradation. *Chem. Rev.* **2007**, *107*, 3904–3951. [CrossRef]

8. Alberti, G.; Casciola, M.; Massinelli, L.; Bauer, B. Polymeric proton conducting membranes for medium temperature fuel cells (110–160°C). *J. Membr. Sci.* **2001**, *185*, 73–81. [CrossRef]
9. Roberti, E.; Carlotti, G.; Cinelli, S.; Onori, G.; Donnadio, A.; Narducci, R.; Casciola, M.; Sganappa, M. Measurement of the Young's modulus of Nafion membranes by Brillouin light scattering. *J. Power Sources* **2010**, *195*, 7761–7764. [CrossRef]
10. Donnadio, A.; Narducci, R.; Casciola, M.; Marmottini, F.; D'Amato, R.; Jazestani, M.; Chiniforoshan, H.; Costantino, F. Mixed Membrane Matrices Based on Nafion/UiO-66/SO_3H-UiO-66 Nano-MOFs: Revealing the Effect of Crystal Size, Sulfonation, and Filler Loading on the Mechanical and Conductivity Properties. *ACS Appl. Mater. Interfaces* **2017**, *9*, 42239–42246. [CrossRef]
11. Escorihuela, J.; Narducci, R.; Compan, V.; Costantino, F. Proton Conductivity of Composite Polyelectrolyte Membranes with Metal-Organic Frameworks for Fuel Cell Applications. *Adv. Mater. Interfaces* **2019**, *6*, 1–30. [CrossRef]
12. Mauritz, K.A.; Moore, R.B. State of Understanding of Nafion. *Chem. Rev.* **2004**, *104*, 4535–4585. [CrossRef] [PubMed]
13. Pabby, A.K.; Rizvi, S.S.H.; Requena, A.M.S. *Handbook of Membrane Separations: Chemical, Pharmaceutical, Food, and Biotechnological Applications*, 2nd ed.; CRC Press: Boca Raton, FL, USA, 2015.
14. Wang, L.; Husar, A.; Zhou, T.H.; Liu, H.T. A parametric study of PEM fuel cell performances. *Int. J. Hydrogen Energy* **2003**, *28*, 1263–1272. [CrossRef]
15. Alberti, G.; Casciola, M.; Capitani, D.; Donnadio, A.; Narducci, R.; Pica, M.; Sganappa, M. Novel Nafion–zirconium phosphate nanocomposite membranes with enhanced stability of proton conductivity at medium temperature and high relative humidity. *Electrochim. Acta* **2007**, *52*, 8125–8132. [CrossRef]
16. Ito, H.; Maeda, T.; Nakano, A.; Takenaka, H. Properties of Nafion membranes under PEM water electrolysis conditions. *Int. J. Hydrogen Energy* **2011**, *36*, 10527–10540. [CrossRef]
17. Siracusano, S.; van Dijk, N.; Payne-Johnson, E.; Baglio, V.; Arico, A.S. Nanosized IrOx and IrRuOx electrocatalysts for the O2 evolution reaction in PEM water electrolysers. *Appl. Catal. B-Environ.* **2018**, *164*, 488–495. [CrossRef]
18. Boaretti, C.; Pasquini, L.; Sood, R.; Giancola, S.; Donnadio, A.; Roso, M.; Modesti, M.; Cavaliere, S. Mechanically stable nanofibrous sPEEK/Aquivion® composite membranes for fuel cell applications. *J. Membr. Sci.* **2018**, *545*, 66–74. [CrossRef]
19. Arico, A.S.; di Blasi, A.; Brunaccini, G.; Sergi, F.; Dispenza, G.; Andaloro, L.; Ferraro, M.; Antonucci, V.; Asher, P.; Buche, S.; et al. High Temperature Operation of a Solid Polymer Electrolyte Fuel Cell Stack Based on a New Ionomer Membrane. *Fuel Cells* **2010**, *10*, 1013–1023. [CrossRef]
20. Skulimowska, A.; Dupont, M.; Zaton, M.; Sunde, S.; Merlo, L.; Jones, D.J.; Roziere, J. Proton exchange membrane water electrolysis with short-side-chain Aquivion® membrane and IrO2 anode catalyst. *Int. J. Hydrogen Energy* **2014**, *39*, 6307–6316. [CrossRef]
21. Giancola, S.; Zaton, M.; Reyes-Carmona, A.; Dupont, M.; Donnadio, A.; Cavaliere, S.; Roziere, J.; Jones, D.J. Composite short side chain PFSA membranes for PEM water electrolysis. *J. Membr. Sci.* **2019**, *570*, 69–76. [CrossRef]
22. Rolfi, A.; Oldani, C.; Merlo, L.; Facchi, D.; Ruffo, R. New perfluorinated ionomer with improved oxygen permeability for application in cathode polymeric electrolyte membrane fuel cell. *J. Power Sources* **2018**, *396*, 95–101. [CrossRef]
23. Kusoglu, A.; Weber, A.Z. New Insights into Perfluorinated Sulfonic-Acid Ionomers. *Chem. Rev.* **2017**, *117*, 987–1104. [CrossRef] [PubMed]
24. Alberti, G.; Narducci, R.; Sganappa, M. Effects of hydrothermal/thermal treatments on the water-uptake of Nafion membranes and relations with changes of conformation, counter-elastic force and tensile modulus of the matrix. *J. Power Sources* **2008**, *178*, 575–583. [CrossRef]
25. Alberti, G.; Narducci, R. Evolution of Permanent Deformations (or Memory) in Nafion 117 Membranes with Changes in Temperature, Relative Humidity and Time, and Its Importance in the Development of Medium Temperature PEMFCs. *Fuel Cells* **2009**, *9*, 410–420. [CrossRef]
26. Alberti, G.; di Vona, M.L.; Narducci, R. New results on the visco-elastic behaviour of ionomer membranes and relations between T–RH plots and proton conductivity decay of Nafion® 117 in the range 50–140 °C. *Int. J. Hydrogen Energy* **2012**, *37*, 6302–6307. [CrossRef]

27. Alberti, G.; Narducci, R.; di Vona, M.L.; Giancola, S. Annealing of Nafion 1100 in the Presence of an Annealing Agent: A Powerful Method for Increasing Ionomer Working Temperature in PEMFCs. *Fuel Cells* **2013**, *13*, 42–47. [CrossRef]
28. Alberti, G.; Narducci, R.; di Vona, M.L.; Giancola, S. More on Nafion Conductivity Decay at Temperatures Higher than 80 °C: Preparation and First Characterization of In-Plane Oriented Layered Morphologies. *Ind. Eng. Chem. Res.* **2013**, *52*, 10418–10424. [CrossRef]
29. Alberti, G.; Narducci, R.; di Vona, M.L.; Giancola, S. Preparation and Nc/T plots of un-crystallized Nafion 1100 and semi-crystalline Nafion 1000. *Int. J. Hydrogen Energy* **2017**, *42*, 15908–15912. [CrossRef]
30. Narducci, R.; di Vona, M.L.; Marrocchi, A.; Baldinelli, G. Stabilized SPEEK Membranes with a High Degree of Sulfonation for Enthalpy Heat Exchangers. *Coatings* **2018**, *8*, 190. [CrossRef]
31. Narducci, R.; Knauth, P.; Chailan, J.F.; di Vona, M.L. How to improve Nafion with tailor made annealing. *Rsc Adv.* **2018**, *8*, 27268–27274. [CrossRef]
32. Alberti, G.; Narducci, R.; di Vona, M.L. *Solid State Proton Conductors: Properties and Applications in Fuel Cells*; Knauth, P., di Vona, M.L., Eds.; Wiley: Hoboken, NJ, USA, 2012; Chapter 8; ISBN 978-0-470-66937-2.
33. Subianto, S.; Pica, M.; Casciola, M.; Cojocaru, P.; Merlo, L.; Hards, G.; Jones, D.J. Physical and chemical modification routes leading to improved mechanical properties of perfluorosulfonic acid membranes for PEM fuel cells. *J. Power Sources* **2013**, *233*, 216–230. [CrossRef]
34. Casciola, M.; Alberti, G.; Sganappa, M.; Narducci, R.J. On the decay of Nafion proton conductivity at high temperature and relative humidity. *Power Sources* **2006**, *162*, 141–145. [CrossRef]
35. Feng, Q.; Yuan, X.Z.; Liu, G.Y.; Wei, B.; Zhang, Z.; Li, H.; Wang, H.J. A review of proton exchange membrane water electrolysis on degradation mechanisms and mitigation strategies. *J. Power Sources* **2017**, *366*, 33–55. [CrossRef]
36. Yin, C.S.; Wang, Z.; Luo, Y.; Li, J.J.; Zhou, Y.W.; Zhang, X.W.; Zhang, H.N.; Fang, P.F.; He, C.Q. Thermal annealing on free volumes, crystallinity and proton conductivity of Nafion membranes. *J. Phys. Chem. Solids* **2018**, *120*, 71–78. [CrossRef]
37. Mugtasimova, K.R.; Melnikov, A.P.; Galitskaya, E.A.; Kashin, A.M.; Dobrovolskiy, Y.A.; Don, G.M.; Likhomanov, V.S.; Sivak, A.V.; Sinitsyn, V.V. Fabrication of Aquivion-type membranes and optimization of their elastic and transport characteristics. *Ionics* **2018**, *24*, 3897–3903. [CrossRef]
38. Lee, K.; Ishihara, A.; Mitsushima, S.; Kamiya, N.; Ota, K. Effect of Recast Temperature on Diffusion and Dissolution of Oxygen and Morphological Properties in Recast Nafion. *J. Electrochem. Soc.* **2004**, *151*, A639–A645. [CrossRef]
39. Casciola, M.; Cojocaru, P.; Donnadio, A.; Giancola, S.; Merlo, L.; Nedellec, Y.; Pica, M.; Subianto, S. Zirconium phosphate reinforced short side chain perflurosulfonic acid membranes for medium temperature proton exchange membrane fuel cell application. *J. Power Sources* **2014**, *262*, 407–413. [CrossRef]
40. Gebel, G.; Aldebert, P.; Pineri, M. Structure and related properties of solution-cast perfluorosulfonated ionomer film. *Macromolecules* **1987**, *20*, 1425–1428. [CrossRef]
41. Casciola, M.; Alberti, G.; Sganappa, M.; Narducci, R. Factors affecting the stability of Nafion conductivity at high temperature and relative humidity. *Desalination* **2006**, *200*, 639–641. [CrossRef]
42. Young, R.J.; Lovell, P.A. *Introduction to Polymers*, 2nd ed.; CRC Press: Boca Raton, FL, USA, 1991.
43. Moukheiber, E.; de Moor, G.; Flandin, L.; Bas, C. Investigation of ionomer structure through its dependence on ion exchange capacity (IEC). *J. Membr. Sci.* **2012**, *389*, 294–304. [CrossRef]
44. Chailan, J.; Khadhraoui, M.; Knauth, P. *Solid State Proton Conductors: Properties and Applications in Fuel Cells*; Knauth, P., di Vona, M.L., Eds.; Wiley: Hoboken, NJ, USA, 2012; Chapter 6; ISBN 978-0-470-66937-2.

membranes

MDPI

Article

Immobilisation and Release of Radical Scavengers on Nanoclays for Chemical Reinforcement of Proton Exchange Membranes

Alia Akrout [1], Aude Delrue [1], Marta Zatoń [1], Fanny Duquet [1], Francesco Spanu [1], Mélanie Taillades-Jacquin [1], Sara Cavaliere [1,2,*], Deborah Jones [1] and Jacques Rozière [1]

1 Institute Charles Gerhardt Montpellier, UMR CNRS 5253, Aggregates Interfaces and Materials for Energy, University of Montpellier, CEDEX 5, 34095 Montpellier, France; alia.akrout@etu.umontpellier.fr (A.A.); delrue.a@live.fr (A.D.); marta.zaton@umontpellier.fr (M.Z.); fanny.duquet@umontpellier.fr (F.D.); francesco.spanu@umontpellier.fr (F.S.); melanie.taillades-jacquin@umontpellier.fr (M.T.-J.); deborah.jones@umontpellier.fr (D.J.); jacques.roziere@umontpellier.fr (J.R.)

2 Institut Universitaire de France (IUF), CEDEX 05, 75231 Paris, France

* Correspondence: sara.cavaliere@umontpellier.fr; Tel.: +33-467-149-098

Received: 17 July 2020; Accepted: 21 August 2020; Published: 28 August 2020

check for updates

Abstract: Mechanical and chemical stability of proton exchange membranes are crucial requirements for the development of fuel cells for durable energy conversion. To tackle this challenge, bi-functional nanoclays grafted with amino groups and with embedded radical scavengers, that is, CeO_2 nanoparticles were incorporated into Aquivion® ionomer. The composite membranes presented high proton conductivity and increased stability to radical attack compared to non-modified Aquivion membranes, demonstrating the effectiveness of the approach based on radical scavenger immobilisation and release from clay nanocontainers.

Keywords: proton exchange membrane fuel cells; radical scavengers; halloysite; cerium oxide

1. Introduction

Despite the significant progress in proton exchange membrane development for fuel cells over the past decade [1–3], improvement of their durability to meet transport application targets [4] is still a great challenge. Advances have also been made in understanding the origins of perfluorosulfonic acid (PFSA) membrane degradation, leading to the development of novel strategies and materials for its mitigation [5–9].

Mechanical degradation of a proton exchange membrane during fuel cell operation occurs as a consequence of dimensional changes due to hydration/dehydration or to the variation in stack compression, leading to crack propagation and pinhole formation [7,10,11]. Mechanical stability has become a greater challenge with the use of thinner membranes to benefit from the advantages of low membrane resistance and improved water transport. Being related to mechanical and chemical degradation of the membrane, failure stress and tear resistance are used as indicators of the durability of membrane-electrode assemblies [12]. Several chemical and physical routes have been employed to increase membrane mechanical resistance, including thermal annealing [13,14] and chemical cross-linking [15] or preparation of composite membranes incorporating polymer reinforcements [16,17], electrospun nanofibres [18–20], carbon nanotubes [21,22], inorganic particles [23–26] and clays [27–30].

Chemical degradation of PFSA membranes is induced by the attack of free radicals (HO•, HOO•) generated in the fuel cell environment [31]. Hydroxyl and hydroperoxyl radicals are the products of decomposition of hydrogen peroxide formed at the cathode [32]. In the presence of traces of iron or other multivalent metal ions (e.g., Cu^{2+} or Ti^{3+}) originated from corrosion of the cell, stack materials or humidifiers, the reaction of H_2O_2 decomposition is catalysed to produce radicals by the Fenton reaction [33,34]. The formed HO• and HOO• attack specific sites on the polymer side or main chains, leading to membrane thinning and pinhole formation due to defragmentation via the unzipping mechanism and fluoride ion release [3,8] and decrease in the ion exchange capacity and consequently in the proton conductivity [6,35]. The former and in particular the fluoride emission rate (FER), that is, the release of fluoride ions per unit area with time from a membrane upon an accelerated stress test (AST) for example, Fenton's reaction, is the conventional indicator of the extent of membrane degradation [36]. To mitigate chemical degradation of membranes the incorporation of radical scavengers, organic (terephthalic acid, λ-tocopherol) as well as inorganic (metal oxides) has been demonstrated to be an effective approach [37–40]. In particular, cerium ions or oxide nanoparticles have been thoroughly investigated due to their faster reaction rate they have with free radicals than that of the free radicals with the polymer membrane [37] and their fast reversible redox reaction in aqueous conditions [41].

Despite the effective stabilisation of the membranes as indicated by strong decrease of the FER upon accelerated stress tests [42,43], the issue of their stability in acidic environment arose [31]. Ce ions can migrate in the operating fuel cell, favoured by concentration gradients and water flow and are leached into exhaust water [44–46] The possibility of immobilising them in the membrane while allowing their controlled release would be a valuable approach to overcome this limitation and make the scavenging effect more durable [31,47–49].

As already mentioned, natural, synthetic and modified clays have been thoroughly used for the preparation of composite membranes, in particular to improve their dimensional and mechanical properties and to provide physical barriers to gas crossover [50,51]. Furthermore, clay materials can be used to immobilise radical scavengers to avoid their elution, playing the role of mechanical as well as chemical stabilisers. Tubular nanoclays and in particular halloysite nanotubes (HNTs) have already been used as containers for the encapsulation and the sustained release of multiple entities (molecules, particles …) especially in the biomedical field [52–54]. HNTs are a naturally occurring aluminosilicate ($Al_2Si_2O_5(OH)_4 \cdot nH_2O$) belonging to the kaolinite group with nanometric dimensions and a hollow tubular structure with an inner *lumen*. The chemical compositions of the inner and outer surfaces of HNTs are different, being formed by Al-OH and Si-O sheets, respectively, a situation that allows regioselective functionalisation [55]. Moreover, halloysite nanotubes have a significant mechanical and thermal reinforcing effect on polymer matrices [56], which makes them promising components of proton exchange membranes.

In this work, HNTs were used as nanocontainers to encapsulate and release cerium oxide nanoparticle radical scavengers to prevent chemical degradation of a PFSA membrane by radical attack. We report the preparation and characterisation of a composite proton exchange membrane based on Aquivion® and the cerium oxide-functionalised HNTs (CeO_2@HNT). Furthermore, amino moieties were grafted on the outer surface of the HNTs using an aminosilane coupling agent in order to improve their compatibility with the acidic ionomer and ensure high dispersion and homogeneity. The bi-functional nanoclays, labelled CeO_2@HNT-NH_2, were characterised by X-ray diffraction (XRD), X-ray fluorescence (XRF), infrared spectroscopy (IR), thermogravimetric analysis (TGA), zeta potential, solid-state ^{29}Si MAS NMR (magic-angle spinning nuclear magnetic resonance) and electron microscopies to evaluate the composition and degree of the inner/outer functionalisation. The composite membranes incorporating such HNTs were characterised for their in-plane proton conductivity and mechanical properties upon strain/stress. They were also submitted to the Fenton reaction to assess the effect of the functionalised clays embedding cerium oxide on the chemical stability, as monitored by the FER. The HNT loading in PFSA was adjusted to find a compromise between chemical stability and proton conductivity adapted for their application in proton exchange membrane fuel cells (PEMFC).

2. Materials and Methods

2.1. Materials

All the following chemicals were purchased from Sigma Aldrich (St Louis, USA) and used as received—halloysite clay nanotubes (HNTs) mined from the High Purity Dragon Mine in Silver City, Utah (USA), oxalic acid solution (1 mol/L), cerium (III) nitrate hexahydrate (99%), anhydrous ethanol, (3-aminopropyl)trimethoxysilane (97%), anhydrous toluene, N,N-dimethylacetamide solution (anhydrous, 99.8%), 2-propanol, hydrogen peroxide (30% volume), ammonium iron(II) sulfate ($(NH_4)_2Fe(SO_4)_2(H_2O)_6$) (99.997%), sulfuric acid (97%), total ionic strength adjustment buffer III solution (TISAB III). Composite membranes were prepared using Aquivion® ionomer (830 EW 24 wt% suspension in water) purchased from Solvay (Brussels, Belgium).

2.2. Preparation of CeO_2 Embedded and Amino Functionalised HNT and Their Characterisation

2.2.1. HNT Pre-Treatment

Pre-treatment with acid was applied as follows to leach the iron ions naturally present in HNTs—1 g of HNTs was dispersed in a 10 mL solution of 0.45 M oxalic acid and the dispersion was stirred for 3 h at 80 °C and then filtered using a 0.22 μm polyvinylidene fluoride (PVDF) filter (Durapore membrane filters). The obtained HNTs were rinsed three times with MilliQ grade water and dispersed in 10 mL of ultra-pure water by ultrasonication (Branson digital sonication) at 5 W for 10 min. The obtained HNTs were dried for 16 h at 80 °C and then characterised by XRF to quantify the amount of the eventual residual iron.

2.2.2. Preparation of CeO_2@HNT

A 3 M solution of Ce $(NO_3)_3$ $6H_2O$ was prepared in EtOH. The treated HNTs were submitted to vacuum-cycling of 0.12 g of HNTs dispersed in 40 mL of this solution. The suspension was kept under vacuum for 1 h in order to remove the air present in the *lumen* and to facilitate insertion of cerium nitrate solution. The suspension was then cycled back to atmospheric pressure. This process was repeated three times. The HNTs embedding cerium nitrate, labelled Ce@HNTs, were separated from the solution by filtration on a PVDF filter (0.22 μm) and rinsed twice with methanol (anhydrous, 99.8%) to ensure the removal of any cerium nitrate that might be present on their surface. Finally, the Ce@HNTs powder was dried at 80 °C for 16 h. In order to ensure the stability of embedded cerium and prevent its leaching, Ce@HNTs were heat treated in air at 300 °C for 5 h to convert the cerium nitrate to cerium oxide nanoparticles [57], embedded within the clay *lumen*, leading to the material labelled CeO_2@HNTs.

2.2.3. Surface Functionalisation of CeO_2@HNT

Organosilane modified CeO_2@HNTs were prepared by adding 0.3 g of CeO_2@HNTs to a solution of 1.5 mL of APTMS in 12.5 mL of dry toluene. The suspension obtained was dispersed using an ultrasonic bath during 30 min and then transferred under reflux at 120 °C for 20 h. A calcium chloride drying tube was used to ensure a dry environment. The suspension was then filtered and washed with toluene six times to afford CeO_2@HNT-NH_2 that was dried at 120 °C for 16 h.

2.2.4. Physico-Chemical Characterisation of HNT

The morphology of halloysite nanotubes (HNT) was investigated by field emission-scanning electron microscopy (FE-SEM) using a Hitachi S-4800 microscope (Hitachi Europe SAS, Velizy, France). Data analysis and particle size distribution of HNT were performed using an image processing software Image J 1.48 v (U. S. National Institutes of Health, Bethesda, MD, USA). Halloysite nanotubes were analysed by transmission electron microscopy (TEM) using a JEOL 2200FS (Source: FEG) microscope operating at 200 kV equipped with a CCD camera Gatan USC (16 MP) (JEOL, Tokyo, Japan).

The different components present were identified by X-ray diffraction (XRD, PANalytical X'Pert in Bragg-Brentano configuration with CuK_α radiation, (Malvern Panalytical, Cambridge, United Kingdom from 10 to 70° with a step angle of 0.032°.

Pristine and functionalised HNTs were characterised by thermogravimetric analysis (TGA, Netzsch, Selb, Germany) with a NETZSCH STA 409 PC from 20 °C to 1200 °C with a ramp of 10 °C min^{-1} under nitrogen.

Wavelength dispersive X-Ray Fluorescence (XRF) spectrometer (Axios max, PANalytical, Cambridge, United Kingdom) was used for quantitative elemental analysis of materials. The X-ray tube in the spectrometer had an Rh anode and operated at a maximum power of 4 kW with a maximum voltage of 60 kV or maximum current of 160 mA. To obtain high resolution fluorescence spectra, eight LiF200 dispersive crystals were used. The measurements were performed under vacuum in fourteen different scans. Each scan covered a range of the expected elements and the peak areas were determined using Spectra Evaluation of SuperQ software. Samples were prepared by grinding 400 mg of sample and pressing it to obtain a pellet of 32 mm diameter. The same protocol was used to prepare the standards to obtain a calibration line. For iron determination, standards were prepared using a mixture containing alumina (prepared in the laboratory) and varying amounts of iron oxide (Sigma Aldrich, Saint Louis, MO, USA, nanopowder, <50 nm particle size—0.1, 0.2, 0.3 and 0.5 wt%). For cerium determination, standards were prepared with halloysite and varying amounts of CeO_2 powder (Sigma Aldrich, Saint Louis, MO, USA, <5 μm, 99.9%: 1, 3, 5, 7 and 10 wt%).

Surface charge of the clays was determined using a Malvern Zetasizer 3000HSa (Malvern Panalytical, Cambridge, United Kingdom) at pH 7 after preparing a dispersion of 1 mg of HNTs in 5 mL of deionised water.

Infrared spectroscopy was used to determine the surface functionalities of the different halloysites upon acidic treatment and functionalisation using a spectrum two spectrometer (Perkin Elmer, Waltham, MA, USA). 5 mg of the samples were analysed in powder form.

The solid-state ^{29}Si NMR spectra of APTMS functionalised clays were recorded on a 300 MHz VARIAN VNMRS300 300 MHz spectrometer (7.05 Tesla "Wide Bore" magnet, LabX, Midland, ON, Canada). A VARIAN T^3 MAS (Magic Angle Spinning, LabX, Midland, ON, Canada) probe with 7.5 mm ZrO_2 rotors was used. The measurements were carried out with the CPMAS technique (non-quantitative) and Single Pulse (quantitative/single pulse ^{29}Si followed by ^1H decoupling). For CPMAS, a $\pi/2$ pulse of 6 μS, a contact time of 3 ms and a recycling time of 3 s were used. For the Single Pulse and to guarantee the quantitative analysis a 2 μs π /6 pulse and a recycling time of 60 s were used. The samples rotated at 5 kHz. The chemical shift values were calibrated using Q8M8H (octakis (dimethylsiloxy) octasilsesquioxane) as a secondary reference (line at −2.25 ppm). The acquisition window was 50 kHz and the filtering (line broadening) 50H. The assignment of the chemical shifts and the identification of the materials were performed based on a model spectrum obtained with the ChemBio Draw software (Ultra 14.0, PerkinElmer, Waltham, MA, USA).

2.3. Membrane Preparation and Characterisation

Composite membranes were prepared by casting Aquivion® 830 EW (10 wt%) dispersion containing different amounts of CeO_2@HNT-NH_2 (2, 4, 5 and 10 wt%). First, the appropriate amount of CeO_2@HNT-NH_2 was sonicated in 2-propanol at 10 W for 5 min. Then, 2.2 g of ionomer dispersion in water were stirred together with 0.5 g of DMAc for 1 h. Finally, CeO_2@HNT-NH_2 suspension in 2-propanol and Aquivion® in DMAc were mixed and ultra-sonicated at 10 W for 5 min. After 10 min, the resulting suspension was cast on a Teflon sheet with a 200 μm blade. The membrane was treated in an oven at 80 °C for 16 h in order to remove the solvent and afterwards annealed at 170 °C for 2 h leading to a composite membrane 15 μm thick (measured with an electronic micrometer with resolution = ±0.001 mm and validated by scanning electron microscopy (SEM, (Hitachi Europe SAS, Velizy, France). For comparison purpose a 15 μm thick reference membrane of Aquivion® 830 EW (10 wt%) was prepared by casting a solution prepared by mixing and stirring for 1 h a solution of 2.2 g of Aquivion® 830 EW with 0.5 g DMAc and 2.3 g of 2-propanol on a Teflon sheet using a 200 μm blade. The PFSA membrane was submitted to the same thermal treatments used for the composite membrane.

2.3 1. Fenton Reaction Protocol

The chemical degradation of the membrane was accelerated by the Fenton reaction. 160 mg of the prepared membranes were immersed in a solution containing 45 mL of hydrogen peroxide (30% volume), 55 mL of ultrapure water, 7.5 μL of concentrated sulfuric acid and 28 mg of $(NH_4)_2Fe(SO_4)_2(H_2O)_6$. Then, the membrane was left in the Fenton reagent for 4 h at 75 °C under reflux and with stirring. Afterwards, the solution was recovered and the TISAB III was added. A selective electrode was used to quantify the amount of fluoride ions released in the medium after the Fenton reaction.

2.3.2. Membrane Characterisation

In-plane proton conductivity of the membranes was determined from resistance measurements made with a BT-512 BekkTech Conductivity Test System including a Keithley 2400 Sourcemeter (TeKtronix, Beaverton, OR, USA). The membrane samples (0.5 cm × 4 cm) were treated in sulfuric acid (1 mol/L) for 1 h and washed 3 times in deionised water for 15 min and left to dry overnight before being placed in the conductivity cell in contact with 4 platinum microelectrodes placed at a distance of 3.5 mm from each other. The temperature and relative humidity were controlled using the BT-201 Temperature Control System. The data were collected and analysed using BekkTech Conductivity Testing (BT512, FuelCellStore, Texas, USA). & LabView Data Analysis Software (8.1, National Instruments, Austin, TX, USA).

The concentration of fluoride ions released in the solution upon Fenton test (fluoride emission rate, FER) was obtained using an ion selective electrode (ISE, Thermo Scientific Orion Star Series Meter—ISO 10359-2: 1994, ThermoFisher Scientific, Waltham, USA). Prior to measurements, the electrode was calibrated using standard solutions at concentrations of 190 ppm, 1900 ppm and 19,000 ppm. The calibration line is validated when the slope is between −54 and −60 mV. The detection limit for this technique is *ca* 100 ppb.

The mechanical properties of the membranes were determined at 22 °C and a relative humidity of 40%. The tensile stress tests were carried out with a Z1.0 testing machine from Zwick Roell (Ulm, Germany), with a 200 N static sensor using at least three repetitions. The membranes were cut into 5 mm × 60 mm rectangles. Data were treated by the TestXpert Master software (11.0, Zwick Roell, Ulm, Germany).

3. Results and Discussion

3.1. Characterisation of HNTs

The morphology of halloysite clays was characterised by scanning and transmission electron microscopies (Figure 1). They presented a tubular structure with length varying with a wide dispersion from 100 nm to 800 nm. The average outer diameter measured 70 nm, while the average inner diameter (*lumen*) was 20 nm.

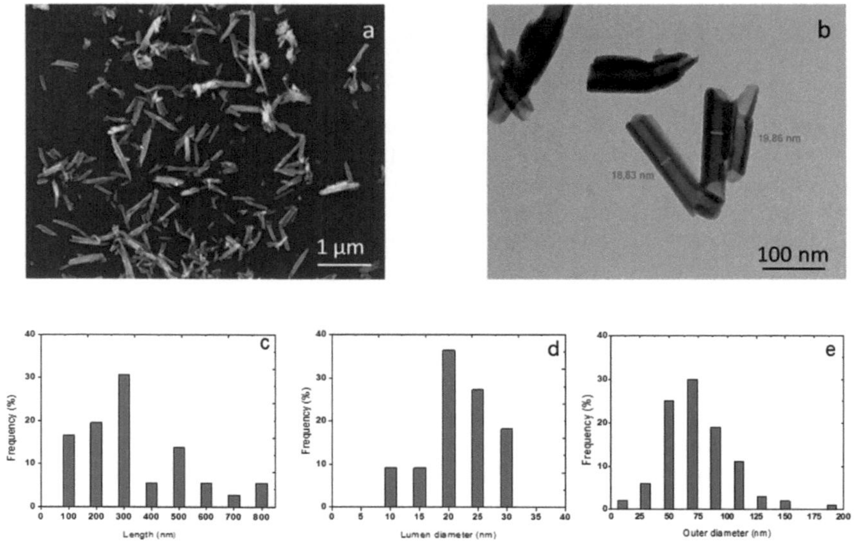

Figure 1. Scanning electron microscopy (SEM) (**a**) and transmission electron microscopy (TEM) micrographs of halloysite nanotubes (HNTs) (**b**) corresponding distribution histograms of length (**c**) *lumen* internal diameter (**d**) and outer diameter (**e**).

The structural analysis of the halloysite nanotubes (HNTs) (Figure S1a, Supplementary Materials) demonstrated peaks typical of halloysite [$Al_2Si_2O_5(OH)_4$] together with SiO_2 and Fe_3O_4 co-crystallised into the kaolinite layers. A sharp peak at 12.1° corresponds to a basal spacing of 0.73 nm confirming the identity of the composite as dehydrated HNTs [58,59]. Other peaks with lower intensity at 20°, 24.5°, 35°, 37.9°, 54.5° and 62.5° also assigned to HNTs according to the standard JCPDS card no 00-029-1487 [60,61].

The chemical composition of HNTs was evaluated by XRF (Table 1) Together with the expected presence of Al and Si in agreement with XRD results, a non-negligible amount of iron was detected (0.34 wt%).

Table 1. Chemical composition of the pristine HNTs, acid-treated HNTs, CeO_2@HNTs and CeO_2@HNT-NH_2 materials determined by X-Ray Fluorescence (XRF) and elemental analysis.

Material	Al (wt %)	Si (wt %)	O (wt %)	N (wt %)	Fe (wt %)	Ce (wt %)
HNT	21.85	22.00	55.15	-	0.34	-
treated HNT	18.39	21.36	56.57	-	0.23	-
CeO_2@HNT	18.39	21.36	56.57	-	0.23	8.0
CeO_2@HNT-NH_2	11.21	16.21	72.40	1.30	0.23	8.0

Iron is naturally present in halloysites. There is a correlation between the amount of iron present in halloysites and their morphology (flat, spherical, tubular) [62,63]. When the isomorphic substitution of Al^{3+} by Fe^{3+} increases, the curvature of the halloysite sheet decreases. Flat halloysites contain the largest amounts of Fe (from 2 to 6 wt%), while tubular halloysites relatively small amounts (from 0 to 3 wt%). Tubular halloysites are the most influenced by iron content with an inverse relationship between tube length and Fe content [55].

The presence of Fe^{x+} in HNTs is critical, as it may catalyse the formation of radicals and thus promote the chemical degradation of the clay-filled membranes. Treatment with oxalic acid, chosen for its acidic, reducing and chelating properties, [64,65] was performed in order to remove or reduce the amount of iron in the HNTs. XRF elemental analysis demonstrated that this treatment led to the reduction of iron content from 0.34 wt% to 0.23 wt% (Table 1) in HNTs without affecting their tubular morphology (Figure S2, Supplementary Materials). The crystal structure of halloysite was also unaffected by oxalic acid treatment. The diffractogram recorded after this treatment (Figure S1b, Supplementary Materials) is identical to that of untreated halloysite. This lack of structural change can be attributed to the use of a relatively low acid concentration for leaching compared to those reported by Panda et al. [63] and Zhang et al. [64]. The peak corresponding to the presence of FeO_x is still present in the diffractogram of the acid-treated HNTs and it is concluded that the remaining iron is structural, that is, occupies the Al^{3+} site in the clay structure [55]. Its possible activity as a catalyst for the Fenton reaction in the composite membrane will be described in paragraph 3.4. All the HNTs characterised and modified in this work were pre-treated with oxalic acid and contain the minimum amount of iron.

3.2. *Preparation and Characterisation of CeO_2@HNT*

In order to increase the chemical stability of the composite membrane embedding HNTs against attack of the hydroxyl HO• and hydroperoxy HOO• free radicals, which may be formed in the presence of multivalent cations leached from the fuel cell or iron from the same HNTs (Section 3.2), the clays were functionalised with cerium oxide nanoparticles.

HNTs have been modified with a range of active agents for their storage and release, especially for biomedical and pharmaceutical applications, either by intercalation between the sheets, adsorption on the external walls or encapsulation in the *lumen* [66–68]. The latter approach offers the highest capacity as well as the possibility of controlled release of the encapsulated entities. The empty HNT *lumen* corresponds to approximately 20 % of the total tube volume, which makes it suitable for loading with approximately 10 to 15 vol% of the active agents. In this work the embedding of CeO_2 nanoparticles in the HNT *lumen* was adapted from a procedure developed by Abdullayev et al. [67] for insertion of the corrosion inhibitor benzotriazole. Firstly, ionic cerium in the form of nitrate was inserted in the inner porosity of HNT forming Ce@HNTs. From XRF analysis, the amount of cerium incorporated as salt was 8 wt%. A previous investigation on the kinetics of the release of the inorganic salt from the halloysite nanotubes evidenced the completion of the process in less than 24 h (data not shown). This result can be explained by the increased mobility and solubility of inorganic ions in the release medium and their weak interaction with the HNT walls [69]. In order to stabilize cerium in the *lumen* and allow its slower and controlled release for a prolonged protection of the membrane from radical attack, Ce@HNTs was converted into cerium oxide nanoparticles (CeO_2@HNTs) by thermal treatment in air [57]. It is known that CeO_2 nanoparticles are formed by thermal decomposition of the hydrated nitrate precursor according to the following mechanism:

$$Ce(NO_3)_3 \,.\, 6\,H_2O \overset{70\text{–}150\,^\circ C}{\rightarrow} Ce(NO_3)_3 \,.\, H_2O + 5\,H_2O$$
$$Ce(NO_3)_3 \,.\, H_2O \overset{150\text{–}225\,^\circ C}{\rightarrow} Ce(NO_3)_3 + H_2O$$
$$Ce(NO_3)_3 \overset{225\text{–}400\,^\circ C}{\rightarrow} CeO_2 + NO_X$$

At 300 °C the encapsulated cerium nitrate is considered fully decomposed to the corresponding oxide The cerium loading was determined by XRF as 8 wt% (Table 1). XRD analysis (Supplementary Materials, Figure S1d) demonstrated the presence of CeO_2 by the appearance of the corresponding peaks at 28.5°, 33°, 47.4° and 56.3° associated with reflections from (111), (200), (220), (311) planes of the fluorite cubic structure according to the JCPDS 01-075-0390 [31]. From the deconvolution of the high intensity peak at 28.5° (Supplementary Materials, Figure S3) and application of the Scherrer equation the size of the ceria nanoparticles was estimated to be 4–5 nm. In reasonable agreement, TEM analysis confirmed the presence of 3.5 nm cerium oxide particles located in the *lumen* (Supplementary Materials, Figure S4).

3.3. Preparation and Characterisation of CeO_2@HNT-NH_2

In order to improve the interface between the PFSA ionomer and the clay and thus the homogeneity of the clay dispersion in the membrane, the surface of the HNT was chemically modified by grafting of with an aminosilane agent (APTMS). Performed in anhydrous medium [70], this reaction gives rise to NH_2 surface groups able to interact with the acidic functionalities of Aquivion®. The amount of APTMS grafted was quantified by comparing TGA traces of functionalised and bare HNT (Figure S5, Supplementary Materials). The two materials showed the same profile of degradation until 200 °C corresponding to the dehydration of the clays and a different mass loss at higher temperature attributed to the degradation of APTMS until 500 °C. The yield at 1200 °C of HNTs was 84.74% which is assigned to the inorganic components of HNTs, while for APTMS-HNTs was 79.08%. From the mass change values it is possible to estimate the amount of the APTMS grafted on the surface as 5.6 ± 0.2 wt%. The amount of nitrogen (from grafted APTMS) was also estimated by elemental analysis as being 1.3 wt%.

The Fourier-transform infrared (FT-IR) spectra of HNTs before and after functionalisation are shown in Figure 2. The absorption peaks at 3624 and 3694 cm^{-1} correspond to the OH stretching of inner surface hydroxyl groups and outer surface hydroxyl groups [56]. Other signals characteristic of HNT are displayed, such as deformation vibrations of Si–O–Si and Al–O–Si at 458 and 522 cm^{-1} and the inner Si–O stretching vibration at 1024 cm^{-1}. The intensity of Al-O–H deformation vibration of the inner hydroxyl groups appears at 907 cm^{-1}. The weak band at 1651 cm^{-1} is ascribed to the O–H deformation vibration of the adsorbed water.

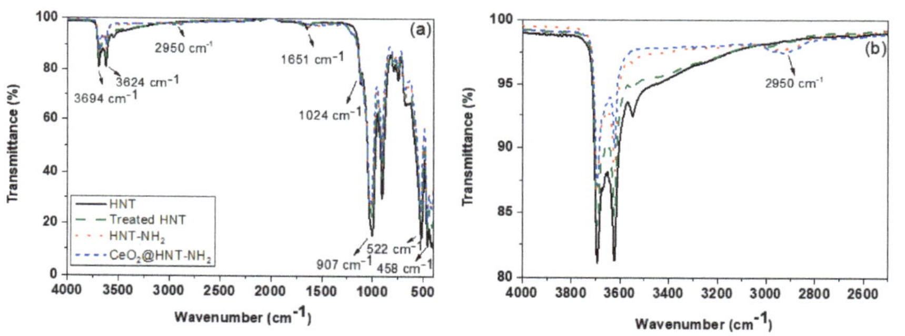

Figure 2. Fourier-transform infrared (FT-IR) spectra of HNTs, acid-treated HNTs, HNT-NH_2-and CeO_2@HNT-NH_2 (**a**). In (**b**) is represented the enlarged 4000–2500 cm^{-1} region.

After the functionalisation of HNTs with APTMS, the peak at 2950 cm^{-1} is assigned to the symmetric stretching vibration of $–CH_2$ and the decrease in the intensity of the hydroxyl groups at 3624 and 3694 cm^{-1} demonstrated the presence of the coupling agent and the external surface functionalisation of HNTs.

Zeta potential measurements were performed at pH 7 after dispersing the HNTs in ultrapure water. Bare HNTs displayed a zeta potential of −26 mV, which corresponds to the charge of the silanol functionalities on the surface. SiO_2 is negatively charged above pH 4. The zeta potential of HNT-NH_2 was 27.2 mV at pH 7. The change of the charge value from a negative to a positive value after reaction with APTMS is in agreement with the effective grafting of NH_2 moieties on the HNT external surface.

In order to gain information on the nature of the coordination at the silicon atoms of the HNT surface, CP/MAS ^{29}Si NMR analysis was carried out on HNT-NH_2 and the spectrum obtained is presented in Figure 3.

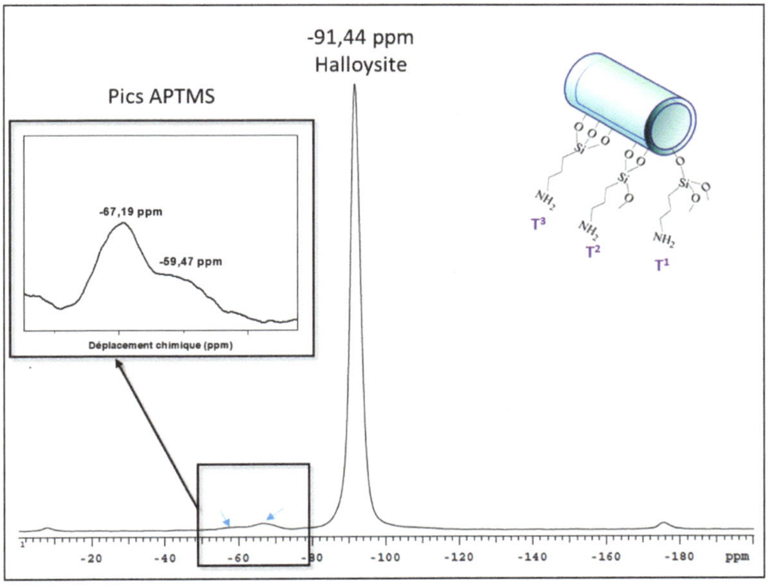

Figure 3. ^{29}Si NMR spectrum of HNT-NH_2.

It displays a high intensity peak at −92 ppm attributed to the presence of silicon Q^3, $Si(OSi)_3(OAl)$ of HNTs. Low intensity signals at −68 ppm and −61 ppm correspond to the tridentate (T^3) and bidentate (T^2) coordination of silicon of the APTMS, respectively. Observation of the bidentate form of Si means that some APTMS species possess one methoxy or hydroxyl group that is not condensed [71].

Transmission electron microscopy observation was performed before and after the functionalisation of HNTs by cerium oxide particles and APTMS (Supplementary Materials, Figure S6). From TEM micrographs, the presence of CeO_2 spherical particles inside the *lumen* with diameter ranging from 2 to 5 nm was demonstrated. Their average diameter of 3.5 nm is in agreement with the cerium oxide domain size determined from XRD (Supplementary Materials, Figures S1d and S3). The cerium oxide content of CeO_2@HNT-NH_2 after the reaction with APTMS (Table 1) was unchanged (8 wt%), demonstrating the stability of CeO_2 nanoparticles in the *lumen*.

The results discussed so far demonstrate the effective formation of cerium oxide nanoparticles in the HNT *lumen* and the grafting of APTMS on the outer surface hydroxyl groups leading to the formation of a bi-functional material CeO_2@HNT-NH_2. The next section will present their incorporation into Aquivion® ionomer to prepare nanocomposite membranes.

3.4. Composite Membrane Characterisation

To investigate the effect of the residual iron in the halloysite nanoclays in the Fenton reaction, membranes of Aquivion® 830 EW containing 20 wt% of as-received and acid-treated HNTs were prepared and compared with a reference membrane of Aquivion® 830 EW. A greater amount of HNTs than that used in this work (20 vs. 10 wt%) was incorporated in the membranes to maximise the iron amount and therefore the release of fluoride ions from the ionomer according to the detection limit of the fluoride electrode.

When investigating the composite membranes no $(NH_4)_2Fe(SO_4)_2(H_2O)_6$ was added to the hydrogen peroxide solution (iron being already present in HNTs). During the study of the reference membrane in the Fenton reaction and for the purpose of comparison with the composite membrane containing the acid-treated HNTs, an amount of $(NH_4)_2Fe(SO_4)_2(H_2O)_6$ corresponding to 0.23 wt% of iron (as in the HNTs after oxalic acid treatment) was added. The reaction was followed by monitoring the amount of fluoride ion released (FER) in the medium after 4 h (Figure 4).

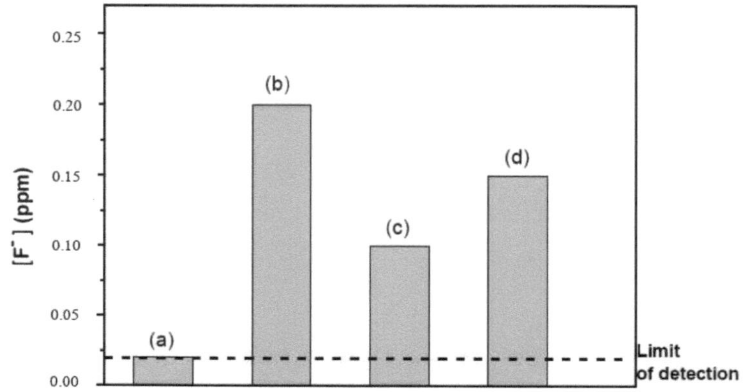

Figure 4. Histogram of the fluoride concentration released after 4 h in the Fenton reaction for the Aquivion® membrane (without added Fe) (a) and composite membranes comprising untreated HNT (0.34 wt% Fe) (b) and acid-treated HNTs (0.23 wt% Fe) (c) and Aquivion® membrane (with 0.23% of Fe added as $(NH_4)_2Fe(SO_4)_2(H_2O)_6$) (d).

The reference Aquivion® membrane immersed in a Fenton solution without Fe, showed no chemical degradation since the concentration of fluoride ions measured in the medium after 4 h was 0.02 ppm, that is, within the detection limits of the ion-selective electrode (ISE). The composite membrane with pristine HNTs (containing 0.34 wt% of iron) released twice the amount of fluoride ions double (0.2 ppm) than the membrane containing the acid-treated HNTs (containing 0.23 wt% of iron) (0.1 ppm), while the ratio of the amounts of iron in the halloysites is less than 2 (1.48). On comparing the FER for the membrane containing the acid-treated HNTs and an Aquivion® membrane in the presence of the same amount of iron (0.23 wt% by $(NH_4)_2Fe(SO_4)_2(H_2O)_6$ addition), it is concluded that the degradation induced by the residual iron in the HNTs is negligible (higher fluoride release for the Aquivion® membrane in the presence of Fe^{2+}). Therefore, structural iron has an effect in the production of radicals in the Fenton reaction but to a lower extent than free iron ions at the same concentration. The acidic treatment procedure was effective and the influence of the residual iron was negligible. From these results, the amount of clay incorporated in the membranes was reduced to 4 wt%, with radical scavenger incorporated into the HNTs, to counter the oxidative effect of structural iron.

The dispersion and aggregation behaviour of halloysite are known to be highly related to the surface charge [72]. To verify this assumption, morphological analysis and proton conductivity determination were performed on the composite membranes embedding functionalised and non-functionalised nanoclays. From the SEM images of the cross section (Figure 5) of the composite membranes comprising 10% acid-treated HNT and CeO_2@HNT-NH_2 it may be seen that the latter presented higher compatibility with the ionomer.

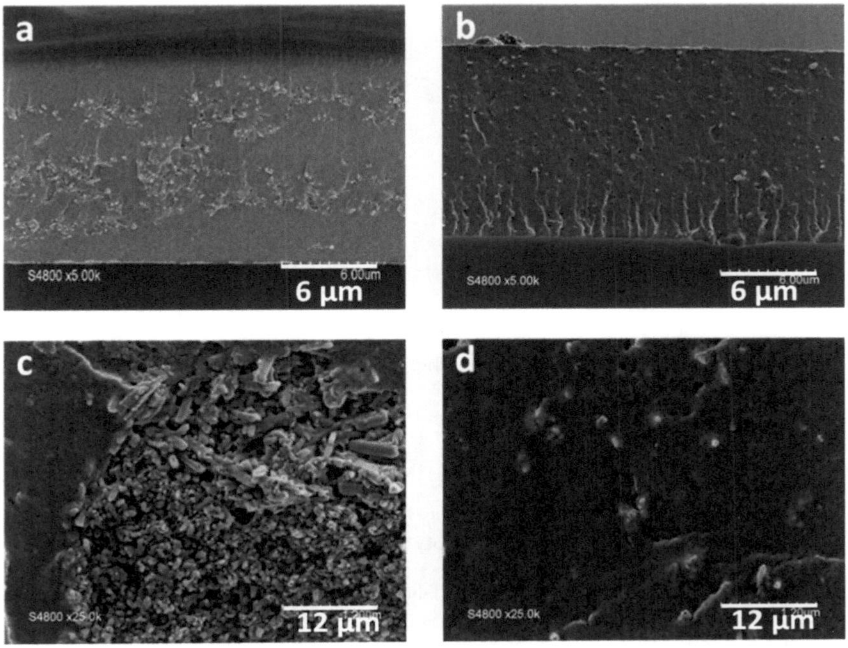

Figure 5. SEM images of the cross section of composite membrane with Aquivion® loaded with 10 wt% HNT (**a**,**c**) and HNT-NH_2 (**b**,**d**).

The homogeneity of the dispersion of the functionalised HNTs is observed throughout the thickness of the membrane (Figure 5b,d), while for the composite membrane comprising the same amount of HNTs, the presence of agglomerates of clays of micrometric size (2–3 µm) (Figure 5a,c) indicates poor interaction with the ionomer. The proton conductivity, Figure 6, of the reference Aquivion® membrane at 90 °C and 95%RH (190 mS/cm) is in agreement with results reported elsewhere [73]. As a direct result of the poor dispersion, the incorporation of (non-functionalised) HNTs in Aquivion® led to a significant decrease in the proton conductivity to 50 mS/cm, under the same conditions. Surface functionalisation of the HNT had a positive effect on the proton transfer in the composite membrane however, since the proton conductivity reaches 154 mS/cm at 90 °C and 95 %RH, only slightly lower than that the reference membrane, due to immobilisation of some of the protons in ionic crosslinking. The acid/basic interaction between the sulfonic groups of Aquivion® and the amine groups of HNT-NH_2 could lead to a decrease in the effective ion exchange capacity of the ionomer with a subsequent decrease in the conductivity of the corresponding membrane.

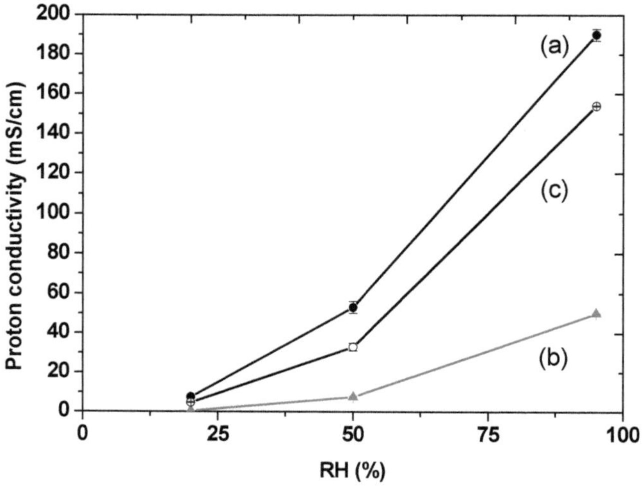

Figure 6. Proton conductivity measurements at 90 °C at different RH of pristine Aquivion® membrane (a), composite membrane with 10 wt% loading of HNT (b) and 10 wt% loading of HNT-NH_2 (c).

The effect of the incorporation of CeO_2@HNT-NH_2 on the proton conduction and chemical stability properties of the corresponding membranes was investigated by preparing composite membranes with different amounts of CeO_2@HNT-NH_2 (the cerium loaded in the *lumen* being constant at 8 wt%). Composite membranes were prepared by casting a solution of Aquivion® 830 EW (10 wt%) and different amounts of CeO_2@HNT-NH_2 from 2 to 10 wt% corresponding to Ce/HSO_3^- mole ratio from 1% to 5 %. The molar ratio of cerium to HSO_3^- groups in Aquivion® PFSA ratio is a straightforward method to compare immobilised radical scavenger. It has been reported that 1% led to a significant (7 times) increase in durability in fuel cell measurement [47].

The chemical stability of the above composite membranes was evaluated in a Fenton test and the quantity of fluoride ions released over time (FER) was measured after every 4 h. The Fenton solution was renewed each time in order to ensure the same conditions throughout the experiment. The time-dependent FER for all composite membranes is depicted in Figure 7. The higher the quantity of cerium incorporated into the membranes, the lower the concentration of F^- measured in the Fenton solution and therefore the higher the chemical stability of the membrane against radical attack. However, higher CeO_2@HNT-NH_2 content in the membrane leads to lower proton conductivity, which could be explained by a release of cerium ions from the nanometric oxide encapsulated in halloysite and its migration in the membrane, blocking proton transfer sites [44–46,74,75].

For further study, membranes containing 4 wt% of CeO_2@HNT-NH_2, corresponding to 2 mol% Ce/HSO_3^- ratio, was selected since it enabled high proton conductivity (160 mS/cm) and significant scavenging activity. This Ce/HSO_3^- ratio corresponds to that considered as optimum in a study on silica-immobilised ceria radical scavenger [47]. The microscopy images of the bi-functional clays and of the cross-section of the corresponding composite membranes are depicted in Figure S6 of Supplementary Materials. The dispersion of the nanomaterials is homogeneous all over the membrane thickness.

Tensile stress/strain measurements were performed on the Aquivion® 830 EW reference membrane and the composite membranes containing 4 and 10 wt% of CeO_2@HNT-NH_2. The stress/strain curves are shown in Figure S7 of the Supplementary Materials and the corresponding calculated mechanical properties (Young modulus, yield stress, breaking strength) are summarised in Table 2. Average mechanical properties derived from stress/strain test conducted at 22 °C and 40 % RH on reference and composite membranes.

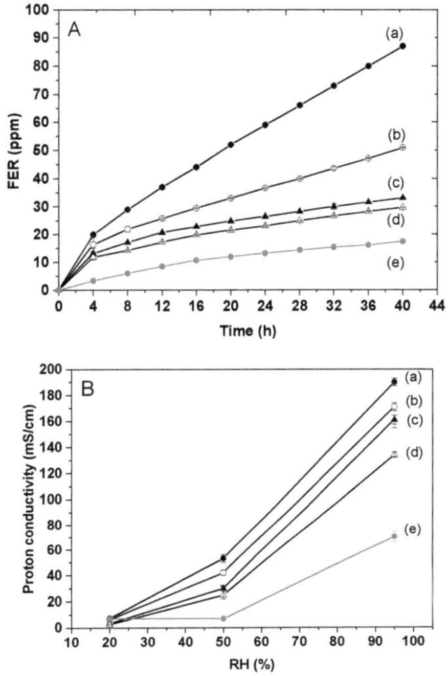

Figure 7. Fluoride Emission Rate (FER) as a function of time (**A**) and in-plane proton conductivity at 90 °C as a function relative humidity (**B**) of composite membranes loaded with different amounts of CeO_2@HNT-NH_2 corresponding to the following Ce/HSO_3^- ratios: 1% (b), 2% (c), 2.5% (d), 5% (e) and comparison with the pristine Aquivion® membrane (a).

Table 2. Mechanical properties of reference membrane Aquivion® 830EW and composite membrane Aquivion® + 4 % CeO_2@HNT-NH_2 and Aquivion® + 10 % CeO_2@HNT-NH_2.

Membrane	Young Modulus (N/mm^2)	Breaking Strength ($N\ mm^2$)	Yield Stress ($N\ mm^2$)
Aquivion® 830 EW	238 ± 7	14 ± 0.7	10 ± 0.3
Aquivion® + 4 % CeO_2@HNT-NH_2	237 ± 5	11 ± 1.3	9 ± 0.6
Aquivion® + 10 % CeO_2@HNT-NH_2	326 ± 9	15 ± 1.5	12 ± 0.3

All membranes have ductile mechanical behaviour. The reference membrane has a higher elongation at break (160%) followed by the membrane loaded with 4 wt% (93%) and 10 wt% (77%) of CeO_2@HNT-NH_2. This result is expected because the incorporation of an inorganic component in a polymer matrix generally leads to increased hardness of the resulting composite material [42,43]. The ductility of the membrane therefore decreases with the rate of incorporated functionalised HNTs. Since the dimensional change of membranes in the plane of the membrane during fuel cell operation is always significantly lower than these stress at break values, more pertinent indicators are the Young's modulus and the yield stress, which both increased significantly (+85% and 26%, respectively) for the composite membrane with 10 wt% CeO_2@HNT-NH_2. The selected composite membrane loaded with 4 wt% of functionalised clays presented very similar values of mechanical property indicators to those of the reference membrane.

It is concluded that the incorporation of 4 wt% CeO_2@HNT-NH_2 in Aquivion® did not lead to significant effect on the membrane mechanical properties. The composite membrane with such composition presented high proton conductivity and significant increase in chemical stability against free radicals, demonstrating the effectiveness of the approach. In situ investigation in a single fuel cell will further validate the strategy of immobilisation and release of radical scavengers in PFSA membranes for their enhanced lifetime.

4. Conclusions

With the aim of enhancing their chemical and mechanical stability, composite proton-exchange membranes incorporating radical scavengers immobilized in nanoclays were prepared and characterized. Bi-functional halloysites, grafted with amino groups and embedding CeO_2 nanoparticles (CeO_2@HNT-NH_2), were used as nanocontainers to immobilise and release the radical scavenger to the Aquivion® ionomer. Composite membranes incorporating 4 wt% CeO_2@HNT-NH_2 presented unchanged tensile properties but high proton conductivity and increased stability to radical attack compared to non-modified Aquivion® membranes, demonstrating the effectiveness of the approach. In situ characterisation in a single fuel cell will further validate it and specific approaches to improve mechanical resistance are currently under investigation.

Supplementary Materials: The following are available online at http://www.mdpi.com/2077-0375/10/9/208/s1, Figure S1: X-ray diffraction of pristine HNT, acid-treated HNT, HNT-NH_2 and CeO_2@HNT materials; Figure S2: SEM micrograph of acid-treated HNTs; Figure S3: Deconvolution of the high intensity peak of CeO_2 in the X-ray diffractogram of CeO_2@HNT; Figure S4. TEM micrographs of HNTs (a), CeO_2@HNT-NH_2 (b) and histogram of diameter size of cerium oxide particles (c); Figure S5. TGA curves of acid-treated HNTs and HNTs-NH_2; Figure S6. SEM micrographs of the bi-functional CeO_2@HNT-NH_2 clays (a) and of the cross-section of the corresponding composite membrane loaded at 4 wt% (2 mol% Ce/HSO_3^-); Figure S7. Stress/strain test curves registered at 22 °C and 40% RH on reference and composite membranes.

Author Contributions: Data curation, F.S.; Formal analysis, A.A. and F.S.; Funding acquisition, D.J. and J.R.; Investigation, A.A., A.D., M.Z. and F.D.; Methodology, M.Z., S.C., D.J. and J.R.; Project administration, S.C., D.J. and J.R.; Supervision, M.T.-J., S.C. and J.R.; Validation, D.J. and J.R.; Writing—original draft, S.C. All authors have read and agreed to the published version of the manuscript.

Funding: This research was funded by French National Research Agency under the COMEHTE project (contract number ANR-15-CE05-0025-01).

Acknowledgments: S.C. acknowledges the financial support from the French IUF funding agency.

Conflicts of Interest: The authors declare no conflict of interest.

References

1. De Bruijn, F.A.; Dam, V.A.T.; Janssen, G.J.M. Review: Durability and degradation issues of PEM fuel cell components. *Fuel Cells* **2008**, *8*, 3–22. [CrossRef]
2. Kusoglu, A.; Weber, A.Z. New insights into perfluorinated Sulfonic-Acid Ionomers. *Chem. Rev.* **2017**, *117*, 987–1104. [CrossRef] [PubMed]
3. Borup, R.; Meyers, J.; Pivovar, B.; Kim, Y.S.; Mukundan, R.; Garland, N.; Myers, D.; Wilson, M.; Garzon, F.; Wood, D.; et al. Scientific aspects of polymer electrolyte fuel cell durability and degradation. *Chem. Rev.* **2007**, *107*, 3904–3951. [CrossRef] [PubMed]
4. Available online: https://www.energy.gov/eere/fuelcells/doe-technical-targets-fuel-cell-systems-and-stacks-transportation-applications (accessed on 15 June 2020).
5. Rodgers, M.P.; Bonville, L.J.; Kunz, H.R.; Slattery, D.K.; Fenton, J.M. Fuel cell perfluorinated sulfonic acid membrane degradation correlating accelerated stress testing and lifetime. *Chem. Rev.* **2012**, *112*, 6075–6103. [CrossRef] [PubMed]
6. Zatoń, M.; Rozière, J.; Jones, D.J. Current understanding of chemical degradation mechanisms of perfluorosulfonic acid membranes and their mitigation strategies: A review. *Sustain. Energy Fuels* **2017**, *1*, 409–438. [CrossRef]

7. Subianto, S.; Pica, M.; Casciola, M.; Cojocaru, P.; Merlo, L.; Hards, G.; Jones, D.J. Physical and chemical modifi cation routes leading to improved mechanical properties of per fl uorosulfonic acid membranes for PEM fuel cells. *J. Power Sources* **2013**, *233*, 216–230. [CrossRef]
8. Curtin, D.E.; Lousenberg, R.D.; Henry, T.J.; Tangeman, P.C.; Tisack, M.E. Advanced materials for improved PEMFC performance and life. *J. Power Sources* **2004**, *131*, 41–48. [CrossRef]
9. Venkatesan, S.v.; Lim, C.; Holdcroft, S.; Kjeang, E. Progression in the morphology of fuel cell membranes upon conjoint chemical and mechanical degradation. *J. Electrochem. Soc.* **2016**, *163*, F637–F643. [CrossRef]
10. Mauritz, K.A.; Moore, R.B. State of understanding of Nafion. *Chem. Rev.* **2004**, *104*, 4535–4585. [CrossRef]
11. Ramaswamy, N.; Hakim, N.; Mukerjee, S. Degradation mechanism study of perfluorinated proton exchange membrane under fuel cell operating conditions. *Electrochim. Acta* **2008**, *53*, 3279–3295. [CrossRef]
12. Huang, X.; Solasi, R.; Zou, Y.U.E.; Feshler, M.; Reifsnider, K.; Condit, D.; Burlatsky, S.; Madden, T. Mechanical endurance of polymer electrolyte membrane and PEM fuel cell durability. *J. Polym. Sci. Part B Polym. Phys.* **2006**, *44*, 2346–2357. [CrossRef]
13. Alberti, G.; Narducci, R.; Di Vona, M.L.; Giancola, S. Annealing of nafion 1100 in the presence of an annealing agent: A powerful method for increasing ionomer working temperature in PEMFCs. *Fuel Cells* **2013**, *13*, 42–47. [CrossRef]
14. Giancola, S.; Arciniegas, R.A.B.; Fahs, A.; Chailan, J.-F.; Di Vona, M.L.; Knauth, P.; Narducci, R. Study of Annealed Aquivion® Ionomers with the INCA Method. *Membranes* **2019**, *9*, 134. [CrossRef] [PubMed]
15. Zatoń, M.; Cavaliere, S.; Jones, D.J.; Rozière, J. Design of Heterogeneities and Interfaces with Nanofibers in Fuel Cell Membranes BT—Handbook of Nanofibers. In *Handbook of Nanofibers*; Barhoum, A., Bechelany, M., Makhlouf, A., Eds.; Springer International Publishing: Cham, Switzerland, 2018; pp. 1–37.
16. Bahar, B.; Hobson, A.R.; Kolde, J.A.; Zuckerbrod, D.; Bahar, B.; Hobson, A.R.; Kolde, J.A.; Zuckerbrod, D. Ultra-Thin Integral Composite Membrane. U.S. Patent No. 5,547,551, 20 August 1996.
17. Xiao, P.; Li, J.S.; Tang, H.L.; Wang, Z.; Pan, M. Physically stable and high performance Aquivion/ePTFE composite membrane for high temperature fuel cell application. *J. Memb. Sci.* **2013**, *442*, 65–71. [CrossRef]
18. Sood, R.; Cavaliere, S.; Jones, D.J.; Rozière, J. Electrospun nanofibre composite polymer electrolyte fuel cell and electrolysis membranes. *Nano Energy* **2016**, *26*, 729–745. [CrossRef]
19. Subianto, S.; Donnadio, A.; Cavaliere, S.; Pica, M.; Casciola, M.; Jones, D.J.; Rozière, J. Reactive coaxial electrospinning of ZrP/ZrO2 nanofibres. *J. Mater. Chem. A* **2014**, *2*, 13359–13365. [CrossRef]
20. Choi, J.; Lee, K.; Wycisk, R.; Pintauro, P.N.; Mather, P.T. Nanofiber composite membranes with low equivalent weight perfluorosulfonic acid polymers. *J. Mater. Chem.* **2010**, *20*, 6282–6290. [CrossRef]
21. Cele, N.P.; Sinha Ray, S.; Pillai, S.K.; Ndwandwe, M.; Nonjola, S.; Sikhwivhilu, L.; Mathe, M.K. Carbon nanotubes based nafion compositemembranes for fuel cell applications. *Fuel Cells* **2010**, *10*, 64–71.
22. Jérôme, R.; Thomassin, J.-M.; Kollar, J.; Caldarella, G.; Germain, A.; Detrembleur, C. Beneficial effect of carbon nanotubes on the performances of Nafion membranes in fuel cell applications. *J. Memb. Sci.* **2007**, *303*, 252–257.
23. Donnadio, A.; Pica, M.; Carbone, A.; Gatto, I.; Posati, T.; Mariangeloni, G.; Casciola, M. Double filler reinforced ionomers: A new approach to the design of composite membranes for fuel cell applications. *J. Mater. Chem. A* **2015**, *3*, 23530–23538. [CrossRef]
24. Guo, C.; Zhou, L.; Lv, J. Effects of expandable graphite and modified ammonium polyphosphate on the flame-retardant and mechanical properties of wood flour-polypropylene composites. *Polym. Polym. Compos.* **2013**, *21*, 449–456. [CrossRef]
25. Di Noto, V.; Gliubizzi, R.; Negro, E.; Pace, G. Effect of SiO2 on relaxation phenomena and mechanism of ion conductivity of [Nafion/(SiO2)x] composite membranes. *J. Phys. Chem. B* **2006**, *110*, 24972–24986. [CrossRef] [PubMed]
26. Thayumanasundaram, S.; Piga, M.; Lavina, S.; Negro, E.; Jeyapandian, M.; Ghassemzadeh, L.; Müller, K.; Di Noto, V. Hybrid inorganic–organic proton conducting membranes based on Nafion, SiO2 and triethylammonium trifluoromethanesulfonate ionic liquid. *Electrochim. Acta* **2010**, *55*, 1355–1365. [CrossRef]
27. Burgaz, E.; Lian, H.; Alonso, R.H.; Estevez, L.; Kelarakis, A.; Giannelis, E.P. Nafion-clay hybrids with a network structure. *Polymer (Guildf)* **2009**, *50*, 2384–2392. [CrossRef]
28. Herrera Alonso, R.; Estevez, L.; Lian, H.; Kelarakis, A.; Giannelis, E.P. Nafion-clay nanocomposite membranes: Morphology and properties. *Polymer (Guildf)* **2009**, *50*, 2402–2410. [CrossRef]

29. Beauger, C.; Lainé, G.; Burr, A.; Taguet, A.; Otazaghine, B.; Rigacci, A. Nafion®-sepiolite composite membranes for improved proton exchange membrane fuel cell performance. *J. Memb. Sci.* **2013**, *430*, 167–179. [CrossRef]
30. Nicotera, I.; Enotiadis, A.; Angjeli, K.; Coppola, L.; Ranieri, G.A.; Gournis, D. Effective improvement of water-retention in nanocomposite membranes using novel organo-modified clays as fillers for high temperature PEMFCs. *J. Phys. Chem. B* **2011**, *115*, 9087–9097. [CrossRef]
31. Zatoń, M.; Rozière, J.; Jones, D.J. Mitigation of PFSA membrane chemical degradation using composite cerium oxide–PFSA nanofibres. *J. Mater. Chem. A* **2017**, *5*, 5390–5401. [CrossRef]
32. Yuan, X.Z.; Li, H.; Zhang, S.; Martin, J.; Wang, H. A review of polymer electrolyte membrane fuel cell durability test protocols. *J. Power Sources* **2011**, *196*, 9107–9116. [CrossRef]
33. Gubler, L.; Dockheer, S.M.; Koppenol, W.H. Radical (HO•, H• and HOO•) formation and ionomer degradation in polymer electrolyte fuel cells. *J. Electrochem. Soc.* **2011**, *158*, B755. [CrossRef]
34. Mittal, V.O.; Kunz, H.R.; Fenton, J.M. Membrane degradation mechanisms in PEMFCs. *J. Electrochem. Soc.* **2007**, *154*, B652. [CrossRef]
35. Peron, J.; Nedellec, Y.; Jones, D.; Rozière, J. The effect of dissolution, migration and precipitation of platinum in Nafion®-based membrane electrode assemblies during fuel cell operation at high potential. *J. Power Sources* **2008**, *185*, 1209–1217. [CrossRef]
36. Healy, J.; Hayden, C.; Xie, T.; Olson, K.; Waldo, R.; Brundage, M.; Gasteiger, H.; Abbott, J. Aspects of the chemical degradation of PFSA ionomers used in PEM fuel cells. *Fuel Cells* **2005**, *5*, 302–308. [CrossRef]
37. Gubler, L.; Koppenol, W.H. Kinetic simulation of the chemical stabilization mechanism in fuel cell membranes using cerium and manganese redox couples. *J. Electrochem. Soc.* **2011**, *159*, B211–B218. [CrossRef]
38. Endoh, E. Development of highly durable PFSA membrane and MEA for PEMFC under high temperature and low humidity conditions. *ECS Trans.* **2008**, *16*, 1229–1240. [CrossRef]
39. Pearman, B.P.; Mohajeri, N.; Brooker, R.P.; Rodgers, M.P.; Slattery, D.K.; Hampton, M.D.; Cullen, D.a.; Seal, S. The degradation mitigation effect of cerium oxide in polymer electrolyte membranes in extended fuel cell durability tests. *J. Power Sources* **2013**, *225*, 75–83. [CrossRef]
40. Danilczuk, M.; Perkowski, A.J.; Schlick, S. Ranking the stability of perfluorinated membranes used in fuel cells to attack by hydroxyl radicals and the effect of Ce(III): A competitive kinetics approach based on spin trapping ESR. *Macromolecules* **2010**, *43*, 3352–3358. [CrossRef]
41. Schlick, S.; Danilczuk, M.; Drews, A.R.; Kukreja, R.S. Scavenging of Hydroxyl Radicals by Ceria Nanoparticles: Effect of Particle Size and Concentration. *J. Phys. Chem. C* **2016**, *120*, 6885–6890. [CrossRef]
42. Lei, M.; Yang, T.Z.; Wang, W.J.; Huang, K.; Zhang, Y.C.; Zhang, R.; Jiao, R.Z.; Fu, X.L.; Yang, H.J.; Wang, Y.G.; et al. One-dimensional manganese oxide nanostructures as radical scavenger to improve membrane electrolyte assembly durability of proton exchange membrane fuel cells. *J. Power Sources* **2013**, *230*, 96–100. [CrossRef]
43. Trogadas, P.; Parrondo, J.; Ramani, V. CeO_2 surface oxygen vacancy concentration governs in situ free radical scavenging efficacy in polymer electrolytes. *ACS Appl. Mater. Interfaces* **2012**, *4*, 5098–5102. [CrossRef]
44. Baker, A.M.; Mukundan, R.; Spernjak, D.; Judge, E.J.; Advani, S.G.; Prasad, A.K.; Borup, R.L. Cerium migration during PEM fuel cell accelerated stress testing. *J. Electrochem. Soc.* **2016**, *163*, 1023–1031. [CrossRef]
45. Stewart, S.M.; Spernjak, D.; Borup, R.; Datye, A.; Garzon, F. Cerium migration through hydrogen fuel cells during accelerated stress testing. *ECS Electrochem. Lett.* **2014**, *3*, F19–F22. [CrossRef]
46. Zatoń, M.; Prélot, B.; Donzel, N.; Rozière, J.; Jones, D.J. Migration of Ce and Mn ions in PEMFC and its impact on PFSA membrane degradation. *J. Electrochem. Soc.* **2018**, *165*, F3281–F3289. [CrossRef]
47. D'Urso, C.; Oldani, C.; Baglio, V.; Merlo, L.; Aricò, A.S. Towards fuel cell membranes with improved lifetime: Aquivion®Perfluorosulfonic Acid membranes containing immobilized radical scavengers. *J. Power Sources* **2014**, *272*, 753–758. [CrossRef]
48. D'Urso, C.; Oldani, C.; Baglio, V.; Merlo, L.; Aricò, A.S. Immobilized transition metal-based radical scavengers and their effect on durability of Aquivion®perfluorosulfonic acid membranes. *J. Power Sources* **2016**, *301*, 317–325. [CrossRef]
49. Park, Y.; Kim, D. Chemical stability enhancement of Nafion membrane by impregnation of a novel organic ·OH radical scavenger, 3,4-dihydroxy-cinnamic acid. *J. Memb. Sci.* **2018**, *566*, 1–7. [CrossRef]

50. Jones, D.J.; Rozière, J. Advances in the development of inorganic–organic membranes for fuel cell applications. In *Fuel Cells I—Advances in Polymer Science*; Springer: Berlin/Heidelberg, Germany, 2008; Volume 215, pp. 219–264.
51. Chua, S.; Fang, R.; Sun, Z.; Wu, M.; Gu, Z.; Wang, Y.; Hart, J.N.; Sharma, N.; Li, F.; Wang, D.W. Hybrid solid polymer electrolytes with two-dimensional inorganic nanofillers. *Chem. A Eur. J.* **2018**, *24*, 18180–18203. [CrossRef] [PubMed]
52. Veerabadran, N.G.; Price, R.; Lvov, Y.M. Clays nanotubes for encapsulation and sustaines release of drugs. *Nano* **2007**, *02*, 115–120. [CrossRef]
53. Yuan, P.; Tan, D.; Annabi-Bergaya, F. Properties and applications of halloysite nanotubes: Recent research advances and future prospects. *Appl. Clay Sci.* **2015**, *112–113*, 75–93. [CrossRef]
54. Kausar, A. Review on Polymer/Halloysite Nanotube nanocofmposite. *Polym. Plast. Technol. Eng.* **2018**, *57*, 548–564. [CrossRef]
55. Joussein, E.; Petit, S.; Churchman, J.; Theng, B.; Righi, D.; Delvaux, B. Halloysite clay minerals—A review. *Clay Miner.* **2005**, *40*, 383–426. [CrossRef]
56. Gaaz, T.; Sulong, A.; Kadhum, A.; Al-Amiery, A.; Nassir, M.; Jaaz, A. The impact of halloysite on the thermo-mechanical properties of polymer composites. *Molecules* **2017**, *22*, 838. [CrossRef] [PubMed]
57. Bhosale, A.K.; Tarwal, N.L.; Shinde, P.S.; Kadam, P.M.; Patil, R.S.; Barman, S.R.; Patil, P.S. Effective utilization of spray pyrolyzed CeO2 as optically passive counter electrode for enhancing optical modulation of WO3. *Solid State Ionics* **2009**, *180*, 1324–1331. [CrossRef]
58. Pasbakhsh, P.; Churchman, G.J.; Keeling, J.L. Characterisation of properties of various halloysites relevant to their use as nanotubes and microfibre fillers. *Appl. Clay Sci.* **2013**, *74*, 47–57. [CrossRef]
59. Zargarian, S.S.; Haddadi-Asl, V.; Hematpour, H. Carboxylic acid functionalization of halloysite nanotubes for sustained release of diphenhydramine hydrochloride. *J. Nanoparticle Res.* **2015**, *17*, 218. [CrossRef]
60. Falcón, J.M.; Sawczen, T.; Aoki, I.V. Dodecylamine-Loaded halloysite nanocontainers for active anticorrosion coatings. *Front. Mater.* **2015**, *2*, 1–13. [CrossRef]
61. Lun, H.; Ouyang, J.; Yang, H. Natural halloysite nanotubes modified as an aspirin carrier. *RSC Adv.* **2014**, *4*, 44197–44202. [CrossRef]
62. Bauluz Lázaro, B. Halloysite and kaolinite: Two clay minerals with geological and technological importance. *Rev. la Acad. Ciencias Exactas Físicas Químicas y Nat. Zaragoza* **2015**, *70*, 7–38.
63. Churchman, G.J.; Lowe, D.J. Alteration, formation, and occurrence of minerals in soils introduction: The role of mineralogy in soil science. In *Handbook of Soil Science*; CRC Press: Boca Raton, FL, USA, 2012; Volume 1, pp. 33–48.
64. Saklar, S.; Yorukoglu, A. Effects of acid leaching on halloysite. *Physicochem. Probl. Miner. Process.* **2015**, *51*, 83–94.
65. Ambikadevi, V.R.; Lalithambika, M. Effect of organic acids on ferric iron removal from iron-stained kaolinite. *Appl. Clay Sci.* **2000**, *16*, 133–145. [CrossRef]
66. Bediako, E.G.; Nyankson, E.; Dodoo-Arhin, D.; Agyei-Tuffour, B.; Łukowiec, D.; Tomiczek, B.; Yaya, A.; Efavi, J.K. Modified halloysite nanoclay as a vehicle for sustained drug delivery. *Heliyon* **2018**, *4*, e00689. [CrossRef] [PubMed]
67. Abdullayev, E.; Price, R.; Shchukin, D.; Lvov, Y. Halloysite tubes as nanocontainers for anticorrosion coating with benzotriazole. *ACS Appl. Mater. Interfaces* **2009**, *1*, 1437–1443. [CrossRef] [PubMed]
68. Shu, Z.; Zhang, Y.; Ouyang, J.; Yang, H. Characterization and synergetic antibacterial properties of ZnO and CeO2 supported by halloysite. *Appl. Surf. Sci.* **2017**, *420*, 833–838. [CrossRef]
69. Korsmeyer, R.W.; Gurny, R.; Doelker, E.; Buri, P.; Peppas, N.A. Mechanisms of solute release from porous hydrophilic polymers. *Int. J. Pharm.* **1983**, *15*, 25–35. [CrossRef]
70. Yuan, P.; Southon, P.D.; Liu, Z.; Kepert, C.J. Organosilane functionalization of halloysite nanotubes for enhanced loading and controlled release. *Nanotechnology* **2012**, *23*, 375705. [CrossRef]
71. Yuan, P.; Southon, P.D.; Liu, Z.; Green, M.E.R.; Hook, J.M.; Antill, S.J.; Kepert, C.J. Functionalization of Halloysite Clay Nanotubes by Grafting with γ-Aminopropyltriethoxysilane. *J. Phys. Chem. C* **2008**, *112*, 15742–15751. [CrossRef]
72. Joo, Y.; Sim, J.H.; Jeon, Y.; Lee, S.U.; Sohn, D. Opening and blocking the inner-pores of halloysite. *Chem. Commun.* **2013**, *49*, 4519–4521. [CrossRef]

73. Giancola, S.; Zatoń, M.; Reyes-Carmona, Á.; Dupont, M.; Donnadio, A.; Cavaliere, S.; Rozière, J.; Jones, D.J. Composite short side chain PFSA membranes for PEM water electrolysis. *J. Memb. Sci.* **2019**, *570–571*, 69–76. [CrossRef]
74. Baker, A.M.; Babu, S.K.; Mukundan, R.; Advani, S.G.; Prasad, A.K.; Spernjak, D.; Borup, R.L. Cerium ion mobility and diffusivity rates in perfluorosulfonic acid membranes measured via hydrogen pump operation. *J. Electrochem. Soc.* **2017**, *164*, F1272–F1278. [CrossRef]
75. Park, J.; Kim, D. Effect of cerium/18-crown-6-ether coordination complex OH quencher on the properties of sulfonated poly(ether ether ketone) fuel cell electrolyte membranes. *J. Memb. Sci.* **2014**, *469*, 238–244. [CrossRef]

membranes

MDPI

Article

Energy Harvesting from Brines by Reverse Electrodialysis Using Nafion Membranes

Ahmet H. Avci [1], Diego A. Messana [1,2], Sergio Santoro [1], Ramato Ashu Tufa [3], Efrem Curcio [1,4], Gianluca Di Profio [2,4] and Enrica Fontananova [2,*]

1 Department of Environmental Engineering, University of Calabria, 87036 Rende (CS), Italy; ahmethalilavci@hotmail.com (A.H.A.); diego.messana.ded@gmail.com (D.A.M.); sergio.santoro@unical.it (S.S.); efrem.curcio@unical.it (E.C.)
2 Institute on Membrane Technology of the National Research Council (ITM-CNR), at University of Calabria, 87036 Rende (CS), Italy; g.diprofio@itm.cnr.it
3 Department of Energy Conversion and Storage, Technical University of Denmark, Building 310, 2800 Kgs. Lyngby, Denmark; rastu@dtu.dk
4 SELIGENDA Membrane Technologies SrL, 87036 Rende (CS), Italy
* Correspondence: e.fontananova@itm.cnr.it; Tel.: +39-0984-492010; Fax: +39-0984-402103

Received: 23 June 2020; Accepted: 24 July 2020; Published: 28 July 2020

Abstract: Ion exchange membranes (IEMs) have consolidated applications in energy conversion and storage systems, like fuel cells and battery separators. Moreover, in the perspective to address the global need for non-carbon-based and renewable energies, salinity-gradient power (SGP) harvesting by reverse electrodialysis (RED) is attracting significant interest in recent years. In particular, brine solutions produced in desalination plants can be used as concentrated streams in a SGP-RED stack, providing a smart solution to the problem of brine disposal. Although Nafion is probably the most prominent commercial cation exchange membrane for electrochemical applications, no study has investigated yet its potential in RED. In this work, Nafion 117 and Nafion 115 membranes were tested for NaCl and NaCl + $MgCl_2$ solutions, in order to measure the gross power density extracted under high salinity gradient and to evaluate the effect of Mg^{2+} (the most abundant divalent cation in natural feeds) on the efficiency in energy conversion. Moreover, performance of commercial CMX (Neosepta) and Fuji-CEM 80050 (Fujifilm) cation exchange membranes, already widely applied for RED applications, were used as a benchmark for Nafion membranes. In addition, complementary characterization (i.e., electrochemical impedance and membrane potential test) was carried out on the membranes with the aim to evaluate the predominance of electrochemical properties in different aqueous solutions. In all tests, Nafion 117 exhibited superior performance when 0.5/4.0 M NaCl fed through 500 μm-thick compartments at a linear velocity 1.5 $cm \cdot s^{-1}$. However, the gross power density of 1.38 $W \cdot m^{-2}$ detected in the case of pure NaCl solutions decreased to 1.08 $W \cdot m^{-2}$ in the presence of magnesium chloride. In particular, the presence of magnesium resulted in a drastic effect on the electrochemical properties of Fuji-CEM-80050, while the impact on other membranes investigated was less severe.

Keywords: reverse electrodialysis; Nafion; brine

---

1. Introduction

The increasing demand for water and energy requires sustainable and environmentally friendly solutions. Therefore, the old-fashioned linear approach ("take, make and dispose") gives way to the circular economy approach in which any waste is potentially considered as a valuable source for another process. In this regard, reverse electrodialysis (RED) is a promising electromembrane-based process that harvests the Gibbs free energy of mixing of solutions with different salinity [1]. For example, although the brine solution coming from seawater desalination is currently considered as a waste, thanks to its high salinity it can be exploited as a valuable source for RED [2–4].

A typical RED unit (Figure 1) is similar to an electrodialysis (ED) unit, a well-established and commercialized technology. However, the operating conditions of RED are different. The inputs to ED are a feed solution and the electrical energy, producing separately a concentrate and a dilute. On the other hand, the inputs to RED are a concentrated solution and a dilute solution, mixed together in a controlled manner to produce spontaneously electrical energy [5]. In a RED stack, alternately arranged cation exchange membranes (CEMs) and anion exchange membranes (AEMs) are separated by spacers and piled up in a repetitive organization. When feeding concentrated and diluted solutions throughout the channels created by spacers, a Nernst potential is generated which drives the ions from high electrochemical potential to low electrochemical potential. However, only counter-ions (oppositely charged ions with respect to fixed charge groups of ion exchange membranes) can diffuse through IEM, while co-ions (having the same charge of IEM) are retained. As a result, a steady ion flux occurs between adjacent compartments. Utilization of the appropriate electrolyte solution and electrode couple at the end of compartments allows the transformation of this ion flux into an electric current [6].

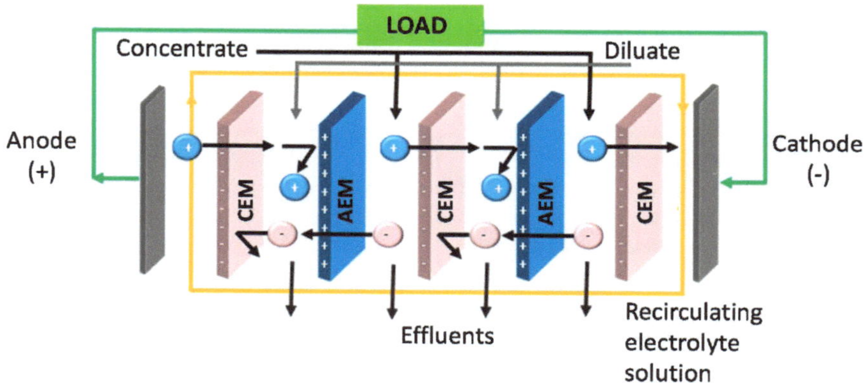

Figure 1. Illustration of the RED process.

IEMs are one of the most important components of a RED stack: in order to maximize generated power, high permselectivity and ion conductivity are essential. Beside these two properties, adequate mechanical strength and low cost are also desired. Moreover, the use of sustainable membrane production protocols for optimizing the green benefits of advanced separation techniques is a key issue of the modern membrane industry [7,8].

So far, numerous researchers readapted IEMs designed for other electrochemical processes (i.e., electrodialysis) to RED process [9]. Due to their high costs, perfluorosulfonic acid polymer electrolyte membranes were not tested before in reverse electrodialysis applications although these membranes are widely used for many applications such as chlor-alkali electrolysis [10,11], water electrolysis [12,13], polymer electrolyte fuel cells [14,15].

The chemical structure of Nafion, one of the most commercially relevant perfluorosulfonic acid polymers, is shown in Figure 2 [16]. It is synthesized by perfluorinated vinyl ether comonomer and tetrafluoroethylene copolymerization. The resulting polymer has outstanding long-term chemical and thermal stability. Beside its stability, previous researchers revealed notable permselectivity and conductivity of Nafion membranes in NaCl solutions [17–19].

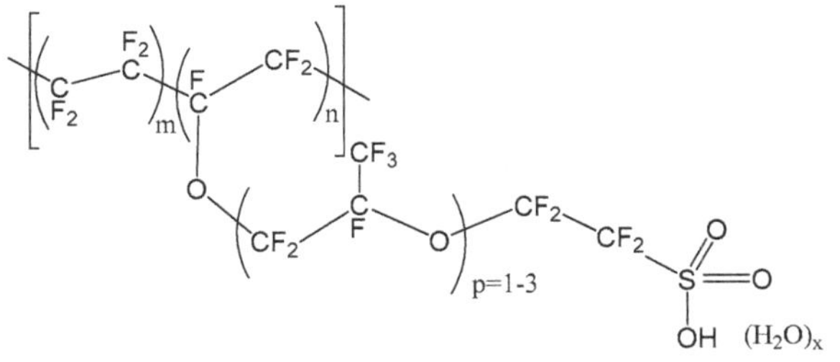

Figure 2. Chemical structure of Nafion.

One of the main disadvantages of Nafion membranes is the high cost. For instance, Nafion 117 price was stated between \$1400/m^2 and \$2200/m^2 [20]; Yee et al. (2012) reported a normalized cost of Nafion 117 and Nafion 115 of \$3800/m^2 and \$3100/m^2, respectively [21]. Processing huge volumes of solutions with different salinity requires a large membrane area in RED. Consequently, an elevated capital cost makes the operation economically infeasible. According to Daniilidis et al. (2014), for a 2.7 W·m^{-2} power-producing RED stack having a competitive levelized cost of electricity (LCE) with conventional renewable technologies, the cost of IEM must be around 4 €·m^{-2} [22]. Although the current price of Nafion is far from this estimation, a reduction is projected for large scale production and technological improvements [23].

Toupin et al. (2016) carried out a study on the cost of Nafion and other perfluorinated sulfonic acid (PFSA) polymer electrolyte membranes to use in fuel cell vehicles; in particular, the cost of membranes was estimated in the case of a different annual production rate for melt blowing and e-PTFE solution cast methods (Figure 3) [24]. Both methods were able to reduce the cost of the membranes by approximately two orders of magnitude while the melt blowing method of production resulted in superior value of 7.7 €·m^{-2} for 5 million m^2 annual production [24]. Moreover, this value is expected to get lower with increasing technological maturation and production rates.

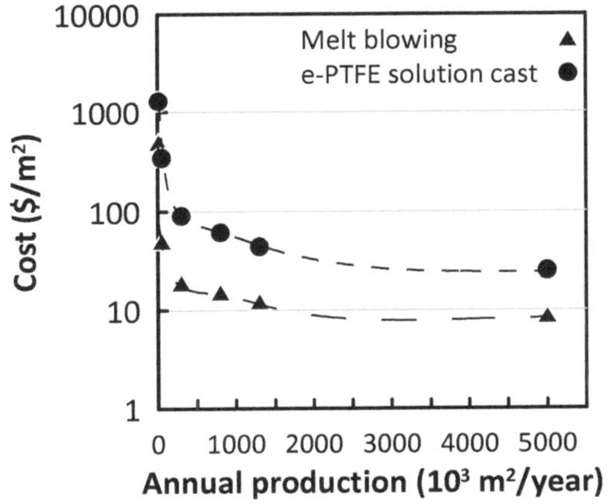

Figure 3. Cost estimation of perfluorinated sulfonic acid polymer electrolyte membrane produced by two different methods as a function of the total yearly production (data from [24]).

Other commercial membranes designed for electrochemical processes that are suitable for electrodialysis or RED are expected to be acquired for a lower price compared to Nafion. In this work, CMX Neosepta from Astom Corporation (Tokyo, Japan) and Fuji-CEM-80050 from Fujifilm Manufacturing Europe B.V (Tilburg, the Netherlands) cation exchange membranes were considered as the benchmark for their frequent use in RED application as cation exchange membranes. Unlike

Nafion, these membranes are non-perflourinated based membranes. Although the information about preparation methods of these commercial membranes is limited in the literature, it is known that Fujifilm cation exchange membranes have an aliphatic polyamide backbone with sulfonic groups reinforced by uncharged polyolefin support [25,26]. On the other hand, CMX membranes are prepared by the so-called "paste method": a paste embedded into polyvinyl chloride fabric contains sulfonated styrene monomer, a crosslink agent (i.e., divinylbenzene), polymerization initiator and polyvinyl chloride [26,27].

Different saline solutions have mixed in RED to produce electricity, mainly: fresh water/seawater [5,22,23], seawater/brine [28,29], seawater/groundwater [30], brackish water/brine [31–33]. Among them, mixing seawater and brine (the waste from the desalination process) solutions have operational, economic, and environmental advantages. Due to low electrolyte concentration in fresh water and brackish water, the conductivity of these solutions limits the efficient transportation of ions within a compartment. Conversely, utilizing seawater in the low concentration compartments reduces the total stack resistance and increases the generated power.

Studies over the past decade have provided important information on seawater/brine mixing by RED. Daniilidis et al. (2014) investigated the performance of Neosepta CMS and ACS in RED for a wide range of NaCl solutions: for 0.5 M/5.0 M NaCl mixing, 1.5 $W \cdot m^{-2}$ gross power density was detected. It is noteworthy that permselectivity was about 20% lower when compared to 0.1/0.5 M NaCl feed, while a fourfold decrease in stack resistance was observed compared to 0.01 M/0.5 M NaCl feed [22]. In a study on an integrated membrane distillation–reverse electrodialysis system, for a RED stack equipped with Fuji-CEM 80050 and AEM 80045 and operated with 0.5 M/4.0 M NaCl feed, Tufa et al. (2015) measured an open circuit voltage (OCV), a stack resistance (R_{stack}) and a gross power density (P_d) of 1.25 V, 7 $\Omega \cdot cm^2$ and 0.9 $W \cdot m^{-2}$, respectively [4].

With a share of ~10%, magnesium is the second most abundant cation in seawater [34]. Despite its importance, the effect of Mg^{2+} on RED performance at high salinity is still poorly investigated [35]. In one of these studies, Avci et al. (2016) observed a 20% and a 60% reduction of OCV and power density, respectively, when 10% molal $MgCl_2$ was present in feed solutions. It was also noted that the reason for significant power loss can be attributed to tripled resistance of Fuji-CEM-80050, while no notable change was observed for Fuji-AEM-80045 [36]. Similarly, Fontananova et al. (2017) compared the electrochemical properties of the abovementioned membranes in analogous operative conditions: 40% loss in permselectivity and 3.5 times higher resistance was observed for CEM, while AEM permselectivity decreased only by 16% with resistance remaining almost stable. Consequently, gross power density reduced from 0.96 to 0.67 $W \cdot m^{-2}$ [37].

The main purpose of this study is to characterize the electrochemical properties of Nafion membranes for RED operations carried out at high salinity gradients and compare them with commercially available non-perflourinated membranes frequently utilized in RED. For this reason, Nafion 117, Nafion 115, CMX and Fuji-CEM-80050 were characterized by electrochemical impedance spectroscopy (EIS) and potential cell for NaCl and NaCl + $MgCl_2$ solutions with ionic strengths mimicking seawater (0.5 $mol \cdot kg^{-1}$) and hypersaline brine (4.3 $mol \cdot kg^{-1}$).

To the best of our knowledge, this is the first work in which Nafion membranes were tested in a RED stack, although numerous works were carried out in different fields such as fuel cells and chlor-alkali processes.

2. Materials and Methods

2.1. Feed and Electrolyte Solutions

Three different solutions are required to operate reverse electrodialysis: a high concentration compartment (HCC) solution, a low concentration compartment (LCC) solution, and an electrolyte compartment solution. Solutions for RED experiments and electrochemical characterization were prepared by dissolving appropriate amounts of NaCl, $MgCl_2 \cdot 6H_2O$, $K_4[Fe(CN)_6] \cdot 3H_2O$ and

$K_3[Fe(CN)_6]$ (supplied by Sigma-Aldrich, Milan, Italy) in deionized water (0.055 μS·cm^{-1}, produced by PURELAB, Elga LabWaters, High Wycombe, United Kingdom). The compositions of the solutions are reported in Table 1.

Table 1. Concentration and ionic strength of feed and electrolyte solutions used in RED.

Compartment	Composition	Ionic Strength (mol·kg^{-1})
LCC	0.5 M NaCl	0.51
HCC	4.0 M NaCl	4.3
LCC	0.34 M NaCl + 0.054 M $MgCl_2$	0.51
HCC	2.7 M NaCl + 0.43 M $MgCl_2$	4.3
Electrolyte	0.3 M $K_4[Fe(CN)_6]$ + 0.3 M $K_3[Fe(CN)_6]$ + 2.5 M NaCl	7.3

The ionic strength I (mol/kg of water) of a solution is calculated as:

$$I = \frac{1}{2}\sum m_i z_i^2 \quad (1)$$

where m_i and z_i are the molality (i.e., moles solute per kg of solvent) and the charge of the i-th ion, respectively.

A 3M KCl (Sigma-Aldrich, Milan, Italy) was prepared to fill Haber-Luggin capillaries in EIS measurements.

2.2. Membranes

Nafion 117 and Nafion 115 were purchased from Quintech (Göppingen, Germany). Fuji CEM 8005C (hereinafter referred to as "Fuji-CEM") was kindly supplied by FujiFilm Manufacturing Europe B.V. ((Tilburg, the Netherlands). Neosepta CMX and AMX were kindly supplied by Eurodia (Pertuis, France).

All dry membrane samples were initially activated in 0.5 M NaCl solution. Additionally, they were conditioned in the specific test solution before use. For example, prior to the electrochemical impedance characterization in 4 M NaCl, membrane samples were immersed in this solution for at least 24 h and the solution was changed at least 3 times during this period.

2.3. Membrane Permselectivity

The membrane potential was measured by using two Ag/AgCl reference electrodes (Gamry Instruments, Warminster, PA, US) as in Figure 4. DC voltage drop across the membrane was recorded by a digital multimeter in the range of 0 to 600 mV (Fluke 117, Fluke Corporation, Everett, WA, US). Membrane potential of CEMs were characterized in two different solution pairs: 0.5/4.0 M NaCl and 0.34 M NaCl + 0.054 M $MgCl_2$/2.72 M NaCl + 0.43 M $MgCl_2$. Test solutions were kept at 25 ± 3 °C and fed to the cell at a flow rate of 1.5 cm·s^{-1}.

After obtaining the membrane potential experimentally (ΔV_{exp}), permselectivity (α) was calculated by taking the ratio to theoretical membrane potential (ΔV_{theo}):

$$\alpha = \frac{\Delta V_{exp}}{\Delta V_{theo}} \quad (2)$$

The theoretical membrane potential was calculated by the Nernst equation [38]:

$$\Delta V_{theo} = \sum \frac{RT}{z_i F} ln \frac{\gamma_i^c c_i^c}{\gamma_i^d c_i^d} \quad (3)$$

where R is the universal gas constant (8.3144 J·K^{-1}mol^{-1}), T is the temperature (K), z is the valence number (−), F is the Faraday constant (96485 C·mol^{-1}), γ is the activity coefficient and c is the molality. Subscript i stands for the component type, while superscripts c and d refer to the concentrated solution and the diluted solution, respectively. The activity coefficients were calculated by the interpolation from experimental values [39].

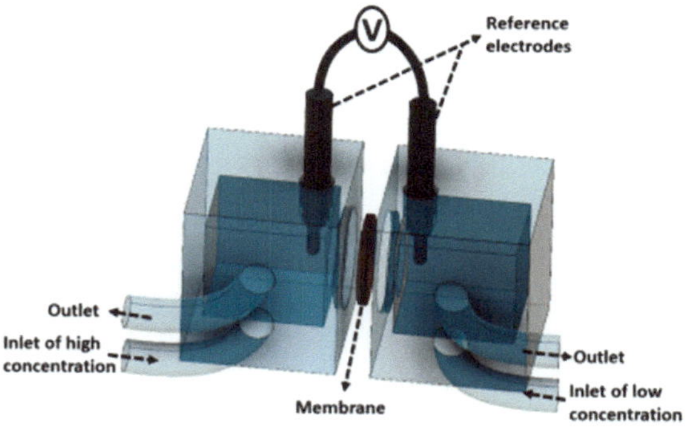

Figure 4. Scheme of the two compartments/two electrodes cell used for membrane potential measurements.

2.4. *Electrochemical Impedance Spectroscopy (EIS)*

In order to characterize ohmic and nonohmic resistance of a membrane-solution system, EIS experiments were carried out with a potentiostat/galvanostat combined with a frequency response analyzer (PGSTAT302N, Metrohm Autolab B.V., Utrecht, the Netherlands). As it is shown in Figure 5, a specifically designed four-electrode configuration was used in the impedance cell with 3.14 cm^2 active membrane area [40]. An alternating current in the frequency range 1000–0.01 Hz, with a signal amplitude of 10 mV, was applied between working and counter electrodes (made of Ag), while the response (voltage drop) was measured by the reference electrodes immersed in the Haber–Luggin capillaries containing 3M KCl solution.

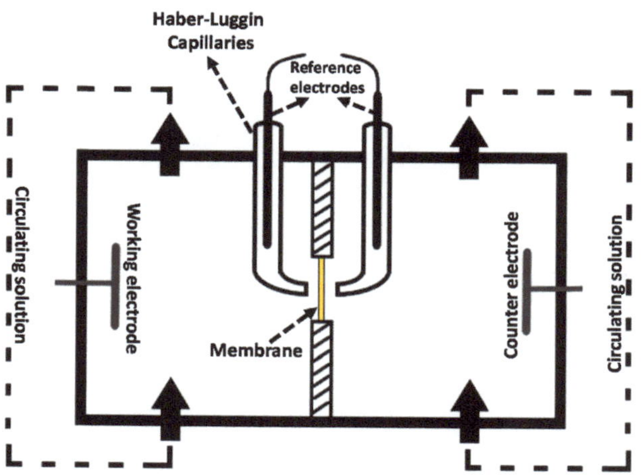

Figure 5. Scheme of the two compartments/four electrodes electrochemical impedance cell.

The response of the membrane solution system was plotted into a Nyquist diagram and fitted to the equivalent circuit shown in Figure 6, generated by the software Nova 1.9.16 (from Metrohm Autolab B.V., Utrecht, the Netherlands). The membrane-solution resistance is an ohmic resistance obtained from the intersection point of the curve and $-Z'' = 0$ at high frequency. In order to calculate stand-alone membrane resistance, repetition of the experiment under the same conditions without the membrane was required; by subtracting solution resistance from the membrane-solution resistance, membrane resistance could be determined. On the other hand, electrical double layer resistance (EDL) and the diffusion boundary layer (DBL) cannot be modelled by only resistance due to their electrochemical nature. As illustrated in Figure 6, EDL consists of a resistance and a capacitance in parallel while DBL consists of a resistance and a constant phase element in parallel. Both of them appeared as a semi-circle in the Nyquist plot at different frequency ranges: EDL at medium frequencies whereas DBL at low frequencies.

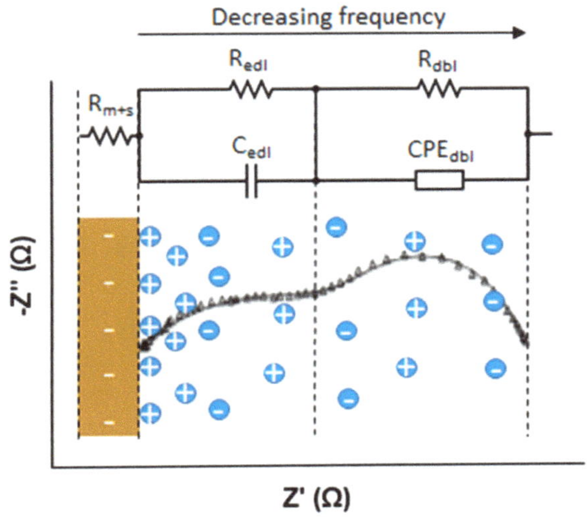

Figure 6. Equivalent circuit of a membrane-solution system impedance on the Nyquist diagram.

The EIS experiments were carried out at 25 °C and 1.5 cm·s^{-1} by circulating the LCC and HCC solutions individually.

2.5. Water Uptake, Ion Exchange Capacity and Fixed Charge Density

Membrane water uptake (WU) was calculated by weighing the membrane swelled in 0.5 M NaCl solution ($w_{swelled}$) and dry membrane (w_{dry});

$$WU\% = \frac{w_{swelled} - w_{dry}}{w_{dry}} \cdot 100 \tag{4}$$

Ion exchange capacity of CEMs were calculated as reported previously [37]. In order to saturate negative fixed charge groups of CEMs, samples were kept in excess 1 M HCl solution overnight; then to remove all uncoupled H^+ present in the surface water, the samples were washed with demi-water. Following this, H^+ ions were exchanged with Na^+ ions by immersing the samples into 40 mL of 2 M NaCl. Finally, the immersed solutions were collected into a beaker and titrated with 0.01 M NaOH. The pH values were monitored with a pH meter (WTW Inolab Terminal Level 3, Weilheim, Germany). The *IEC* (meq·g dry membrane^{-1}) was calculated by using the following equation:

$$IEC = \frac{V_{NaOH} \cdot M_{NaOH}}{m_{dry}} \tag{5}$$

in which V_{NaOH} is the volume of NaOH titrant (l), M_{NaOH} is the molarity of NaOH titrant (mol·L^{-1}) and m_{dry} is the dry weight of the sample (g) after washing with water and leaving in an oven at 70 °C overnight.

The fixed charge density (C_{fix}) was calculated by using water uptake, *IEC* values and water density at 25 °C (d_w):

$$C_{fix} = \frac{IEC \cdot d_w}{wu\%} \cdot 100 \quad (6)$$

2.6. Reverse Electrodialysis

The lab-scale electrodialysis cell PCCell 200, provided by PCCell GmbH (Heusweiler, Germany), was used in reverse electrodialysis mode to characterize electrochemical performance of the stack equipped with the aforementioned CEMs paired with AMX Neosepta. CEMs were cut into 26.2 × 12.5 cm^2 pieces to fit 500 µm thick spacers for 207 cm^2 total active area. The electrode compartments included anode and cathode made of inert Pt/Ir-coated titanium mesh. The electrode compartments were separated from the central compartments by CMX membranes. Between the central membrane and these CMX membranes, AMX membranes were utilized as anion exchange membranes.

The performance of the RED unit was investigated at 25 °C and the linear flow velocity of the concentrated and diluted compartments was 1.5 cm·s^{-1}. Flowrate of electrolyte solution was fixed to 30 L·h^{-1}. Solutions were fed by Masterflex L/S digital peristaltic pumps (Cole-Palmer, Vernon Hills, IL, US) and conditioned to the desired temperature by a refrigerated/heated circulating bath (PolyScience, Niles, IL, US) before entering the stack. Two different salinity gradients were tested: 0.5 M/4.0 M NaCl and 0.34 M NaCl + 0.054 M $MgCl_2$/2.7 M NaCl + 0.43 M $MgCl_2$ Table 1.

The current (I) versus voltage (V) curve, that is linear coherently with Ohmic law, was plotted by applying DC current by Methrom Autolab in the range of 0–32 A/m^2. Open circuit voltage (OCV) was obtained from both fitted data (at I = 0 A) and experimental measurements, while stack resistance (R_{stack}) was calculated from the slope of I-V curve. Then, gross power density (P_d, W·m^2) and current density (A_d, I·m^2) were determined and fitted as a parabola.

In line with the ohmic behavior of RED, gross power density $P_{d,max}$ is proportional to the OCV^2 and reversely proportional to R_{stack}:

$$P_{d,max} = \frac{OCV^2}{4N \cdot R_{stack}} \quad (7)$$

The maximum power density ($P_{d,max}$) was calculated from the maximum of parabola.

3. Results

Ion exchange membranes have a great importance for energy conversion from salinity gradients by reverse electrodialysis [41]: the power potential of a RED unit, estimated from OCV and R_{stack}, is strictly related to permselectivity and electrical membrane resistance. In turn, these properties are interrelated to other characteristics, i.e., thickness, ion exchange capacity (IEC), water uptake (WU), and fixed charge density (C_{fix}) (Table 2). It is difficult to have a straightforward comment on the effect of a single IEM property due to strong interconnections and counteractions among all of them. For example, high IEC is a way to reduce the resistance. However, since water uptake increases with increasing IEC, the concentration of fixed charged groups attached to the polymeric matrix decreases, thus reducing permselectivity. A significant increase of IEC also results in swollen and mechanically weak membranes.

Table 2. Relevant physical and electrochemical properties of membranes at 25 °C.

Membrane	Thickness (µm)	IEC (meq·g^{-1})	Water Uptake (%)	Charge Density (mol·L^{-1})
Nafion 115	139 ± 8	0.90 *	11.2 ± 0.02	8.0
Nafion 117	201 ± 4	0.90 *	11.7 ± 0.01	7.7
Fuji-CEM-80050	114 ± 2	1.1 ± 0.1	34.0 ± 0.00	3.2
CMX	166 ± 1	1.61 ± 0.03	25.5 ± 0.1	6.3 ± 0.23

* From the manufacturer.

The thickness of CEMs used in this study ranges between 114–201 μm, which is typical for CEMs used previously for RED [42]. Even though the thickness and the ionic resistance are proportional, thinner membrane does not necessarily perform better. Tedesco et al. (2018) carried out experiments with FAS and FKS Fumasep (FUMATECH BWT GmbH, Bietigheim-Bissingen, Germany) membranes with varying thickness between 14–90 μm, and concluded having thinner membranes was not beneficial for maximum power density [43].

Referring to Table 2, although Nafion 115 and 117 membranes exhibited relatively lower IEC, low water uptake made C_{fix} superior compared to the investigated benchmark membranes. Conversely, high WU and moderate IEC of the Fuji-CEM membrane resulted in the lowest C_{fix}.

3.1. Electrochemical Properties of CEMs

3.1.1 Permselectivity

For ion exchange membranes, the permselectivity is an indication of the ability to selectively transport counter-ions over co-ions. To be able to control the mixing of ions in a preferred direction during a RED process, a permselectivity higher than 0.95 is desired [44]. Most of the reported commercial CEMs have acceptable permselectivity in this regard. However, generally, permselectivity characterization is carried out in 0.1/0.5 M NaCl or KCl, which is not representative for high concentration or complex solutions with multivalent ions used in real cases. Therefore, in this study, permselectivity of Nafion 115, Nafion 117, CMX and Fuji-CEM were characterized for concentrated (ionic strength > 0.5 mol·kg^{-1}) and multicomponent solutions (Table 1).

Figure 7 compares the permselectivity (α) of the membranes at 25 °C. In standard 0.1/0.5 M NaCl test solution pairs, all membranes performed satisfactorily enough for a RED application; Nafion membranes characterized as ideal (1.0) while CMX resulted in almost ideal (0.99) and Fuji-CEM had sufficient permselectivity (0.94). Having C_{fix} around 8 mol·L^{-1}, Nafion membranes exhibited high co-ion exclusion with pure NaCl solutions, even when one side of the membrane was in touch with 4.0 M NaCl (α = 0.88) whereas the permselectivity of CMX and Fuji-CEM membranes was 8% and 10% lower, respectively. This deviation from unity is in accordance with the previous literature data [28,37]. The co-ion equilibrium in an ion exchange membrane for ideal monovalent electrolyte can be expressed as the following equation:

$$C_{co}^{m} = \frac{C_{co}^{2}}{C_{fix}^{m}} \tag{8}$$

where C is the concentration, m stands for membrane, subscript "*co*" and "*fix*" are co-ion and fixed charge, respectively. From Equation (8), it can be deduced that a low fixed charge concentration leads to a lack of co-ion exclusion when the membrane is exposed to a high concentration of electrolyte.

A more detrimental effect on permselectivity was observed with the introduction of $MgCl_2$ to the electrolyte solution. The losses of permselectivity were recorded between 32–38%; the lowest permselectivity (α = 0.49) was measured for Fuji-CEM. In general, this drastic reduction can be explained by investigating the binding affinity of counter-ions [45]. With an increasing binding affinity of a counter ion/fixed charge group, the possibility of condensation of the counter-ion increases. Therefore, counter-ion concentration in the ionic state decreases as long as the neutralization of the fixed charge groups occurs. Consequently, co-ion transport across the membrane increases. According to Luo et al. (2018), the divalent cations affinity to sulfonic groups is significantly higher than the Na^+ [46].

3.1.2. Electrochemical Impedance

The total RED stack resistance consists of CEMs, AEMs, HCC, LCC and electrolyte compartment resistances [47]. When RED is operated with seawater and river water, LCC resistance dominates the total resistance [48]. However when using high concentrated feed solutions, the contribution of ionic membrane resistance becomes critical in understanding the RED performance [49]. Electrochemical impedance test is a powerful technique to quantify not only the membrane resistance, but also the boundary layer resistance at the membrane solution interface. Figure 8 illustrates the resistance of four CEMs against four different concentrations. For 0.5 M NaCl, the lowest resistance (1.50 $\Omega \cdot cm^2$) was measured for Nafion 115, while 15% higher resistance was obtained for Nafion 117. The resistance of CMX and Fuji-CEM were 2.20 and 2.41 $\Omega \cdot cm^2$, respectively. These findings are comparable with those of Fontananova et al. [40] and Galama et al. (2016) [49] who measured Fuji-CEM and CMX resistance as 2.97 and 2.58 $\Omega \cdot cm^2$, respectively. Although thickness and water uptake values are in favor of Fuji-CEM compared to the others, the low charge density determines relatively high resistance.

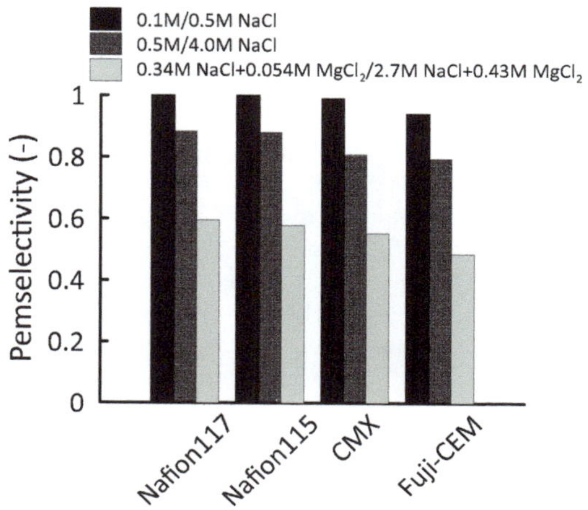

Figure 7. Permselectivity of Nafion 117, Nafion 115, Neosepta CMX, Fuji-CEM-80050 in −0.1/0.5 M NaCl, 0.5/4.0 M NaCl and 0.34 M NaCl + 0.054 M $MgCl_2$/2.7 M NaCl + 0.43 M $MgCl_2$ at 25 °C.

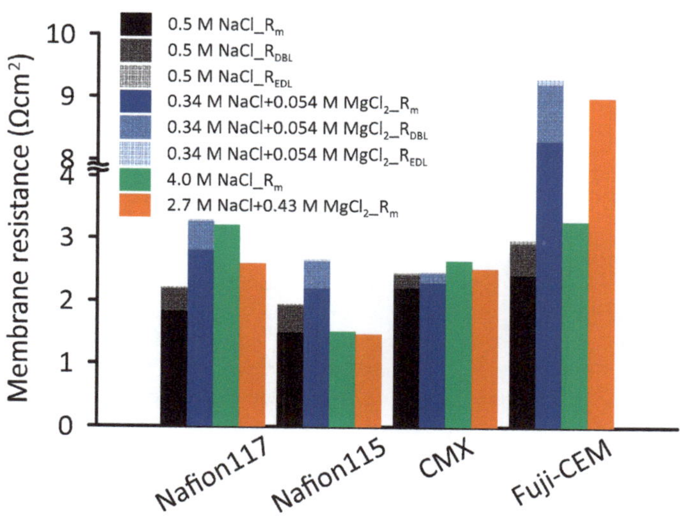

Figure 8. Ionic resistance of Nafion 117, Nafion 115, CMX, Fuji-CEM-80050 in 0.5 M NaCl, 4.0 M NaCl, 0.34 M NaCl + 0.054 M $MgCl_2$ and 2.7 M NaCl + 0.43 M $MgCl_2$ at 25 °C.

Interestingly, at 4.0 M NaCl, the resistance of Nafion 115 remained constant compared to the resistance at 0.5 M NaCl, whereas other CEM resistances increased 20–75%. It seems possible that these results are due to the decreasing water content in CEMs [37]: with decreasing water uptake, the interstitial channels through the membrane cross section get narrower, so the ionic transfer is impeded.

Due to the significant amount of magnesium in seawater, characterizing the resistance of CEMs in presence of Mg^{2+} ions by EIS provides deeper understanding on RED performance. EIS tests were carried out in NaCl + $MgCl_2$ solutions having ionic strength equal to 0.5 M and 4.0 M pure NaCl solutions (Table 1). As shown in Figure 8, the presence of magnesium in the test solution significantly affected the Fuji-CEM conductivity: the resistance increased by 3.4 and 2.7 times when the ionic strength was 0.51 mol·kg^{-1} and 4.3 mol·kg^{-1}, respectively. The ionic conductivity of Nafion membranes was halved while no significant effect was observed on CMX at ionic strength of 0.51 mol·kg^{-1}. The observed increase in resistance could be attributed to the binding affinity of Mg^{2+}, as in the case of permselectivity. As discussed by Cassady et al. (2016), counter-ions in the membrane lattice can exist as a solvated pair or as a condensed salt, having prevalence of the latter form when binding affinity is higher. Consequently, a fixed charge in a condensed salt is electrically neutralized, does not facilitate the counter ion transport anymore, and IEM conductivity reduces [45].

EIS results revealed that, at high salinity, the R_{EDL} and R_{DBL} were insignificant compared to R_m. Conversely, for 0.5 M NaCl, interfacial (nonohmic) resistances contributed by 10–23% to the total resistance, while in the presence of magnesium at equivalent ionic strength of 0.51 mol·kg^{-1}, the contribution varied from 7% to 17%.

At lower concentrations, the nonohmic resistance is more significant and its contribution can reach 50% [37,50] For example, in 0.1M NaCl solution, the total resistance of Fuji-CEM 80050 resulted in around 4.6 Ω·cm^2 in which approximately 2.4 Ω·cm^2 was contributed by the diffusion boundary layer and electrical double layer resistances [50].

With respect to the extent of nonohmic resistances, CMX was found less prone under the investigated conditions. In general, R_{DBL} was the dominant nonohmic resistance with more than 90% for all CEMs.

In order to diminish the effect of the diffusion boundary layer, several studies were focused on enhancing the fluid mixing in feed compartments. In one of these studies, Guler et al. (2014) prepared microstructured membranes in order to eliminate the usage of spacers: increasing flow rate from 2 to 40 mL·min^{-1} resulted in minimal nonohmic resistance [51]. However, it should be noted that increasing the flow rate leads to a reduction of net power density. Vermaas et al. (2011) investigated the net power density of a RED stack equipped with FKS and FAS (Fumatech) membranes by using different spacer thickness (60, 100, 200 and 485 μm). Each spacer resulted in its maximum at a different Reynold number; for example when 100 μm-thick spacers were used, maximum net power density was measured for $R_e = 0.5$ while maximum gross power density was obtained for $R_e \sim 2.0$ [48].

3.1.3. Reverse Electrodialysis Performance

Figure 9 illustrates the comparison from electrochemical tests of four different commercial CEMs utilized in the RED stack for solutions at two different compositions as detailed in Table 1.

From the current-voltage curve (Figure 9a,b), the OCV was in the range of 0.167–0.171 V when using pure NaCl solution; the addition of $MgCl_2$ resulted in a slight narrowing of this range, i.e., 0.160–0.164 V. In both cases, the decreasing order of OCV was Nafion 117 > Nafion 115 > CMX > Fuji-CEM.

Using a single membrane pair to test RED is the main reason for such a slight variation of OCV; a higher number of membrane pairs enhances the voltage drop across the stack and makes this difference explicit. Even so, the OCV of the RED stack was in line with the permselectivity of CEMs.

The total stack resistance consists of individual resistances that constitute the RED system:

$$R_{stack} = N(R_{CEM} + R_{AEM} + R_{HCC} + R_{LCC}) + R_{EL} \quad (9)$$

where N is the number of the membrane pair and subscript EL stands for electrolyte. In most cases, when large numbers of membrane pairs are used, R_{EL} is neglected; in this study, being RED operated for a single cell ($N = 1$), this effect has to be considered. Therefore, R_{stack} lined up very close for all CEMs when feeding 0.5 M/4.0 M NaCl to RED: the lowest and the highest measured R_{stack} were 0.1709 and 0.1818 Ω for Nafion 117 and Fuji-CEM, respectively. On the other hand, when feeding 0.34 M NaCl + 0.054 M $MgCl_2$/2.7 M NaCl + 0.43 M $MgCl_2$ solutions, a 35% increase in the R_{stack} of Fuji-CEM was detected; this increase was limited to 8–15% for the other investigated membranes. This finding corroborates the results obtained in EIS characterization.

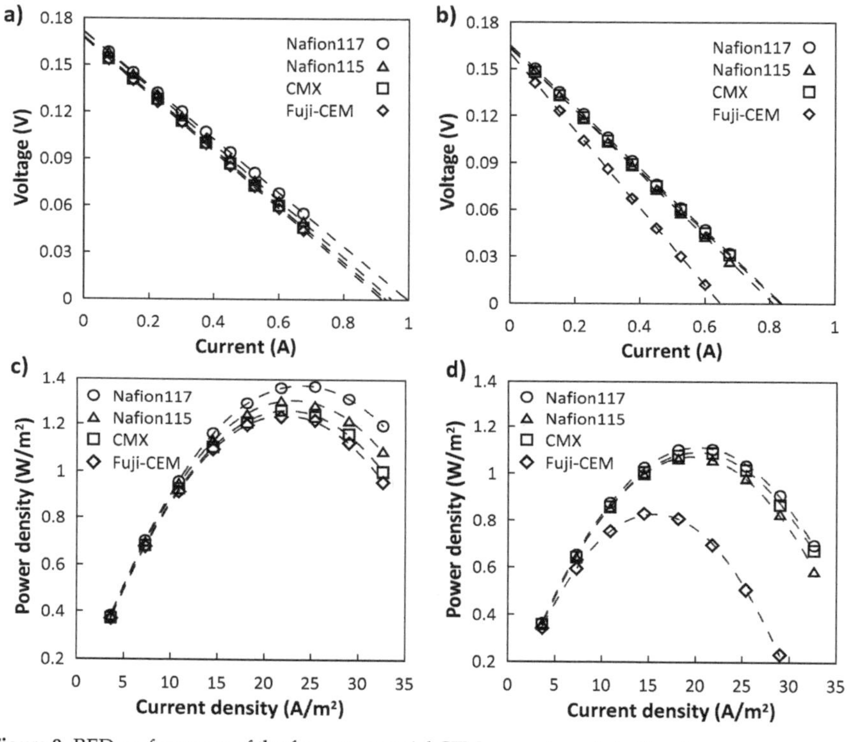

Figure 9. RED performance of the four commercial CEMs investigated: (**a**,**b**) voltage versus current; (**c**,**d**) gross power density versus current density. For (**a**,**c**) test in 0.5/4.0 M NaCl; for (**b**,**d**) test with 0.34 M NaCl + 0.054 M $MgCl_2$/2.7 M NaCl + 0.43 M $MgCl_2$.

Figure 9c,d show that, whether magnesium was present or not, the best and worst performing membranes in terms of maximum power density were Nafion 117 and Fuji-CEM, respectively. When the RED stack was equipped with Nafion 117, gross power density of 1.38 and 1.08 W·m^{-2} were measured for NaCl and NaCl + $MgCl_2$ solutions; correspondingly, 1.24 and 0.824 W·m^2 were obtained with Fuji-CEM. The significant P_d reduction is attributed to loss in both permselectivity and conductivity of the membrane. Moreover, the main reason why Fuji-CEM differed from the others can be explained by analyzing Equation (8). The low fixed charge density of Fuji-CEM (3.2 mol·L^{-1}) makes it vulnerable to high salinities whereas having high fixed charge density helps maintain the exclusion capacity, as in the example of N117 (8.0 mol·L^{-1}).

These results are coherent with our previous studies on the effect of Mg^{2+}. Avci et al. (2016) performed experiments using similar salinity gradients in a RED stack equipped with 25 cell pairs of Fuji AEM 80045 and Fuji CEM 80050: with a 20% reduction in OCV and 60% increase in R_{stack}, gross power density was more than halved. Specific investigations revealed that the power loss was substantially due to the critical effect of Mg^{2+} on the performance of Fuji CEM [36]. Similarly, Fontananova et al. (2017) reported a 30% decrease in $P_{d,max}$ for the same concentration of feed solution used in this study when the stack was equipped with 25 pairs of AEM 80045 and Fuji CEM 80050 [37].

Nafion and CMX membranes exhibited very similar performance.

For a better comparison of electrochemical performance of the commercial CEMs tested, permselectivity, membrane resistance from EIS tests, stack resistance, open circuit voltage from RED experiments and resulting maximum gross power density are reported in Table 3.

Table 3. Electrochemical characterization data from single-cell RED experiments.

	EIS Tests						Single-Cell RED Tests					
	Permselec. (-)		**Membrane Resist. ($\Omega \cdot cm^2$)**				**OCV (V)**		**R_{stack} (Ω)**		**$P_{d,max}$ ($W \cdot m^{-2}$)**	
CEM	A/B	C/D	A *	B *	C *	D *	A/B	C/D	A/B	C/D	A/B	C/D
Nafion 115	0.88	0.60	1.5	1.5	2.2	1.5	0.171	0.164	0.18	0.20	1.30	1.08
Nafion 117	0.88	0.58	1.8	3.2	2.8	2.6	0.171	0.165	0.17	0.20	1.38	1.11
CMX	0.81	0.55	2.2	2.6	2.3	2.5	0.168	0.163	0.18	0.20	1.26	1.09
Fuji-CEM	0.79	0.49	2.4	3.3	8.3	9.0	0.167	0.160	0.18	0.25	1.24	0.82

* A: 0.5 M NaCl; B: 4.0 M NaCl; C: 0.34 M NaCl + 0.054 M $MgCl_2$ and D: 2.7 M NaCl + 0.43 M $MgCl_2$.

Regarding the permselectivity, CEMs suffered from the presence of magnesium showing a very clear and sharp decrease. Consequently, OCV values declined since it is proportional to average permselectivity of AEM and CEM.

An increase in R_{stack} by 11–18% was observed for the stacks equipped with Nafion 115, Nafion 117 and CMX, whereas R_{stack} increased by about 40% when using Fuji-CEM.

An increase of both R_{HCC} and R_{LCC} contributed to the general enhancement of R_{stack}. In fact, the conductivity of LCC solution reduced from 47.9 to 40.2 $mS \cdot cm^{-1}$ when the feed solution was changed from 0.5 M NaCl to 0.34 M NaCl + 0.054 M $MgCl_2$, respectively. Likewise, HCC conductivity reduced from 270.7 to 200.3 $mS \cdot cm^{-1}$ when the feed solution was changed from 4.0 M NaCl to 2.7 M NaCl + 0.43 M $MgCl_2$, respectively.

Furthermore, for Fuji-CEM, an additional relevant contribution was associated to the drastic increase in membrane resistance (Figure 8), rising from 2.4 to 8.3 $\Omega \cdot cm^2$ when changing LCC solution from 0.5 M NaCl to 0.34 M NaCl + 0.054 M $MgCl_2$ and, analogously, rising from 3.3 to 9.0 $\Omega \cdot cm^2$ when changing LCC solution from 4.0 M NaCl to 2.7 M NaCl + 0.43 M $MgCl_2$.

Therefore, considering the Nernst Equation (3), stack resistance Equation (9) and maximum gross power density Equation (7), it can be concluded that the experimental single-cell RED parameters were in line with CEMs characterization.

Although this study aims to compare the potential of Nafion-based membranes with typically used CEMs for RED, further optimization of stack components can boost the generated energy. For example, reducing spacer thickness allowed enhanced gross power density to be obtained by decreasing stack resistance [48]. Similarly, a favorable response with less impact to spacer thickness reduction would be expected in this study as well, since feed salinity is high enough to provide required conductivity.

4. Conclusions

The present study was designed to elucidate the possible utilization of perfluorosulfonic acid based Nafion in reverse electrodialysis under a high salinity gradient. Additionally, by investigating the effect of magnesium ions, we extended this study from paradigmatic NaCl solutions to multicomponent NaCl + $MgCl_2$ solutions in view of a more realistic approach to RED operations in natural environments. In this regard, single cell RED experiments were carried out by using Nafion 117, Nafion 115, CMX and Fuji-CEM-80050 as cation exchange membranes. When operating with 0.5 M/4.0 M NaCl solutions, Nafion membranes resulted in the highest $P_{d,max}$ thanks to their outstanding permselectivity compared to other CEMs. In the presence of magnesium, 17 and 20% $P_{d,max}$ reductions were recorded for Nafion 115 and Nafion 117, respectively; both membranes maintained their low resistance, while a significant loss in permselectivity was measured. Even so, Nafion membranes outperformed other commercial membranes such as CMX and Fuji-CEM-80050.

Although Nafion membranes exhibited better performance than CMX and Fuji-CEM-80050, their use is limited by high cost, and a significant reduction of membrane price is required for affordable RED applications.

Author Contributions: Conceptualization, E.C., G.D.P. and E.F.; Data curation, E.F.; Formal analysis, E.F.; Funding acquisition, E.F.; Investigation, A.H.A., D.A.M., S.S., R.A.T., E.C., G.D.P. and E.F.; Methodology, E.F.; Project administration, E.F.; Supervision, E.F.; Writing–original draft, A.H.A., G.D.P. and E.F. All authors have read and agreed to the published version of the manuscript.

Funding: This work was partially supported by the Italian Ministry of Education University and Research (prot. MIUR no. 10912, 06/06/2016; concession grant decree no. 3366, 12/18/2018) within the project "Development of a solar powered, zero liquid discharge Integrated desalination membrane system to address the needs for water of the Mediterranean region", (IDEA-ERANETMED2–72-357).

Conflicts of Interest: The authors declare no conflict of interest.

References

1. Lacey, R.E. Energy by reverse electrodialysis. *Ocean Eng.* **1980**, *7*, 1–47. [CrossRef]
2. Long, R.; Li, B.; Liu, Z.; Liu, W. Hybrid membrane distillation-reverse electrodialysis electricity generation system to harvest low-grade thermal energy. *J. Memb. Sci.* **2017**, *525*, 107–115. [CrossRef]
3. Tufa, R.A.; Noviello, Y.; Di Profio, G.; Macedonio, F.; Ali, A.; Drioli, E.; Fontananova, E.; Bouzek, K.; Curcio, E. Integrated membrane distillation-reverse electrodialysis system for energy-efficient seawater desalination. *Appl. Energy* **2019**, *253*, 113551. [CrossRef]
4. Tufa, R.A.; Curcio, E.; Brauns, E.; Van Baak, W.; Fontananova, E.; Di Profio, G. Membrane Distillation and Reverse Electrodialysis for Near-Zero Liquid Discharge and low energy seawater desalination. *J. Memb. Sci.* **2015**, *496*, 325–333. [CrossRef]
5. Benneker, A.M.; Rijnaarts, T.; Lammertink, R.G.H.; Wood, J.A. Effect of temperature gradients in (reverse) electrodialysis in the Ohmic regime. *J. Memb. Sci.* **2018**, *548*, 421–428. [CrossRef]
6. Avci, A.H.; Tufa, R.A.; Fontananova, E.; Di Profio, G.; Curcio, E. Reverse Electrodialysis for energy production from natural river water and seawater. *Energy* **2018**, *165*, 512–521. [CrossRef]
7. Ran, F.; Xu, X.; Pan, D.; Liu, Y.; Bai, Y.; Shao, L. Ultrathin 2D Metal–Organic Framework Nanosheets In situ Interpenetrated by Functional CNTs for Hybrid Energy Storage Device. *Nano-Micro Lett.* **2020**, *12*, 1–13. [CrossRef]
8. Zhang, Y.; Cheng, X.; Jiang, X.; Urban, J.J.; Lau, C.H.; Liu, S.; Shao, L. Robust natural nanocomposites realizing unprecedented ultrafast precise molecular separations. *Mater. Today* **2020**, *36*, 40–47. [CrossRef]
9. Dlugolecki, P.; Nijmeijer, K.; Metz, S.; Wessling, M. Current status of ion exchange membranes for power generation from salinity gradients. *J. Memb. Sci.* **2008**, *319*, 214–222. [CrossRef]
10. Gronowski, A.A.; Yeager, H.L. Factors Which Affect the Permselectivity of Nafion® Membranes in Chlor-Alkali Electrolysis, I. *J. Electroanal. Chem.* **1991**, *138*, 2690. [CrossRef]
11. Kruissink, C.A. The effect of electro-osmotic water transport on current efficiency and cell performance in chlor-alkali membrane electrodialysis. *J. Memb. Sci.* **1983**, *14*, 331–366. [CrossRef]
12. Ito, H.; Maeda, T.; Nakano, A.; Takenaka, H. Properties of Nafion membranes under PEM water electrolysis conditions. *Int. J. Hydrogen Energy* **2011**, *36*, 10527–10540. [CrossRef]
13. Yeo, R.S.; McBreen, J.; Kissel, G.; Kulesa, F.; Srinivasan, S. Perfluorosulphonic acid (Nafion) membrane as a separator for an advanced alkaline water electrolyser *. *J. Appl. Electrochem.* **1980**, *10*, 741–747. [CrossRef]
14. Passalacqua, E.; Lufrano, F.; Squadrito, G.; Patti, A.; Giorgi, L. Nafion content in the catalyst layer of polymer electrolyte fuel cells: Effects on structure and performance. *Electrochim. Acta* **2001**, *46*, 799–805. [CrossRef]
15. Samms, S.R.; Wasmus, S.; Savineil, R.F. Thermal Stability of Nafion ® in Simulated Fuel Cell Environments. *J. Electrochem. Soc.* **1996**, *143*, 1498–1504. [CrossRef]
16. Mauritz, K.A.; Moore, R.B. State of understanding of Nafion. *Chem. Rev.* **2004**, *104*, 4535–4585. [CrossRef] [PubMed]

17. Lehmani, A.; Turq, P.; Michelle, P.; Jacques, P.; Simonin, J.-P. Ion transport in Nation 117 membrane. *J. Electroanal. Chem.* **1997**, *428*, 81–89. [CrossRef]
18. Lindheimer, A.; Molenat, J.; Gavach, C. A study of the superselectivity of Nafion perfluorosulfonic membranes. *J. Electroanal. Chem.* **1987**, *216*, 71–88. [CrossRef]
19. Izquierdo-Gil, M.A.; Barragan, V.M.; Villaluenga, J.P.G.; Godino, M.P. Water uptake and salt transport through Nafion cation-exchange membranes with different thicknesses. *Chem. Eng. Sci.* **2012**, *72*, 1–9. [CrossRef]
20. Hernández-Flores, G.; Poggi-Varaldo, H.M.; Solorza-Feria, O. Comparison of alternative membranes to replace high cost Nafion ones in microbial fuel cells. *Int. J. Hydrogen Energy* **2016**, *41*, 23354–23362. [CrossRef]
21. Yee, R.S.L.; Rozendal, R.A.; Zhang, K.; Ladewig, B.P. Cost effective cation exchange membranes: A review. *Chem. Eng. Res. Des.* **2012**, *90*, 950–959. [CrossRef]
22. Daniilidis, A.; Vermaas, D.A.; Herber, R.; Nijmeijer, K. Experimentally obtainable energy from mixing river water, seawater or brines with reverse electrodialysis. *Renew. Energy* **2014**, *64*, 123–131. [CrossRef]
23. Banerjee, S.; Curtin, D.E. Nafion perfluorinated membranes in fuel cells. *J. Flourine Chem.* **2004**, *125*, 1211–1216. [CrossRef]
24. Toupin, M.; Malek, K.; Mokrini, A. Techno-economics of a new high throughput process for proton exchange membranes manufacturing. *World Electr. Veh. J.* **2016**, *8*, 431. [CrossRef]
25. Antheunis, H.; Hessing, J.; Van Berchum, B. Curable Compositions and Membranes 2015. U.S. Patent US 8968965 B2, 5 November 2015.
26. Sarapulova, V.; Shkorkina, I.; Mareev, S.; Pismenskaya, N.; Kononenko, N.; Larchet, C.; Dammak, L.; Nikonenko, V. Transport characteristics of fujifilm ion-exchange membranes as compared to homogeneous membranes AMX and CMX and to heterogeneous membranes MK-40 and MA-41. *Membranes* **2019**, *9*, 84. [CrossRef]
27. Hori, Y.; Nakatani, T.; Mizutani, Y. Morphology of ion exchange membranes. *J. Electron Microsc. (Tokyo)* **1986**, *35*, 220–226.
28. Avci, A.H.; Rijnaarts, T.; Fontananova, E.; Di Profio, G.; Vankelecom, I.F.V.; De Vos, W.M.; Curcio, E. Sulfonated polyethersulfone based cation exchange membranes for reverse electrodialysis under high salinity gradients. *J. Memb. Sci.* **2019**, *595*, 117585. [CrossRef]
29. Jagur-Grodzinski, J.; Kramer, R. Novel process for direct conversion of free energy of mixing into electric power. *Ind. Eng. Chem. Process Des. Dev.* **1986**, *25*, 443–449. [CrossRef]
30. Farrell, E.; Hassan, M.I.; Tufa, R.A.; Tuomiranta, A.; Avci, A.H.; Politano, A.; Curcio, E.; Arafat, H.A. Reverse electrodialysis powered greenhouse concept for water- and energy-self-sufficient agriculture. *Appl. Energy* **2017**, *187*, 390–409. [CrossRef]
31. Tufa, R.A.; Curcio, E.; Van Baak, W.; Veerman, J.; Grasman, S.; Fontananova, E.; Di Profio, G. Potential of brackish water and brine for energy generation by salinity gradient power-reverse electrodialysis (SGP-RE). *RSC Adv.* **2014**, *4*, 42617–42623. [CrossRef]
32. Tedesco, M.; Cipollina, A.; Tamburini, A.; Micale, G. Towards 1 kW power production in a reverse electrodialysis pilot plant with saline waters and concentrated brines. *J. Memb. Sci.* **2017**, *522*, 226–236. [CrossRef]
33. Tedesco, M.; Scalici, C.; Vaccari, D.; Cipollina, A.; Tamburini, A.; Micale, G. Performance of the first Reverse Electrodialysis pilot plant for power production from saline waters and concentrated brines. *J. Memb. Sci.* **2016**, *500*, 33–45. [CrossRef]
34. Pitzer, K.S. *Activity Coefficients in Electrolyte Solutions*, 2nd ed.; CRC Press: Boca Raton, FL, USA, 2018; ISBN 9781315890371.
35. Besha, A.T.; Tsehaye, M.T.; Aili, D.; Zhang, W.; Tufa, R.A. Design of monovalent ion selective membranes for reducing the impacts of multivalent ions in reverse electrodialysis. *Membranes* **2020**, *10*, 7. [CrossRef]
36. Avci, A.H.; Sarkar, P.; Tufa, R.A.; Messana, D.; Argurio, P.; Fontananova, E.; Di Profio, G.; Curcio, E. Effect of Mg2+ ions on energy generation by Reverse Electrodialysis. *J. Memb. Sci.* **2016**, *520*, 499–506. [CrossRef]
37. Fontananova, E.; Messana, D.; Tufa, R.A.; Nicotera, I.; Kosma, V.; Curcio, E.; Van Baak, W.; Drioli, E.; Di Profio, G. Effect of solution concentration and composition on the electrochemical properties of ion exchange membranes for energy conversion. *J. Power Sources* **2017**, *340*, 282–293. [CrossRef]
38. Strathmann, H. *Ion-Exchange Membrane Separation Processes*; Elsevier B.V.: Amsterdam, the Netherlands, 2004; ISBN 044450236X.

39. Haynes, W.M. *CRC Handbook of Chemistry and Physics*, 96th ed.; CRC Press/Taylor and Francis: Boca Raton, FL, USA, 2016.
40. Fontananova, E.; Zhang, W.; Nicotera, I.; Simari, C.; van Baak, W.; Di Profio, G.; Curcio, E.; Drioli, E. Probing membrane and interface properties in concentrated electrolyte solutions. *J. Memb. Sci.* **2014**, *459*, 177–189. [CrossRef]
41. Yip, N.Y.; Elimelech, M. Comparison of Energy Efficiency and Power Density in Pressure Retarded Osmosis and Reverse Electrodialysis. *Environ. Sci. Technol.* **2014**, *48*, 11002–11012. [CrossRef]
42. Tufa, R.A.; Pawlowski, S.; Veerman, J.; Bouzek, K.; Fontananova, E.; di Profio, G.; Velizarov, S.; Goulão Crespo, J.; Nijmeijer, K.; Curcio, E. Progress and prospects in reverse electrodialysis for salinity gradient energy conversion and storage. *Appl. Energy* **2018**, *225*, 290–331. [CrossRef]
43. Tedesco, M.; Hamelers, H.V.M.; Biesheuvel, P.M. Nernst-Planck transport theory for (reverse) electrodialysis: III. Optimal membrane thickness for enhanced process performance. *J. Memb. Sci.* **2018**, *565*, 480–487. [CrossRef]
44. Post, J.W.; Goeting, C.H.; Valk, J.; Goinga, S.; Veerman, J.; Hamelers, H.V.M.; Hack, P.J.F.M. Towards implementation of reverse electrodialysis for power generation from salinity gradients. *Desalin. Water Treat.* **2010**, *16*, 182–193. [CrossRef]
45. Cassady, H.J.; Cimino, E.C.; Kumar, M.; Hickner, M.A. Specific ion effects on the permselectivity of sulfonated poly (ether sulfone) cation exchange membranes. *J. Memb. Sci.* **2016**, *508*, 146–152. [CrossRef]
46. Luo, T.; Abdu, S.; Wessling, M. Selectivity of ion exchange membranes: A review. *J. Memb. Sci.* **2018**, *555*, 429–454. [CrossRef]
47. Geise, M.; Curtis, A.J.; Hatzell, M.C.; Hickner, M.A.; Logan, B.E. Salt Concentration Differences Alter Membrane Resistance in Reverse Electrodialysis Stacks. *Environ. Sci. Technol. Lett.* **2014**, *1*, 36–39. [CrossRef]
48. Vermaas, D.A.; Saakes, M.; Nijmeijer, K. Doubled power density from salinity gradients at reduced intermembrane distance. *Environ. Sci. Technol.* **2011**, *45*, 7089–7095. [CrossRef] [PubMed]
49. Galama, A.H.; Hoog, N.A.; Yntema, D.R. Method for determining ion exchange membrane resistance for electrodialysis systems. *Desalination* **2016**, *380*, 1–11. [CrossRef]
50. Galama, A.H.; Vermaas, D.A.; Veerman, J.; Saakes, M.; Rijnaarts, H.H.M.; Post, J.W.; Nijmeijer, K. Membrane resistance: The effect of salinity gradients over a cation exchange membrane. *J. Memb. Sci.* **2014**, *467*, 279–291. [CrossRef]
51. Güler, E.; Elizen, R.; Saakes, M.; Nijmeijer, K. Micro-structured membranes for electricity generation by reverse electrodialysis. *J. Memb. Sci.* **2014**, *458*, 136–148. [CrossRef]

membranes

MDPI

Article

Chemically Crosslinked Sulfonated Polyphenylsulfone (CSPPSU) Membranes for PEM Fuel Cells

Je-Deok Kim [1,*], Akihiro Ohira [2] and Hidenobu Nakao [3]

1 Hydrogen Production Materials Group, Center for Green Research on Energy and Environmental Materials, National Institute for Materials Science (NIMS), 1-1 Namiki, Tsukuba, Ibaraki 305-0044, Japan
2 Energy Storage Technology Group, Research Institute for Energy Conservation, National Institute of Advanced Industrial Science and Technology (AIST), 1-1-1 Higashi, Tsukuba, Ibaraki 305-8565, Japan; a-oohira@aist.go.jp
3 Hydrogen Materials Engineering Group, Research Institute for Energy Conservation, National Institute of Advanced Industrial Science and Technology (AIST), 1-1-1 Higashi, Tsukuba, Ibaraki 305-8565, Japan; nakao.hidenobu@nims.go.jp
* Correspondence: kim.jedeok@nims.go.jp

Received: 22 January 2020; Accepted: 14 February 2020; Published: 18 February 2020

check for updates

Abstract: Sulfonated polyphenylsulfone (SPPSU) with a high ion exchange capacity (IEC) was synthesized using commercially available polyphenylsulfone (PPSU), and a large-area (16 × 18 cm^2) crosslinked sulfonated polyphenylsulfone (CSPPSU) membrane was prepared. In addition, we developed an activation process in which the membrane was treated with alkaline and acidic solutions to remove sulfur dioxide (SO_2), which forms as a byproduct during heat treatment. CSPPSU membranes obtained using this activation method had high thermal, mechanical and chemical stabilities. In I-$V_{iR\ free}$ studies for fuel cell evaluation, high performances similar to those using Nafion were obtained. In addition, from the hydrogen (H_2) gas crossover characteristics, the durability is much better than that of a Nafion212 membrane. In the studies evaluating the long-term stabilities by using a constant current method, a stability of 4000 h was obtained for the first time. These results indicate that the CSPPSU membrane obtained by using our activation method is promising as a polymer electrolyte membrane.

Keywords: PPSU; High IEC, CSPPSU; activation; PEMFCs

1. Introduction

A low-carbon society in which safe, highly efficient, renewable and sustainable energy sources are used to sustain economic growth, environmental protection and energy security has become important. Proton exchange membrane fuel cells (PEMFCs) are among the most promising electrochemical devices for a low-carbon society and highly efficient power generation. Although PEMFCs using perfluorosulfonic acid polymer membranes have been commercialized, to increase their performances, more development and new material components, such as catalyst electrodes and polymer membranes, are required. Although it is important to find non-platinum catalyst materials for use as the catalyst electrode [1–3], non-fluorine proton exchange membranes are also necessary. The most commonly used proton exchange membranes for PEMFCs are poly(perfluorosulfonic acid) (PFSA) copolymers such as Nafion, which have high hydrolytic and oxidative stability and excellent proton conductivities [4,5]. However, the glass-transition temperatures (T_g), mechanical stabilities, and gas permeabilities must be improved. Moreover, perfluorinated polymers have high production costs and environmental incompatibilities. The drawbacks of perfluorinated membranes have prompted research into alternative membranes.

For example, several aromatic polymer ionomer membranes, such as polybenzimidazole (PBI) [6,7], sulfonated polyphenylene oxide (SPPO) [8], sulfonated polyimide [9–11], sulfonated polyethersulfone (SPES) [12,13], sulfonated polyphenylene (SPP) [14], sulfonated polyphenylene sulfone (SPPS) [15], sulfonated polysulfone (SPSU) [16–21], sulfonated polyetheretherketone (SPEEK) [22–33], and sulfonated polyphenylsulfone (SPPSU) [34–49], are being actively investigated. The rigid molecular chain structures of aromatic polymers provide high thermal stabilities. However, their conductivities are generally lower than those of PFSA also because the structures can make it difficult to form proton conducting channels. However, increasing the ion exchange capacity (IEC) of the aromatic polymer to increase the conductivity results in weaker mechanical and chemical properties of the polymer. In other words, there is a trade-off relationship between the increase in conductivity and the mechanical and chemical properties. Thus, further development of hydrocarbon polymers without fluorinated moieties is necessary.

Polyphenylsulfone (PPSU), which has an excellent thermal stability, high chemical resistance and low cost, has been studied widely. The sulfonation of PPSU to balance the IEC values and membrane stability is a crucial focus. High IEC values are desirable for good proton conductivities but result in excess swelling, which causes mechanical and chemical instabilities. These mechanical and chemical properties are big obstacles for applying SPPSUs in PEMFCs. We are developing a crosslinked sulfonated polyphenylsulfone (CSPPSU) membrane having good thermal and chemical properties with good proton conductivities [34,36,37]. Organic solvent-free CSPPSU membranes have been reported [34,37]. Moreover, an SPPSU with a high sulfonation degree from bis(4-fluorophenyl)sulfone (FPS) monomer has been used to prepare a CSPPSU membrane [36]. The membranes have high proton conductivities (>0.1 S/cm at 90% RH, 80 °C) [34,36,37]. However, the cell performances are not stable over a long period. We thought that a byproduct such as sulfur dioxide (SO_2) produced during the crosslinking process remaining in the membrane was the cause. In this paper, we report an activation treatment for the CSPPSU membrane to achieve good thermal, chemical, and long-term stabilities. The nanostructures, conductivities, cell performances, and thermal, mechanical and long-term stabilities of the CSPPSU membranes after our activation treatment were investigated for the first time and compared with those of Nafion212 membranes.

2. Experimental

2.1. Materials

PPSU (Solvay Radel R-5000 NT) (Mn = 26,000; Mw = 50,000; Mw/Mn = 1.9) was provided by Solvay Specialty Polymers Japan K.K. (glass transition temperature (T_g) = 220 °C). A DuPont™ Nafion® PFSA membrane (NR-212) was purchased from DuPont (USA). The chemical compounds were purchased from commercially available sources and were used as received. Hydrogen peroxide (H_2O_2), sodium chloride (NaCl), sodium hydroxide (NaOH), and sulfuric acid (H_2SO_4) were purchased from Nacalai Tesque, Inc, Japan. Iron (II) chloride tetrahydrate ($FeCl_2{\cdot}4H_2O$) was purchased from Wako Pure Chemical Industries, Ltd., Japan. A dialysis tubing cellulose membrane (molecular weight cut-off; MWCO = 14,000) and dimethyl sulfoxide (DMSO) were purchased from Sigma-Aldrich Co., Ltd. Deionized (DI) H_2O was obtained using a PURELAB® Option-R 7 ELGA LabWater at 15 Mohm cm and 25 °C.

2.2. Synthesis and Characterization of SPPSU

SPPSU was synthesized using an electrophilic aromatic substitution reaction of PPSU with H_2SO_4. The synthesis and properties of SPPSU have been described in detail in a previous paper [34]. The SPPSU polymers had IEC values of 3.68 meq/g, and the degree of sulfonation (D.S.) values were 2.3, which are close to the theoretical IEC value of 3.57 for D.S. = 2.0. The yield was 76%.

2.3. Preparation of Chemically Crosslinked SPPSU (CSPPSU) Membranes

SPPSU (5 g) was dissolved in DMSO 20 mL with stirring at 60 °C to prepare a 20 wt% solution. The solution was casted at a speed of 4.0 mm/min with an applicator (blade) (Tester Sangyo CO.,

Ltd., Japan; 15 cm, gap = 0.5 mm) on a glass plate (27 × 30 cm^2) heated at 80 °C by an automatic film-coating apparatus (KIPAE, KP-3000VH), and then dried for 24 h at 80 °C. Next, the film coated on the glass plate was moved to an oven and annealed in air at 120 °C (24 h), 160 °C (24 h), and 180 °C (24 h). After that, the membrane was peeled off the glass plate with water and activated to remove the remaining sulfur dioxide (SO_2), which is a byproduct from the annealing process, and to cure the nanostructure of the CSPPSU membrane. Activation was performed using the following procedure: heating in 0.5 M NaOH at 80 °C overnight, boiling in DI H_2O for 2 h, heating at 1M H_2SO_4 at 80 °C for 2 h, and boiling in DI H_2O for 2 h. Finally, the CSPPSU membranes were dried at room temperature before using. The CSPPSU membranes were very flexible and light brown.

2.4. Pretreatment of Nafion212 Membranes

Nafion212 membranes were prepared for a comparison with the CSPPSU membranes. Before using, the Nafion212 membranes were pretreated as follows: they were boiled in DI H_2O for 2 h, heated in 1M H_2O_2 at 80 °C for 2 h, followed by heating in 1M H_2SO_4 at 80 °C for 2 h, and then boiled in DI H_2O for 2 h. The membranes were kept in DI H_2O until use.

2.5. IEC, D.S., Water-Uptake (W.U.), λ, and Crosslink Rates ($D_{crosslink}$)

The ion exchange capacity (IEC) was defined as milliequivalents of sulfonic groups per gram of dried sample. A piece of membrane was soaked in 20 mL of a 2 M NaCl solution and equilibrated for more than 24 h to replace the protons with sodium ions. The solution was then titrated with a 0.01 M NaOH solution. The IEC was calculated using the following equation: IEC (meq/g) = cv/W_{dry}, where c (mmol/L) is the concentration of the standardized NaOH aqueous solution used for titration (0.01mol/L), v (L) is the volume of the standardized NaOH aqueous solution used for titration, and W_{dry} (g) is the mass of the dry membrane. The degree of sulfonation (D.S.) of the membranes was calculated using the following equation: D.S. (Sulfonic acid group/repeating unit; R.U.) = [IEC/1000 × F_w (R.U.)]/[1 − (IEC/1000 × F_w (SO_3))], where F_w (R.U.) = 400.45 and F_w (SO_3) = 80.06.

The water-uptake (W.U.) of the membranes at room temperature was calculated using the following: W.U. (%) = $[(W_{wet} - W_{dry})/W_{dry}] \times 100$, where W_{wet} is the mass of the wet membrane. The membranes were cut into 10 mm × 10 mm squares and dried for >24 h at 80 °C in a dry oven. The membranes were immersed in boiling DI H_2O for >1 h before the measurements.

The hydration number (λ) for the membranes was calculated using the following: λ ($[H_2O]/[SO_3H]$) = $[1000(W_{wet} - W_{dry})]/18\ W_{dry}IEC$.

The degree of crosslinking (crosslink rate, $D_{crosslink}$) in the membranes was calculated using the following: $D_{crosslink}$ (%) = $[(IEC_{before\ annealing} - IEC_{after\ annealing})/IEC_{before\ annealing}] \times 100$.

2.6. Oxidative Stability (Fenton's Test)

The oxidative stabilities of the membranes were evaluated by immersing a small piece of sample into Fenton's reagent [3 wt% H_2O_2 and 2 ppm Fe(II) (added as $FeCl_2{\cdot}4H_2O$)] at 80 °C for 1 h while stirring. The samples were dried at 60 °C in a vacuum oven before the measurements. The membranes were washed with DI H_2O repeatedly and dried in a vacuum oven at 60 °C overnight after the reaction. The oxidative stabilities were determined as follows: 100[(mass of residual membrane after the test)/(initial mass of membrane)].

2.7. Thermal Behavior

The thermal and mass properties of the membranes were investigated using thermogravimetric and mass analysis with a Discovery TGA/MS (TA instruments). The samples were heated from room temperature (RT) to 600 °C at 20 °C/min in an He atmosphere.

The dynamic elastic modulus and dissipation (tan δ) of the membranes were measured by DMA Q800 (TA Instruments) at 1 Hz in the temperature range of 20–200 °C at a heating rate of 5 °C/min.

2.8. Mechanical Behavior

The stress–strain tests on the membranes were performed using a Tension Test Machine (Shimazu, EZ-S) at room temperature with a constant crosshead speed of 5 mm/min. The samples were cut using the super dumbbell cutter SDMP-1000 (Dumbbell Co., Japan). The membrane thicknesses were measured using a Mitutoyo 547-401 ABSOLUTE Digimatic Thickness Gauge, Japan.

2.9. Membrane Nanostructure

Small angle X-ray scattering (SAXS) measurements were performed in the beamline BL15A2 of the Photon Factory in KEK (Tsukuba, Japan). The X-ray beam was monochromatized to 1.2 Å with a curved monochromator (triangular Ge (111) crystal with the asymmetric angle of 8.0 degrees). The membrane samples were put on the stage of a rotational autochanger, and the measurements were performed on the membranes under wet conditions, for which a small amount of water was added dropwise. Scattered photons were detected with a two-dimensional semiconductor detector (PILATUS3 1M, 981 × 1043 pixels, pixel size: 0.172 mm × 0.172 mm, Dectris, Switzerland) with a camera distance of 170 cm, and the signals were accumulated for 30 s. The image was processed by using the SAngler program. The image was calibrated with the diffraction image of silver behenate and then averaged over azimuthal angle to obtain a one-dimensional profile with the subtraction of SAXS profile of air.

2.10. Conductivity Measurements

The proton conductivities of the membranes were determined using four-point probe impedance spectroscopy. Through-plane proton conductivities of the membranes using the MTS cell head were measured as a function of the relative humidity (RH, 20–90%) in the temperature range of 40–120 °C, using a 740 membrane test system (MTS, Scribner Associates, Inc.) with a phase sensitive multimeter (model PSM1735, Newtons4th Ltd.) combined with an impedance analysis interface. A frequency range of 1 Hz–1 MHz and a peak-to-peak voltage of 10 mV were used during the impedance measurements. The samples were equilibrated in a temperature and humidity chamber at specified temperatures and relative humidities (RHs) for 30 min before the measurements. The chamber was pressurized at 130 kPa for the measurements in the temperature range of 100–120 °C.

2.11. Membrane, Gas Diffusion Layer, and Catalyst Electrode Information for MEA

The thickness of the membranes was about 50 µm. A Pt/C/ionomer (ionomer/carbon = 1) catalyst electrode containing 0.3 mg/cm^2 of Pt on a GDL electrode (Sigaracet® GDL 25BC of SGL Group Co. Ltd., Japan) was purchased from EIWA Corporation.

2.12. Preparation of Membrane Electrode Assembly (MEA)

The effective electrode area of the single cell was 4 cm^2. A hot press machine (MODEL A-010D, FC-R&D Company, Sagamihara, Japan) was used. The MEA was prepared on Cu plates. Polyimide films were used between the Cu plates to facilitate peeling of the MEA. The MEA was obtained by inserting a membrane between the anode and cathode and hot-pressing at 130 °C at ~9.8 kN for 20 min.

2.13. Single Fuel-Cell Performance and Durability

A single cell (active area: 20 × 20 mm^2) was purchased from Ulimeng Eng Co., Ltd., Korea. The fuel cell system was evaluated using AutoPEM of Toyo Corporation, Japan. The current-voltage (I-V) performance was measured in relation to the amount of hydrogen and oxygen at the anode and cathode, respectively, at 80 °C, 100% and 60% RH, and ambient pressure. The gas utilizations at the anode and cathode were 56% and 14%, respectively. The I-V characteristics of a single cell were obtained by activating at 1 A for 20 h, increasing the current from 0.02 to 6 A, while maintaining a constant current value for 5 min, and sweeping 5 times. Next, the cell was purged for 10 h with nitrogen gas, and cyclic voltammetry (CV) and linear sweep voltammetry (LSV) were performed in the potential range of 0.02–0.9 V at a scan rate of 50 mV/s and 0.02–0.5 V at 2 mV/s, respectively. During the measurements, 100 mL/min humidified nitrogen and 100 mL/min hydrogen were fed to the working electrode and counter electrode, respectively. The effective catalytic area (ECA) from the results (oxidation of adsorbed H) of the CV was calculated using the following equation: ECA ([cm^2/g] = (charge density, μC/cm^2)/[210 (μC/cm^2) × electrode loading, (g/cm^2)]. A long-term test using a constant current method was performed at 80 °C and 100% RH using a PEMTest8900 from Toyo Corporation.

3. Results and Discussion

3.1. Activation of the CSPPSU Membranes

Figure 1 shows an image (16 × 18 cm^2) (Figure 1a) and the chemical structure (Figure 1b) of the chemically crosslinked SPPSU (CSPPSU) membranes. CSPPSU membranes were prepared by using a heat treatment method as previously reported [34,36,37]. Crosslinking between the sulfo groups (–SO_3H) by heat treatment can produce undesirable byproducts, such as sulfur dioxide (SO_2), which do not completely evaporate from the membrane. The remaining SO_2 was found to decrease the original membrane characteristics obtained from the MEA process, causing bad fuel cell performances. We removed SO_2 by activating the membranes with an alkaline solution. In addition, it was found that the mechanical and chemical properties and the durabilities of the membranes were dramatically stabilized after our activation process.

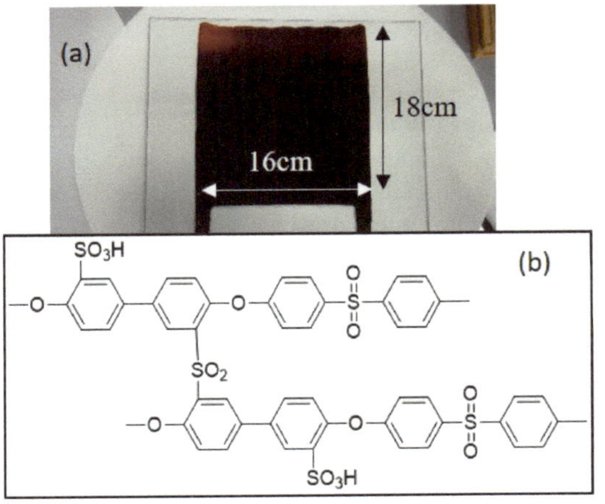

Figure 1. (**a**) Image of a crosslinked sulfonated polyphenylsulfone (CSPPSU) membrane on a glass plate and (**b**) chemical structure of a CSPPSU.

Figure 2 shows the thermal stabilities of synthesized SPPSU, CSPPSU after heat treatment at 180 °C, and CSPPSU membranes activated without and with an NaOH solution. The thermal stabilities of the CSPPSU membranes after heat treatment at 180 °C were better than those of SPPSU. There was no significant difference in the thermal stabilities of the CSPPSU membranes after the heat treatment at 180 °C and the CSPPSU membrane activated without NaOH solution. On the other hand, the stability of the CSPPSU membranes was much higher after activation treatment with NaOH. Mass analysis indicated that the stability of the CSPPSU membrane was due to the removal of the SO_2 trapped in the membrane. From these results, it is suggested that activation with an alkaline solution is necessary when crosslinked membranes are obtained using DMSO as a solvent.

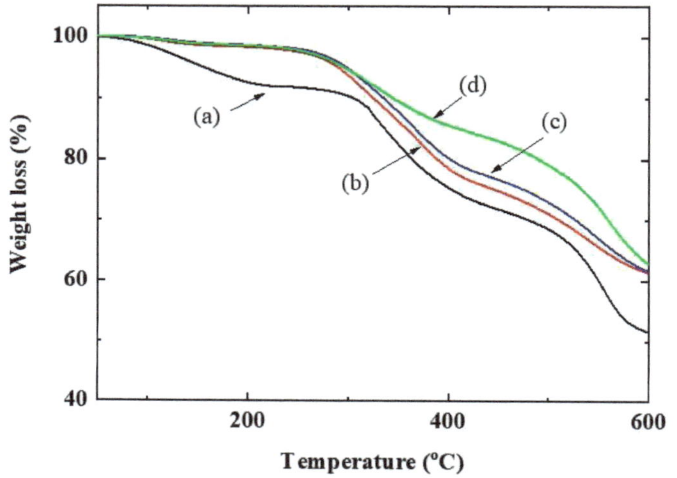

Figure 2. Thermal stabilities in a helium gas flow at a heating rate of 20 °C/min of (**a**) synthesized sulfonated polyphenylsulfone (SPPSU), (**b**) CSPPSU after the heat treatment at 180 °C, and CSPPSU membranes activated (**c**) without and (**d**) with NaOH solution process.

3.2. Mechanical Properties of the CSPPSU Membranes

Figure 3 shows the dynamic elastic modulus and Tan δ properties of CSPPSU and Nafion212 membranes with respect to temperature. The dynamic elastic modulus of CSPPSU membranes was constant up to 200 °C, whereas that of the Nafion212 membrane decreased due to changes in the molecular structure with an increase in the temperature. From Tan δ, the glass transition temperature of the Nafion212 membrane was around 100 °C, and that of the CSPPSU membrane was >200 °C. This result suggests that the CSPPSU membrane is suitable for application as a high-temperature electrolyte membrane over 100 °C.

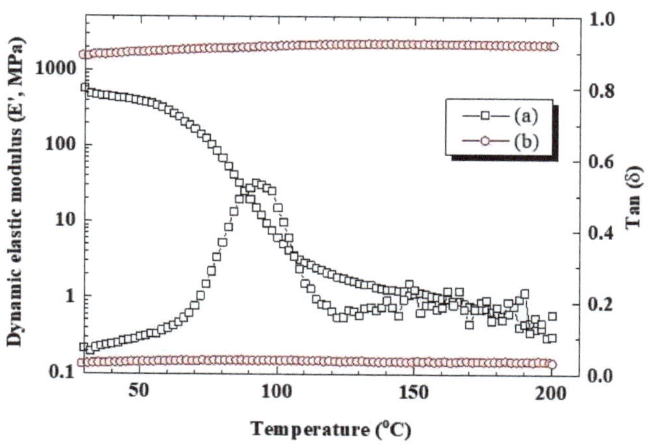

Figure 3. Dynamic elastic modulus and Tan δ of (**a**) Nafion212 and (**b**) CSPPSU membranes at 1 Hz.

Figure 4 shows the tensile strengths and elongation results for the CSPPSU and Nafion212 membranes at room temperature. The Nafion212 membrane had a low tensile strength but very high tensile elongation. The tensile strength of the CSPPSU membranes was higher than that of the Nafion212, but the tensile elongation was lower than that of the Nafion212. On the other hand, the tensile strength and elongation of the CSPPSU membranes were similar or better than those of other hydrocarbon-based polymer electrolyte membranes [26,28,34,37,50]. However, an increase in the tensile elongation (high flexibility) is required to obtain good contact, which can reduce the contact resistance and proton transfer in the membrane at the interface between the membrane and the electrode. The mechanical and thermal properties of PPSU, CSPPSU and Nafion212 membranes are summarized in Table 1.

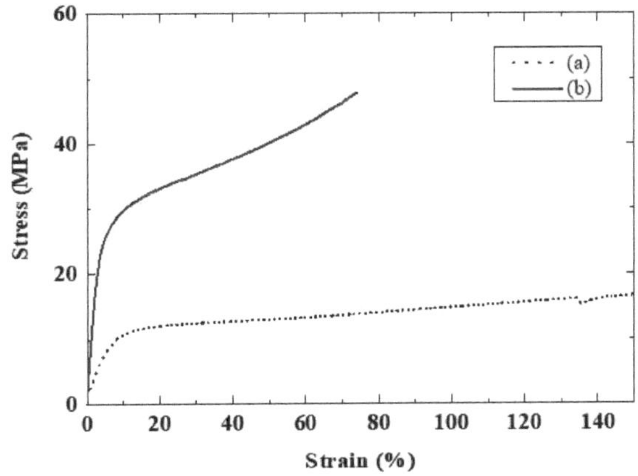

Figure 4. Stress–strain properties of (**a**) Nafion212 (t = 0.050 mm) and (**b**) CSPPSU (t = 0.066 mm) membranes at room temperature.

Table 1. Mechanical and thermal properties of PPSU, CSPPSU and Nafion212 membranes.

Polymer Membrane	PPSU (Solvay) *	CSPPSU	Nafion212
Tensile modulus (MPa)	2340	2000	~500
Tensile strength (MPa)	69.6	48	16 at 150%
Tensile elongation (%) (break)	60–120	74	>150
Flexural modulus (MPa)	2410	757	155
Glass transition (°C)	220	>200	~100
Decomposition (°C)	540	~300	~300

* From Solvay PPSU R-5000_Datasheet.

3.3. *Nanostructures of the CSPPSU Membranes*

The nanostructures of wet CSPPSU membranes were characterized by using small angle X-ray scattering (SAXS) (Figure 5) [4]. An intensity maximum was observed at $q = 1.6\ nm^{-1}$ ($q = 4\pi \sin\theta/\lambda$). The average spacing, L, of the scattering objects was calculated from the peak position by using the Bragg equation ($L = 2\pi/\theta$) to be 3.9 nm. This value was lower than the L value for Nafion (5 nm) [4,51,52]. In addition, the peak intensity of the CSPPSU membrane was lower than that of Nafion. In general, in a Nafion membrane, there is good separation between hydrophilic and hydrophobic phases and it has a good ion conduction path in a humidified state. On the other hand, it is known that hydrocarbon-based membranes, such as SPEEK, have a longer distance between sulfone groups and less continuity of the conduction paths than Nafion membranes do [4]. From these results, we believe that the CSPPSU membranes have good conduction paths and that the IEC is high (Table 2). However, the volume density of the ion clusters in the nanostructures is low, and the SAXS peak intensity is low.

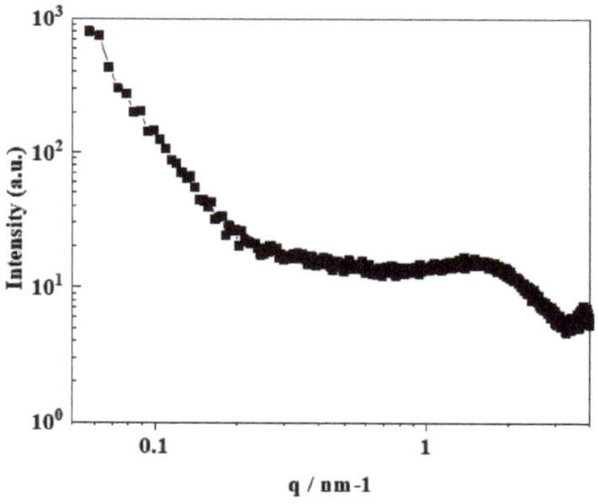

Figure 5. Small angle X-ray scattering (SAXS) profile of a wet-CSPPSU membrane at room temperature.

3.4. Proton Conductivities of the CSPPSU Membranes

Since SPPSU was crosslinked by using heat treatment, it was difficult to precisely control the crosslinking in the nanostructure. However, using this method, stable and macroscopically reproducible CSPPSU membranes were obtained. Figure 6 shows the proton conductivities of CSPPSU membranes obtained from the average value of the conductivities of four different membranes in relation to the temperature and relative humidity. These conductivities were lower than those obtained in our previous studies [34,36,37] due to different synthetic and activation conditions. The conductivities of the CSPPSU membranes were lower than those of the Nafion212 membranes [5,37,52] and were lower by one order of magnitude under low relative humidification conditions (Table 2). The reasons for the difference include the following: the ion cluster size of the CSPPSU membrane is small, the distance between the sulfone groups due to the backbone structure of PPSU is long, and the number of continuous conduction paths is small, as described previously. The design of CSPPSU membranes with high conductivities and high mechanical stabilities is required.

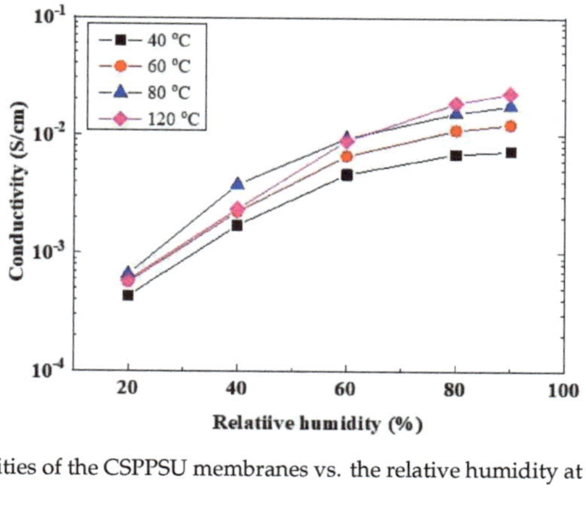

Figure 6. Proton conductivities of the CSPPSU membranes vs. the relative humidity at the specified temperatures.

Table 2. Physicochemical and conductivity properties of CSPPSU and Nafion212 membranes.

Membrane			CSPPSU	Nafion212
IEC (meq/g)			~2	0.9
W.U. (%)			43	50
λ			12	31
$D_{crosslink}$ (%)			47.3	-
$R_{oxidation}$ (%)			91.3–99.4	100
Cluster size (nm)			3.9	5.0
Conductivity (mS/cm)	80 °C	20% RH	0.7	7.3
		90% RH	18	80
	120 °C	20% RH	0.6	6.6
		90% RH	22	105

3.5. IEC, W.U., λ, and $D_{crosslink}$ Values of CSPPSU Membranes

In Table 2, the characteristics of the CSPPSU membranes are summarized in comparison with Nafion212 membranes. The IEC of the CSPPSU membranes was higher than that of Nafion212, but W.U. and λ were lower than those of Nafion212. This is significant for hydrocarbon-based electrolyte membranes because high IECs tend to cause swelling of the membrane in water and thus deteriorate the stability of the membrane. The CSPPSU membranes had a crosslink rate ($D_{crosslink}$) of 47.3% and high mechanical stabilities with a high IEC and reasonable W.U. values. These results indicate that the crosslinking method is effective for increasing the mechanical and chemical stabilities of hydrocarbon electrolyte membranes.

3.6. Oxidative Stability of the CSPPSU Membranes

The degradation phenomena of the electrolyte membrane in a fuel cell are thought to be caused by the following [47]. One is degradation on the anode side due to H radicals and O_2 diffused from the cathode side through the electrolyte membrane. The other is degradation on the cathode side, where HO or HO_2 radicals from O_2 reduction diffused on the cathode side. However, the HO and HO_2 radicals on the cathode side are the main causes of electrolyte membrane degradation. The oxidative stabilities of the CSPPSU membranes were evaluated by using an ex situ method wherein they were treated with Fenton's reagent (3 wt% H_2O_2 + 2 ppm Fe (II)) at 80 °C for 1 h. The CSPPSU membranes had stabilities in the range of 91.3–99.4% (Table 2). However, the membranes broke when rubbed by hand after the Fenton tests. A method for improving the oxidative stabilities while improving the proton conductivities of the CSPPSU membrane is currently under study and will be reported later.

3.7. Fuel Cell Properties Using CSPPSU Membranes

Evaluation of a fuel cell is complicated because the membrane, catalyst electrodes containing an ionomer, the interfaces between a membrane and an electrode and between the electrode and the separator, and gas supply affect the power generation reaction. However, the current-voltage (I-V), cyclic voltammetry (CV), and hydrogen gas crossover characteristics of a single cell were evaluated under similar conditions but with different electrolyte membranes. Figures 7 and 8 show the I-$V_{iR\ free}$ and CV characteristics of the CSPPSU membranes and the Nafion212 membrane evaluated at a cell temperature of 80 °C and 100% and 60% RH, respectively. The voltage differences using the CSPPSU and Nafion212 membranes were 37 mV at 1 A/cm^2 and 100% RH and 33 mV at 0.5 A/cm^2 and 60% RH. On the other hand, the cell resistances using the CSPPSU and Nafion212 membranes were 73 mohm and 24 mohm at 100% RH and 161 mohm and 42 mohm at 60% RH, respectively (Table 3). The I-$V_{iR\ free}$ results show that the difference in the conductivities of the CSPPSU and Nafion212 membranes was large (Table 2), whereas the difference in the fuel cell was not so large. However, on the basis of the resistance value of the entire cell, the difference between the CSPPSU and the Nafion212 membranes was large, especially when RH was 60%. This means that the interface resistance between the CSPPSU membrane and the catalyst electrodes was large and that the resistance due to phase separation between the hydrocarbon-based CSPPSU membrane and the fluorine-based Nafion ionomer was large. Moreover, the CV characteristics show that the catalytic activity on the anode side and the cathode side of the cell containing the CSPPSU membrane was lower than that of the cell containing the Nafion212 membrane (Table 3). It is possible that the catalyst has been poisoned by the CSPPSU membrane.

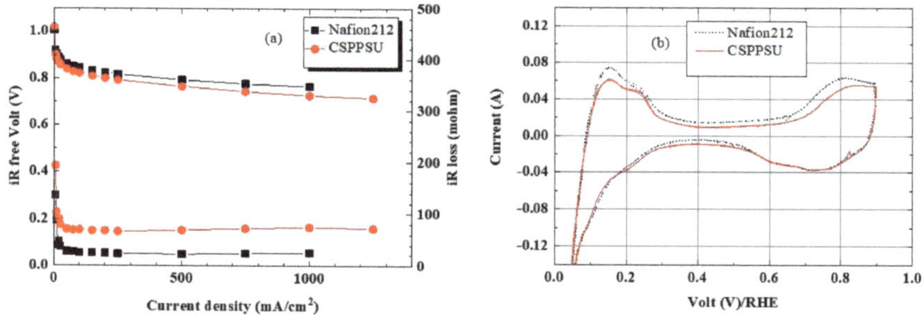

Figure 7. (**a**) I-$V_{iR\ free}$ and (**b**) CV properties of CSPPSU and Nafion212 membranes at 80 °C and 100% RH.

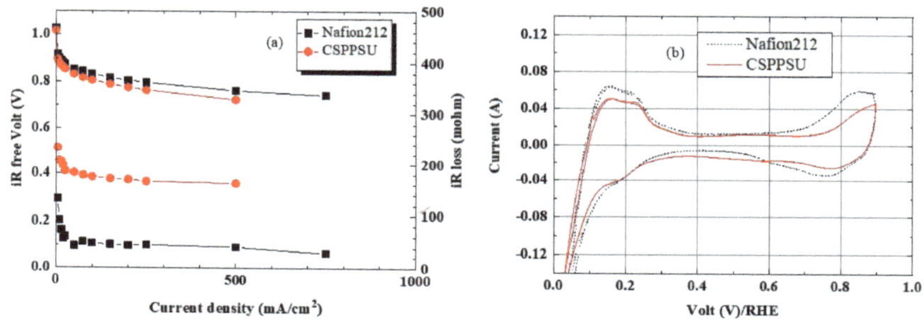

Figure 8. (**a**) I-$V_{iR\ free}$ and (**b**) CV properties of CSPPSU and Nafion212 membranes at 80 °C and 60% RH.

Table 3. I-V and H_2 crossover data for single cells using CSPPSU and Nafion212 membranes.

Membrane	80 °C, 100% RH			80 °C, 60% RH	
	iR Loss (mohm) at 1 A/cm^2	ECA (m^2/g)	H_2 Crossover (mA/cm^2) at 0.4 V	iR Loss (mohm) at 1 A/cm^2	ECA (m^2/g)
CSPPSU	73	82	0.085	161	73
Nafion212	24	99	1.24	42	92

3.8. H_2 Crossover of the CSPPSU Membranes

Figure 9 shows hydrogen gas crossover characteristics for MEA using the CSPPSU and Nafion212 membranes obtained using LSV [53]. The current densities of the MEA using the CSPPSU and the Nafion212 membranes at 0.4 V were 0.08 and 1.24 mA/cm^2, respectively (Table 3). From the literature [53], gas crossover occurs when the current density is >2 mA/cm^2, and short circuit occurs when the current density is >5.5 mA/cm^2. The current density of the Nafion212 membrane was in good agreement with literature values [53]. On the other hand, the current density of the CSPPSU membranes is 15 times smaller than that of the Nafion212 membrane. The current density of the CSPPSU membranes was lower than those using other hydrocarbon membranes [40], indicating that the CSPPSU membrane has high durability against gas crossover.

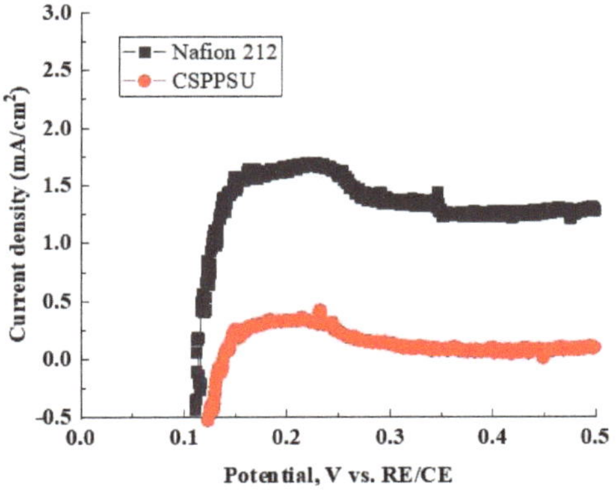

Figure 9. Hydrogen crossover properties of CSPPSU and Nafion212 membranes at 80 °C and 100% RH.

3.9. Durability Using CSPPSU Membranes

Figure 10 shows a plot of the voltage vs. time for MEAs using CSPPSU and Nafion212 membranes with an applied current of 1 A at a cell temperature of 80 °C and 100% RH. The MEA cell using the Nafion212 membrane was very stable during the measurement. On the other hand, the MEA cell using the CSPPSU membrane was a little unstable in the initial stage but became stable by 4000 h. However, the voltage gradually decreased after 4000 h. The single cell was separated, and the MEA was inspected. There were no cracks or pinholes in the membrane where no electrode was attached, and the appearance was the same as that before the test. However, the portion between the membrane and the electrode could not be observed because the membrane and the electrode were firmly attached to each other. However, there is a possibility that the voltage may have decreased due to mechanical deterioration, such as a crack or pinhole, between the membrane and the electrode. Moreover, it is conceivable that HO or HO_2 radicals formed during the reduction of oxygen on the cathode side may chemically degrade the CSPPSU membrane. There are few reports on the long-term stabilities of fuel cells using hydrocarbon-based electrolyte membranes, and we believe this is the first time that the long-term stability of a cell using a CSPPSU membrane has been evaluated for 4000 h. We are currently conducting more detailed evaluations of MEA cells and developing CSPPSU membranes with long-term stabilities like Nafion membranes.

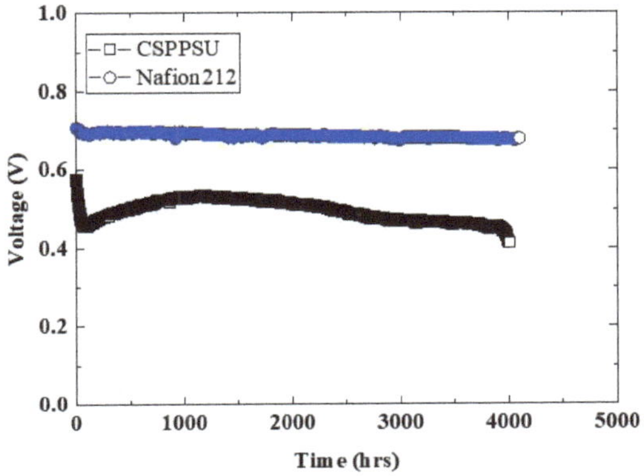

Figure 10. Durability of CSPPSU and Nafion212 membranes at 80 °C, 100% RH, and 1 A.

4. Conclusions

High-IEC SPPSU was synthesized using commercially available PPSU, and a large-area crosslinked membrane was produced. In addition, we developed an activation process using alkaline and acidic solutions to remove the remaining sulfur dioxide (SO_2) formed as a byproduct during the heat treatment. CSPPSU membranes activated with our method were compared with a Nafion212 membrane. The proton conductivities of the CSPPSU membranes were lower than that of the Nafion212 membrane, but the glass transition temperature and tensile strength were higher. This indicated that the CSPPSU membrane could be used even at high temperatures. From SAXS analysis, the CSPPSU membrane had an ion cluster structure that could act as a proton conduction path. The I-$V_{iR\ free}$ characteristics of MEA using the CSPPSU membrane were very good and similar to those when a Nafion212 membrane was used. The durability towards hydrogen gas crossover using the CSPPSU membranes was better than that using the Nafion212 membrane. In the long-term evaluation by using a constant current method, a stability of 4000 h was obtained for the first time. On the other hand, from CV, the resistance between the membrane and the electrode and the poisoning of the catalyst were larger than those when the Nafion212 membrane was used. Thus, further analyses and higher performances of MEAs are required.

Author Contributions: Conceptualization, J.-D.K.; methodology, J.-D.K.; validation, J.-D.K.; formal analysis, J.-D.K., A.O.; investigation, J.-D.K., A.O., H.N.; data curation, J.-D.K.; writing—Original draft preparation, J.-D.K.; writing—Review and editing, J.-D.K. All authors have read and agreed to the published version of the manuscript.

Funding: This research received no external funding.

Conflicts of Interest: The authors declare no conflict of interest.

References

1. Gasteiger, H.A.; Kocha, S.S.; Sompalli, B.; Wagner, F.T. Activity benchmarks and requirements for Pt, Pt-alloy, and non-Pt oxygen reduction catalysts for PEMFCs. *Appl. Catal. B Environ.* **2005**, *56*, 9–35. [CrossRef]
2. Higgins, D.C.; Chen, Z. Recent progress in non-precious metal catalysts for PEM fuel cell applications. *Can. J. Chem. Eng.* **2013**, *91*, 1881–1895. [CrossRef]
3. Sharma, S.; Pollet, B.G. Support materials for PEMFC and DMFC electrocatalysts—A review. *J. Power Sources* **2012**, *208*, 96–119. [CrossRef]
4. Mauritz, K.A.; Moore, R.B. State of understanding of Nafion. *Chem. Rev.* **2004**, *104*, 4535–4585. [CrossRef]
5. Yandrasits, M.; Lindell, M.; Schaberg, M.; Kurkowski, M. Increasing fuel cell efficiency by using ultra-low equivalent weight ionomers. *Electrochem. Soc. Interface* **2017**, *26*, 49–53. [CrossRef]
6. Savinell, R.F.; Litt, M.H. Proton Conducting Polymers Used as Membranes. US Patent Number 5525436, 11 June 1996.
7. Kerres, J.; Ulrich, A.; Meier, F.; Haring, T. Synthesis and characterization of novel acid—Base polymer blends for application in membrane fuel cells. *Solid State Ion.* **1999**, *125*, 243–249. [CrossRef]
8. Wu, D.; Wu, L.; Woo, J.-J.; Yun, S.-H.; Seo, S.-J.; Xu, T.; Moon, S.-H. A simple heat treatment to prepare covalently crosslinked membranes from sulfonated poly(2,6-dimethyl-1,4-phenylene oxide) for application in fuel cells. *J. Membr. Sci.* **2010**, *348*, 167–173. [CrossRef]
9. Yin, Y.; Suto, Y.; Sakabe, T.; Chen, S.; Hayashi, S.; Mishima, T.; Yamada, O.; Tanaka, K.; Kita, H.; Okamoto, K.-I. Water stability of sulfonated polyimide membranes. *Macromolecules* **2006**, *39*, 1189–1198. [CrossRef]
10. Asano, N.; Aoki, M.; Suzuki, S.; Miyatake, K.; Uchida, H.; Watanabe, M. Aliphatic/aromatic polyimide ionomers as a proton conductive membrane for fuel cell applications. *J. Am. Chem. Soc.* **2006**, *128*, 1762–1769. [CrossRef]
11. Aoki, M.; Asano, N.; Miyatake, K.; Uchida, H.; Watanabe, M. Durability of sulfonated polyimide membrane evaluated by long-term polymer electrolyte fuel cell operation. *J. Electrochem. Soc.* **2006**, *153*, A1154–A1158. [CrossRef]
12. Donnadio, A.; Casciola, M.; Di Vona, M.L.; Tamilvanan, M. Conductivity and hydration of sulfonated polyethersulfone in the range 70–120 °C: Effect of temperature and relative humidity cycling. *J. Power Sources* **2012**, *205*, 145–150. [CrossRef]

13. Kim, J.-D.; Donnadio, A.; Jun, M.-S.; Di Vona, M.L. Crosslinked SPES-SPPSU membranes for high temperature PEMFCs. *Int. J. Hydrogen Energy* **2013**, *38*, 1517–1523. [CrossRef]
14. Miyake, J.; Taki, R.; Mocizuki, T.; Shimizu, R.; Akiyama, R.; Uchida, M.; Miyatake, K. Design of flexible polyphenylene proton-conducting membrane for next-generation fuel cells. *Sci. Adv.* **2017**, *3*, 1–8. [CrossRef] [PubMed]
15. Schuster, M.; Kreuer, K.D.; Andersen, H.T.; Maier, J. Sulfonated poly(phenylene sulfone) polymers as hydrolytically and thermooxidatively stable proton conducting ionomers. *Macromolecules* **2007**, *40*, 598–607. [CrossRef]
16. Ding, X.; Liu, Z.; Hua, M.; Kang, T.; Li, X.; Zhang, Y. Poly(ethylene glycol) crosslinked sulfonated polysulfone composite membranes for forward osmosis. *J. Appl. Polym. Sci.* **2016**, *133*, 43941. [CrossRef]
17. Yu, J.; Dong, C.; Liu, J.; Li, C.; Fang, J.; Guan, R. Crosslinked sulfonated poly (bis-A)-sulfones as proton exchange membrane for PEM fuel cell application. *J. Mater. Sci.* **2010**, *45*, 1017–1024. [CrossRef]
18. Lafitte, B.; Karlsson, L.E.; Jannash, P. Sulfophenylation of polysulfones for proton-conducting fuel cell membranes. *Macromol. Rapid Commun.* **2002**, *23*, 896–900. [CrossRef]
19. Lufrano, F.; Gatto, I.; Staiti, P.; Antonucci, V.; Passalacua, E. Sulfonated polysulfone ionomer membranes for fuel cells. *Solid State Ion.* **2001**, *145*, 47–51. [CrossRef]
20. Kerres, J.A. Development of ionomer membranes for fuel cells. *J. Membr. Sci.* **2001**, *185*, 3–27. [CrossRef]
21. Lufrano, F.; Squadrito, G.; Patti, A.; Passalacqua, E. Sulfonated polysulfone as promising membranes for polymer electrolyte fuel cells. *J. Appl. Polym. Sci.* **2000**, *77*, 1250–1257. [CrossRef]
22. Di Vona, M.L.; Alberti, G.; Sgreccia, E.; Casciola, M.; Knauth, P. High performance sulfonated aromatic ionomers by solvothermal macromolecular synthesis. *Int. J. Hydrogen Energy* **2012**, *37*, 8672–8680. [CrossRef]
23. Chen, J.; Zhai, M.; Asano, M.; Huang, L.; Maekawa, Y. Long-term performance of polyetheretherketone-based polymer electrolyte membrane in fuel cells at 95 °C. *J. Mater. Sci.* **2009**, *44*, 3674–3681. [CrossRef]
24. Goto, K.; Rozhanskii, I.; Yamakawa, Y.; Ohtsuki, T.; Naito, Y. Development of aromatic polymer electrolyte membrane with high conductivity and durability for fuel cell. *JSR Tech. Rev.* **2009**, *116*, 1–11. [CrossRef]
25. Di Vona, M.L.; Sgreccia, E.; Licoccia, S.; Alberti, G.; Tortet, L.; Knauth, P. Analysis of temperature-promoted and solvent-assisted cross-linking in sulfonated poly (ether ether ketone)(SPEEK) proton-conducting membranes. *J. Phys. Chem. B* **2009**, *113*, 7505–7512. [CrossRef] [PubMed]
26. Di Vona, M.L.; Sgreccia, E.; Licoccia, S.; Khadhraoui, M.; Denoyel, R.; Knauth, P. Composite proton-conducting hybrid polymers: Water sorption isotherms and mechanical properties of blends of sulfonated PEEK and substituted PPSU. *Chem. Mater.* **2008**, *20*, 4327–4334. [CrossRef]
27. Di Vona, M.L.; Marani, D.; D'Ottavi, C.; Trombetta, M.; Traversa, E.; Beaurroies, I.; Knauth, P.; Lioccia, S. A simple new route to covalent organic/inorganic hybrid proton exchange polymeric membranes. *Chem. Mater.* **2006**, *18*, 69–75. [CrossRef]
28. Harrison, W.L.; Hickner, M.A.; Kim, Y.S.; McGrath, J.E. Poly(arylene ether sulfone) copolymers and related systems from disulfonated monomer building blocks: Synthesis, characterization, and performance—A topical review. *Fuel Cells* **2005**, *5*, 201–212. [CrossRef]
29. Xing, D.M.; Yi, B.L.; Liu, F.Q.; Fu, Y.Z.; Zhang, H.M. Characterization of sulfonated poly(etheretherketone)/polytetrafluoroethylene composite membranes for fuel cell applications. *Fuel Cells* **2005**, *5*, 406–411. [CrossRef]
30. Xing, P.; Robertson, G.P.; Guiver, M.D.; Mikhailenko, S.D.; Wang, K.; Kaliaguine, S. Synthesis and characterization of sulfonated poly(etheretherketone) for proton exchange membranes. *J. Membr. Sci.* **2004**, *229*, 95–106. [CrossRef]
31. Mikhailenko, S.D.; Wang, K.; Kaliaguine, S.; Xing, P.; Robertson, G.P.; Guiver, M.D. Proton conducting membranes based on cross-linked sulfonated poly(etheretherketone) (SPEEK). *J. Membr. Sci.* **2004**, *233*, 93–99. [CrossRef]
32. Li, L.; Zhang, J.; Wang, Y. Sulfonated poly(etheretherketon) membranes for direct methanol fuel cell. *J. Membr. Sci.* **2003**, *226*, 159–167. [CrossRef]
33. Feng, S.; Pang, J.; Yu, X.; Wang, G.; Manthiram, A. High-performance semicrystalline poly(ether ketone)-based proton exchange membrane. *ACS Appl. Mater. Interfaces* **2017**, *9*, 24527–24537. [CrossRef]
34. Matsushita, S.; Kim, J.-D. Organic solvent-free preparation of electrolyte membranes with high proton conductivity using aromatic hydrocarbon polymers and small cross-linker molecules. *Solid State Ion.* **2018**, *316*, 102–109. [CrossRef]

35. Tashvigh, A.A.; Luo, L.; Chung, T.S.; Weber, M.; Maletzko, C. A novel ionically cross-linked sulfonated polyphenylsulfone (SPPSU) membrane for organic solvent nanofiltration (OSN). *J. Membr. Sci.* **2018**, *545*, 221–228. [CrossRef]
36. Zhang, Y.; Kim, J.-D.; Miyatake, K. Effect of thermal crosslinking on the properties of sulfonated poly(phenylene sulfone)s as proton conductive membranes. *J. Appl. Polym. Sci.* **2016**, *44218*, 1–8. [CrossRef]
37. Kim, J.-D.; Ghil, L.-J. Annealing effect of highly sulfonated polyphenylsulfone polymer. *Int. J. Hydrogen Energy* **2016**, *41*, 11794–11800. [CrossRef]
38. Lee, H.; Han, M.; Choi, Y.W.; Bae, B. Hydrocarbon-based polymer electrolyte cerium composite membranes for improved proton exchange membrane fuel cell durability. *J. Power Sources* **2015**, *295*, 221–227. [CrossRef]
39. Takamuku, S.; Wohlfarth, A.; Manhart, A.; Rader, P.; Jannasch, P. Hypersulfonated polyelectrolytes: Preparation, stability and conductivity. *Polym. Chem.* **2015**, *6*, 1267–1274. [CrossRef]
40. Krishnan, N.N.; Henkensmeier, D.; Jang, J.H.; Hink, S.; Kim, H.-J.; Nam, S.-W.; Lim, T.-H. Locally confined membrane modification of sulfonated membranes for fuel cell application. *J. Membr. Sci.* **2014**, *454*, 174–183. [CrossRef]
41. Yu, D.M.; Yoon, S.; Kim, T.-H.; Lee, J.Y.; Lee, J.; Hong, Y.T. Properties of sulfonated poly(arylene ether sulfone)/electrospun nonwonen polyacrylonitrile composite membrane for proton exchange membrane fuel cells. *J. Membr. Sci.* **2013**, *446*, 212–219. [CrossRef]
42. Takamuku, S.; Jannasch, P. Properties and degradation of hydrocarbon fuel cell membranes: A comparative study of sulfonated poly(arylene ether sulfone)s with different positions of the acid groups. *Polym. Chem.* **2012**, *3*, 1202–1214. [CrossRef]
43. Wen, P.; Zhong, Z.; Li, L.; Zhang, A.; Li, X.-D.; Lee, M.-H. Photocrosslinking of sulfonated poly(arylene ether sulfone) in a swollen state. *J. Mater. Chem.* **2012**, *22*, 22242–22249. [CrossRef]
44. Chen, S.; Zhang, X.; Chen, K.; Endo, N.; Higa, M.; Okamoto, K.-I.; Wang, L. Cross-linked miscible blend membranes of sulfonated poly(arylene ether sulfone) and sulfonated polyimide for polymer electrolyte fuel cell applications. *J. Power Sources* **2011**, *196*, 9946–9954. [CrossRef]
45. Feng, S.; Shang, Y.; Xie, X.; Wang, Y.; Xu, J. Synthesis and characterization of crosslinked sulfonated poly(arylene ether sulfone) membranes for DMFC applications. *J. Membr. Sci.* **2009**, *335*, 13–20. [CrossRef]
46. Di Vona, M.L.; Luchetti, L.; Spera, G.P.; Sgreccia, E.; Knauth, P. Synthetic strategies for the preparation of proton-conducting hybrid polymers based on PEEK and PPSU for PEM fuel cells. *C. R. Chim.* **2008**, *11*, 1074–1081. [CrossRef]
47. Xing, D.; Kerres, J. Improved performance of sulfonated polyarylene ethers for proton exchange membrane fuel cells. *Polym. Adv. Technol.* **2006**, *17*, 591–597. [CrossRef]
48. Karlsson, L.E.; Jannasch, P. Polysulfone ionomers for proton-conducting fuel cell membranes 2. Sulfophenylated polysulfones and polyphenylsulfones. *Electrochim. Acta* **2005**, *50*, 1939–1946. [CrossRef]
49. Dyck, A.; Fritsch, D.; Nunes, S.P. Proton-conductive membranes of sulfonated polyphenylsulfone. *J. Appl. Polym. Sci.* **2002**, *86*, 2820–2827. [CrossRef]
50. Nguyen, M.D.T.; Kim, D. Cross-linked poly(arylene ether ketone) proton exchange membranes sulfonated on polymer backbone, pendant, and cross-linked sites for enhanced proton conductivity. *Solid State Ion.* **2015**, *270*, 66–72. [CrossRef]
51. Kim, J.-D.; Ohnuma, M.; Nishimura, C.; Mori, T.; Kucernak, A. Small-angle X-ray scattering and proton conductivity of anhydrous Nafion-Benzimidazole blend membranes. *J. Electrochem. Soc.* **2009**, *156*, B729–B734. [CrossRef]
52. Kim, J.-D.; Ghil, L.-J.; Ohira, A. Annealing effect of Nafion-1,2,3-triazole membrane by autoclave solution processing. *ECS Trans.* **2018**, *85*, 943–959. [CrossRef]
53. Cooper, K.R. In situ PEMFC fuel crossover & electrical short circuit measurement. *Fuel Cell Mag.* **2008**, *8–9*, 1–2.

membranes

Article

The Effects of Temperature and Humidity on the Microstructure of Sulfonated Syndiotactic–polystyrene Ionic Membranes

Maria-Maddalena Schiavone [1], David Hermann Lamparelli [2], Yue Zhao [3], Fengfeng Zhu [1], Zsolt Revay [4] and Aurel Radulescu [1,*]

1 Forschungszentrum Jülich GmbH, Jülich Centre for Neutron Science (JCNS) at Heinz Maier-Leibnitz Zentrum (MLZ), 85747 Garching, Germany; schiavonemariamaddalena@yahoo.it (M.-M.S.); f.zhu@fz-juelich.de (F.Z.)
2 Dipartimento di Chimica e Biologia "Adolfo Zambelli", Università di Salerno, I-84084 Fisciano, Italy; dlamparelli@unisa.it
3 Department of Advanced Functional Materials Research, Takasaki Advanced Radiation Research Institute, National Institutes for Quantum and Radiological Science and Technology (QST), Watanuki-machi 1233, Takasaki 370-1292, Japan; zhao.yue@qst.go.jp
4 Technische Universität Müchen, Forschungsneutronenquelle Heinz Maier-Leibnitz FRM II, Heinz Maier-Leibnitz Zentrum (MLZ), 85747 Garching, Germany; Zsolt.Revay@frm2.tum.de
* Correspondence: a.radulescu@fz-juelich.de; Tel.: +49-89-289-10712

Received: 21 July 2020; Accepted: 12 August 2020; Published: 14 August 2020

Abstract: Polymeric membranes based on the semi-crystalline syndiotactic–polystyrene (sPS) become hydrophilic, and therefore conductive, following the functionalization of the amorphous phase by the solid-state sulfonation procedure. Because the crystallinity of the material, and thus the mechanical strength of the membranes, is maintained and the resistance to oxidation decomposition can be improved by doping the membranes with fullerenes, the sPS becomes attractive for proton-exchange membranes fuel cells (PEMFC) and energy storage applications. In the current work we report the micro-structural characterization by small-angle neutron scattering (SANS) method of sulfonated sPS films and sPS–fullerene composite membranes at different temperatures between 20 °C and 80 °C, under the relative humidity (RH) level from 10% to 70%. Complementary characterization of membranes was carried out by FTIR, UV-Vis spectroscopy and prompt-γ neutron activation analysis in terms of composition, following the specific preparation and functionalization procedure, and by XRD with respect to crystallinity. The hydrated ionic clusters are formed in the hydrated membrane and shrink slightly with the increasing temperature, which leads to a slight desorption of water at high temperatures. However, it seems that the conductive properties of the membranes do not deteriorate with the increasing temperature and that all membranes equilibrated in liquid water show an increased conductivity at 80 °C compared to the room temperature. The presence of fullerenes in the composite membrane induces a tremendous increase in the conductivity at high temperatures compared to fullerenes-free membranes. Apparently, the observed effects may be related to the formation of additional hydrated pathways in the composite membrane in conjunction with changes in the dynamics of water and polymer.

Keywords: proton exchange membranes; semi-crystalline polymers; small-angle neutron scattering

1. Introduction

Polymer electrolyte materials used as proton exchange membranes (PEM) for fuel cells applications (PEMFC) are characterized by a nanoscale phase separation into hydrophilic domains and hydrophobic regions, which is a combination that enables a high, water-mediated, proton conductivity and provides a good chemical and mechanical stability, and thus membrane durability [1–3]. Although it is characterized by excellent conductive properties, the Nafion, which was established as a benchmark for such applications [4], presents also some drawbacks such as high cost, lack of safety, and the requirement of supporting equipment during manufacturing and use [5]. Furthermore, it shows limitations under operating conditions at high temperature (> 80 °C) and low relative humidity (RH), when a decrease in conductivity appears due to dehydration of the membrane at the anode side [5]. Moreover, free radicals such as hydroxyl and hydroperoxyl are produced during the operation of the PEMFC as a result of the reaction of hydrogen and oxygen on the electrodes or the decomposition of hydrogen peroxide with metal contaminants in the membrane. These radicals initiate processes of chemical degradation that affect the durability and the lifetime of the PEM [6,7]. Therefore, improvements in the Nafion properties by incorporating inorganic fillers (such as titania, zirconia, and silica) or carbon-based nanomaterials (carbon nanotubes, fullerenes, etc.) into the membranes [7–14] and alternative low-cost materials that present similar conductive and chemo-mechanical properties, such as the Nafion membranes [14–25], are continuously searched for. Given the recent developments, which enable a controlled sulfonation of only the amorphous phase [26], preserving thus the crystallinity of the material, and an improved resistance to oxidation decomposition when fullerenes are added [27], the sulfonated syndiotactic polystyrene (s–sPS) is a good potential candidate for some PEMFC applications, as it presents a high proton conductivity comparable to Nafion [28], high chemical and thermomechanical stability, and a low cost. The preparation of a s–sPS membrane should start from the δ-form (clathrate with guest molecules), which enables the functionalization of the phenyl groups in the amorphous region and can be subsequently transformed into the thermodynamically stable β-form by high temperature annealing procedures. The s–sPS membranes present a strong dependence of the proton conductivity on sulfonation level, temperature and hydration conditions [29,30].

The proton conduction in PEMs depends on water and is governed by the water behavior at different length scales: at the molecular scale: dissociation of protons and formation of the ion-pair with water, at the nanoscale: the transport through the hydrated domains, at the mesoscale: the long-range mobility within the water network [4]. Therefore, in order to understand the transport properties in different conditions, one should first of all learn about the morphology of hydrated domains at different length scales as a function of hydration level and temperature. The microstructure of PEM materials and the elucidation of the conductive paths at the microscopic level are highly debated topics. In previous works [31,32] we reported a detailed microstructural characterization of s–sPS membranes with δ–clathrate co–crystalline form, which was carried out by extended Q-range small-angle neutron scattering, SANS, where $Q = 4\pi\lambda^{-1}\sin(\theta/2)$ is the modulus of the scattering vector Q, with λ the incident neutron wavelength and θ the scattering angle. Membranes with different sulfonation degrees were investigated in dry and hydrated states at room temperature. A humidity chamber (Anton Paar, Graz, Austria) was used to control the relative humidity (RH) degree between 5% and 95%. The use of uniaxially deformed film samples and neutron contrast variation allowed for the identification and characterization of different structural levels with sizes between nm and μm, which form and evolve with the variation of the hydration level. The neutron scattering length density (SLD) of the crystalline regions was varied using different toluene isotopologues incorporated as guest molecules into the sPS lattice [33], while the variation of the scattering properties of the hydrated amorphous regions was achieved using different H_2O/D_2O mixtures. Deuterated s–sPS films were used in the investigation to obtain a low incoherent background. According to our qualitative and quantitative analysis of the SANS data in terms of structural models, the hydration water is taken-up at low RH in clusters formed around the sulfonic groups. The clusters in the bulk amorphous region grow in size with increasing hydration level, favored by the increased flexibility of the sPS chains in these domains. At very high hydration level, towards RH = 100%, the clusters become interconnected one with another, which gives rise to the formation of cylindrical channels morphology. Observations made by cryo-TEM on fully hydrated films support the SANS conclusions [31]. Moreover, the neutron contrast variation measurements revealed that the hydrated ionic clusters promoting the conductivity of the membranes include the segments of the sPS chains that are affected by sulfonation, hence the phenyl-groups too. The extended Q-range SANS data have also shown that the crystallinity, and hence the robustness of the membrane, is preserved during drying-hydration processes: the 010 crystalline reflection, which is indicative of the crystalline δ–form, was always observed in the scattering pattern, no matter which contrast or hydration conditions were used, while the scattering pattern that is characteristic of a dry membrane was always recovered after exposing the membrane to different hydration procedures.

In this paper, we discuss the temperature effect on the microstructure of the s–sPS membranes at different hydration levels, well below the full hydration state, based on results of a SANS investigation done on cast and uniaxially deformed s–sPS semi-crystalline films. The films consist of the δ–co–crystalline phase of sPS with deuterated toluene and an amorphous phase in which C60 fullerenes have been incorporated. A detailed qualitative analysis was carried out on the cast films, which were studied at different temperatures between 20 °C and 80 °C, with the variation of the hydration level at every temperature between RH = 10% and RH = 80%. Quantitative structural and water content information were extracted from the analysis of the scattering data from the uniaxially deformed films in terms of structural models, which were also used before [31,32]. The use of uniaxially deformed films enabled the separation of the scattering contribution from different components of the hydrated morphologies on the two-dimensional SANS detector while covering different angular scattering ranges, as detailed in our earlier works. Usually, for a thorough characterization of complex multiphase systems such as polymeric membranes used in energy and biomedical applications [32,34], a combination of experimental techniques is necessary and enables the exploration of the microstructure, morphology and composition properties in various chemical and thermodynamic conditions and their relationship with macroscopic properties of interest for particular applications. Complementary information about the composition and crystallinity of the sPS-based membranes were delivered by Fourier-transform infrared spectroscopy (FTIR), X-ray diffraction (XRD) and prompt-gamma neutron activation analysis (PGAA). Finally, the conductivity of membranes was measured on the cast films equilibrated in liquid water at different temperatures. First insights were obtained about the structural changes induced by the variation of temperature and hydration in the membrane morphology. On the other hand, the incorporation of fullerenes in the composite membrane induces a tremendous increase in membrane conductivity at high temperatures compared to fullerenes-free membranes. No evidence to explain this effect was obtained from the micro-structural analysis at this level. Apparently, this effect may be related to the formation of additional hydrated pathways in the composite membrane in conjunction with changes in the dynamics of water and polymer.

2. Materials and Methods

The preparation and subsequent treatment—clathration, sulfonation, and guest-exchange in the crystalline region of uni-axially-oriented deuterated syndiotactic polystyrene films were described elsewhere [31]. For the SANS experiment performed in this study, the exchange of the guest molecules in the polymer clathrate form from d–chloroform, which was loaded during the sulfonation procedure, to d-toluene, was achieved by dipping the films for 1 day in the new solvent, followed by drying at 40 °C under vacuum for a couple of hours.

To prepare the sPS, C60 composite membranes required amounts of newly synthesized deuterated sPS [31], and commercially achieved C60 fullerenes (Merck, Darmstadt, Germany) and h–toluene (ARMAR Chemicals, Döttingen, Switzerland) were taken in hermetically sealed tubes and heated at 170 °C for several minutes, until the solution became homogeneous. The resultant solutions were cast uniformly on quartz substrates to form sPS–C60 composite membranes. Membranes were prepared with a different C60 content in the initial solution, ranging from 0.05 wt% to 1 wt%. The sPS–C60 composite membranes were subsequently functionalized by soaking the films in acyl sulfate solution in d–chloroform (ARMAR Chemicals, Döttingen, Switzerland) at 40 °C. The sulfonation agent was prepared according to the procedure described in [26,28]. Afterwards, the samples were removed from the solution and quickly dipped in acetone for a few minutes to remove traces of impurities (due to the possible remnants of the sulfonation procedures) and dried under the fume-hood for 24 h. The guest exchange from d–chloroform, which replaced the h–toluene, the original guest into the sPS cavities, during the sulfonation procedure, to d–toluene was carried out, to simplify the neutron contrast conditions for the SANS experiments, as discussed in [32].

As proceeded in our previous studies [31,32], the membranes were further on analyzed by FTIR, UV-Vis spectroscopy and prompt–γ neutron activation analysis in terms of composition following the specific preparation and functionalization procedure, and by XRD with respect to crystallinity.

The degree of sulfonation, expressed as S atoms/styrene units × 100 mol% and further indicated as S, was checked at the neutron prompt-gamma activation analysis (PGAA) instrument at the Heinz Maier-Leibnitz Zentrum (MLZ, Garching, Germany). Full descriptions of the experimental method and data interpretation can be found in [35].

Qualitative analysis of the sulfonation was checked by FTIR spectroscopy using a JASCO VIR-200 spectrometer (JASCO Deutschland, Pfungstadt, Germany) in a wavenumber range of 400–4000 cm^{-1}. The film samples were placed in special holders equipped with ZnSe windows.

To check the incorporation of C60 fullerenes in the membranes, UV-Vis analysis was carried out at a Cary 100 SCAN UV-Vis Varian spectrometer (Agilent, Santa Clara, CA, USA) with the film samples placed in a specific holder equipped with quartz windows. The spectra were collected in the range 200–800 nm at a resolution of 100 nm/min.

WAXD analysis of films was done in the range of 2θ between 5° and 35° by means of an X-ray powder diffractometer Brucker 2nd Gen-D2 Phaser (Cu-source) (Brucker, Karlsruhe, Germany). The degree of crystallinity was determined as 100 $A_c/(A_c + A_a)$, where A_c and A_a are the areas determined by resolving the diffraction pattern according to [36] and can be considered proportional to the crystalline and amorphous fractions of the polymer.

The water uptake capacity of the membranes was determined as following the preparation procedure described in [31]. The water uptake was calculated as the percentage increase in mass over the "dry" weight:

$$W_{uptake}(w\%) = \left[\left(W_{wet} - W_{dry}\right)/W_{dry}\right] \times 100\% \quad (1)$$

where Wwet and Wdry are the wet and dry weight of the membrane, respectively.

The conductivity of the membranes was measured in the plane direction at 100 kHz using four-point probe alternating current electrochemical impedance spectroscopy (EIS) with an electrode system connected to an LCR meter (HIOKI 3522 LCR HiTESTER, Nagano, Japan). Membranes were placed between two platinum electrodes and equilibrated in liquid water. The membrane's conductivity was measured at room temperature and 80 °C. The conductivity σ (mS/cm) was calculated from the obtained resistance R (Ω) according to the following equation.

$$\sigma\ (mS/cm) = L/(S \times R) \times 10^3 \quad (2)$$

where L (cm) is the distance between two electrodes, and S (cm^2) is the cross-sectional area of the membrane obtained by multiplying the membrane thickness by the membrane width.

The SANS measurements were carried out at the KWS-2 high intensity/extended-Q range pinhole SANS diffractometer (Forschungszentrum Jülich GmbH, Jülich, Germany) of JCNS at MLZ [37]. A Q-range between 0.003 and 0.6 $Å^{-1}$ was covered by using three sample-to-detector distances, L_D = 2 m, 4 m and 20 m, and a neutron wavelength λ = 5 Å. The film samples were exposed to in-situ controlled hydration and temperature variation by using a humidity chamber (Anton Paar, Graz, Austria). The temperature on the sample was varied between 20 °C and 80 °C, while the relative humidity at a fixed temperature value was varied within the range RH = 10% to 70%. The data treatment and interpretation according to structural models was done according to models and calculations, which are described in details in [32].

Finally, a uniaxially deformed s–sPS membrane loaded with C60 fullerenes, which was characterized by SANS in a previous study [32], was exposed to Fenton's test conditions [4,14,27] for 1 h at 60 °C and then briefly re-investigated with SANS (at KWS-2 diffractometer, Jülich, Germany) to get a first insight about the effect of the peroxyl/hydroxyl radicals, formed from the decomposition of H_2O_2 in an aqueous solution catalyzed by Fe^{2+}/Fe^{3+}, on the crystalline-amorphous morphology of the membrane. The Fenton's test represents a useful method for a preliminary assessment of the oxidative stability of PEMs. Our interest in the current work was limited only to the observation of possible changes in the scattering patterns, as a consequence of an eventual degradation of the membrane after the application of the test.

3. Results

3.1. Composition Characterization

Two sPS-based membranes were characterized prior to their investigation by SANS: sample A—a uniaxially deformed s–sPS film, and sample B—a composite sPS–C60 undeformed sulfonated film, which was prepared at a 0.5 wt% C60 content from the initial common solution with sPS in h–toluene. The membranes are based on deuterated sPS and the clathrate form in the investigated samples consisted of δ–co–crystals of sPS with d–toluene. The PGAA analysis [30] delivered a sulfonation degree of about S = 45% and 41% for the samples A and B, respectively.

The FTIR spectra from a membrane prepared in a similar way to sample A can be found in [31]. The FTIR observations done at different stages of the preparation procedure indicated the successful treatment and functionalization of the membranes of this type. The FTIR spectra from sample B are shown in Figure 1 as collected at different stages of preparation. The low spectrum (black line) was acquired from the sPS–C60 composite membrane after casting from the common solution in h–toluene. Typical spectral features of δ–form of deuterated sPS [38] can be observed, such as the ring stretching and bending modes and backbone C–D and C–D_2 stretching in the region of 2300 to 2150 cm^{-1} (the range indicated by the green horizontal arrows) and the C=C stretching at around 1570 cm^{-1} [39]; while at short wavenumbers, between 500 and 550 cm^{-1}, the bands characteristics of the chain conformation of deuterated sPS in the δ–crystalline form are displayed. These features appear in all further spectra regardless of treatment procedure. As mentioned in our previous work [32], due to the multitude of characteristic bands of sPS, it is very difficult to observe the IR bands of fullerenes. Experimental FTIR characterization of fullerenes in bulk or functionalized polymers can be found in [40] for C60 or [41] for C70, while theoretical calculations were done in [42]. The bands observed at around 1500 cm^{-1} (C–C stretching in the aromatic ring) and around 740 cm^{-1} (out-of-plane C–H bending) that are indicated by the black vertical arrows can be ascribed to vibrations characteristic of the h–toluene molecules, which are trapped in the cavities between the sPS helices (δ–clathrate form). The middle spectrum in Figure 1 (blue) was collected from the same sample after its sulfonation and washing with acetone. The two very broad features observed in the wavenumber ranges of 2500–3700 cm^{-1} and 1000–1250 cm^{-1} (marked with the red horizontal arrows) are indicative of the sulfonation of the sample. On the other hand, during the sulfonation procedure the h–toluene was replaced by the d–chloroform in the clathrate form. Accordingly, the IR band characteristics of h–toluene are not visible anymore in the spectrum, while those from d–chloroform cannot be easily distinguished from those of the deuterated sPS. Additional sharp spectral features due to h–acetone were observed (indicated by the blue vertical arrows). The IR bands in the region 3000 to 2800 cm^{-1} can be ascribed to the C–H stretching modes, while the strong feature at around 1700 cm^{-1} is due the C=O stretching.

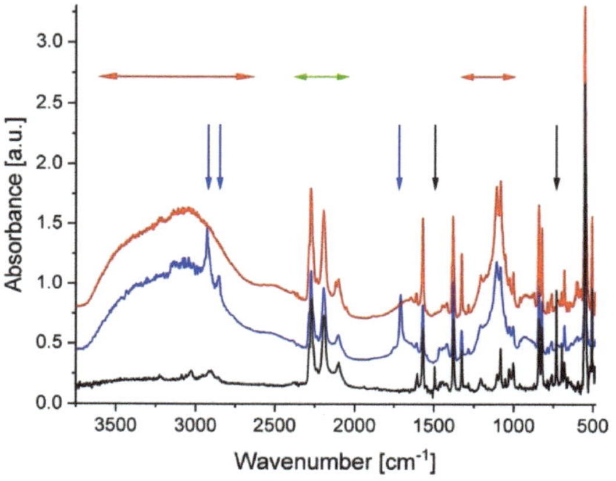

Figure 1. FTIR spectra from film sample B after different steps of the preparation and treatment procedure: black—after casting; blue—after functionalization (sulfonation); red—after dipping in d–toluene (to provide the exchange of guest molecules in the clathrate form). The arrows indicate the regions of interest for IR bands characteristic of different molecular groups, as discussed in the text. The spectra are shifted vertically for clarity.

The washed membrane, which was dipped in d–toluene subsequently to its sulfonation, delivered the IR spectrum shown in red in Figure 1. While the spectral features, which are indicative of the membrane sulfonation, are clearly visible, besides those from the deuterated sPS, the bands from the acetone or other washing agents are not present anymore. The spectral features from d–toluene are not distinguishable from the deuterated sPS bands.

The UV-Vis absorption spectra of the samples A and B are shown in Figure 2a. The UV-Vis analysis was done on sample B prior (blue curve) and after (red curve) its sulfonation. The characteristic absorption features of s–sPS (sample A) occur below 300 nm while above this value the absorbance falls quickly off. Pure C60 exhibits two maxima at 335 and 408 nm [43]. These features are well displayed by the spectrum from the sPS–C60 composite membrane after casting (blue curve). This is indicative of the incorporation of fullerenes in the membrane. Although still indicating a strong absorption above 300 nm, the spectrum from the same membrane after its functionalization (red curve) presents less pronounced and rather broad features compared to its state before sulfonation. Thus, the band at 335 nm, which is known to be affected by the environment of fullerenes [44,45], is broader and less intense, while the small peak at 408 nm completely disappears, and a new broad shoulder-like absorption between 400 and 500 nm appears. The excess absorption above 400 nm seems to be due to agglomeration of the C60 fullerenes. As reported in [44], there is a propensity shown by fullerenes to aggregation that depends on sample preparation and treatment procedures. It seems that the effect observed in the UV-Vis spectra from functionalized sPS/C60 composite membranes, which looks similar to those observed from fullerenes-doped s–sPS membranes [32], is a consequence of the sulfonation of the membrane. However, a clear understanding of this effect requires further investigation.

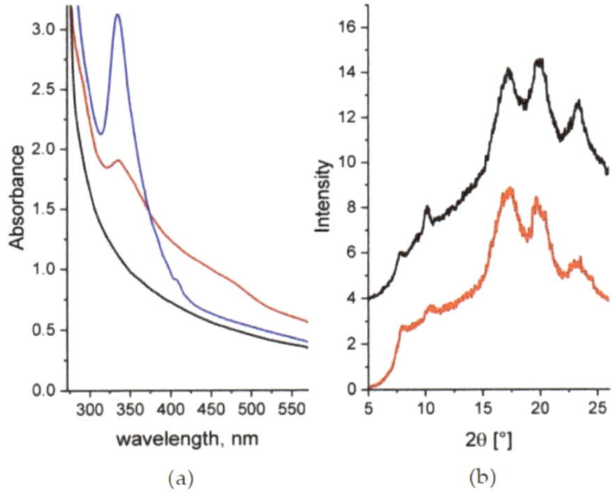

Figure 2. UV-Vis spectra (**a**) and XRD patterns (**b**) from samples A and B. In both panels the black (sample A) and the red (sample B) data are from samples in the final stage of treatment (sulfonated and containing d–toluene in the clathrate form), while the blue curve in the Figure 2a is from the sample B as cast (before functionalization). The XRD patterns are shifted vertically for clarity.

The WAXD spectra from the samples A and B in their final treatment stage, after functionalization and exchange of the guest molecules in the clathrate form to d–toluene, are presented in Figure 2b. The pair of peaks at around 8° and 10.5° in 2θ is indicative of the formation of the crystalline δ–form of the sPS clathrates with guest molecules [46,47]. The presence of these peaks in the patterns collected from both samples indicates that the sPS crystalline habit is preserved, regardless of treatment. The positions of the diffraction peaks are the same in all WAXD patterns, which indicates that the addition of fullerenes does not change the parameters of the polymer crystalline lattice. Evidences about the incorporation of C60 within the crystalline lattice of sPS were reported in [47] following the analysis of the XRD patterns from sPS–C60 composite samples with higher C60 content than our membranes. The additional peaks that were observed besides those from the δ–form of the sPS were attributed to an FCC arrangement of C60. Although we did not observe any additional peaks in the XRD patterns from our film samples, incorporation of some of the C60 within the polymer crystalline lattice cannot be completely ruled out. Although in this case fullerenes may compete with the d–toluene molecules in occupying the cavities between the sPS helices in the clathrate form, however, because of the rather similar neutron scattering length density (SLD) of C60 and d–toluene [32], no change in the neutron scattering properties of the crystalline regions is expected for the SANS experiments. Finally, from the analysis of the WAXD patterns in terms of the A_c (peaks) and A_a (background) areas [36], the crystallinity of the samples A and B was about 33% and 31%, respectively.

3.2. Water Uptake and Conductivity

The results obtained from the analysis of the water uptake by the membranes equilibrated in liquid water at room temperature (Table 1) show that the sulfonated membranes we prepared based on deuterated sPS present similar hydration properties with those made of hydrogenated sPS with a comparable sulfonation degree, which were reported in [28].

Table 1. The characteristics of the sPS-based films after the sulfonation and hydration procedures.

Parameter	Sample A	Sample B
Sulfonation degree (%)	46.3	41.5
Crystallinity (%)	33	31
Water uptake at 25 °C (%)	87.5	76.3
Conductivity 25 °C (mS/cm)	128.2	99.9
Conductivity 80 °C (mS/cm)	160.3	219.8

The determination of the water uptake capacity is important for assessing the membrane functionality and performance in terms of the water and ion-transport, and the swelling and mechanical integrity. For this purpose, it is more suitable to express the amount of water that is taken up by the membrane as the number of water molecules per number of sulfonic acid sites, which defines the water content parameter, or the hydration number, $\lambda \equiv n(H_2O)/n(SO_3H)$ [13]. The hydration number λ relates to the water uptake W_{uptake} (Equation (1)) using the equivalent weight of the membrane (EW, grams of dry polymer per ionic group), which is inversely proportional to the membrane ion-exchange capacity (IEC): $\lambda = W_{uptake} \times EW/M(H_2O)$, with $M(H_2O)$—the molecular weight of water (18 g mol^{-1}). Assuming that the IEC of the s–sPS membranes is about 1.23 meq/g, similar to what we determined in [31] using the titration method, hydration numbers $\lambda = 39.3$ and 34.5 were obtained for our membranes A and B in fully hydrated state at room temperature, respectively. As a comparison, experiments provided values of λ in the range 20 to 30 for Nafion in the same hydration and temperature conditions [4]. Such high values of λ correspond to a bulk-like water regime that is reached as a consequence of the growth and connectivity of the hydrophilic domains when the membrane is equilibrated in liquid water. In this regime, the water molecules move freely, although still confined within nanodomains. The swelling of the hydrophilic domains as a consequence of water uptake is a multistep process [4], which starts with the formation of a hydration shell around the sulfonic acid groups and dissociation of protons, which become solvated and mobile at low hydration numbers ($\lambda = 1$ to 2)—the strongly bound water regime and continues with the formation of multiple solvation shells and water domains and the water percolation, in increasing the hydration level (λ up to 5 to 6). A further increase in hydration level leads to the growth and interconnecting hydrophilic domains and the transition from bound to free water ($\lambda > 6$).

Usually the proton conductivity depends on the water uptake capacity of the membrane [4], which depends on its sulfonation degree [28]. The measured conductivity values of samples A and B at room temperature and at 80 °C are also reported in Table 1. At room temperature, the sample A, which has a higher sulfonation degree, is characterized by a higher conductivity than sample B. A detailed analysis of the dependence of proton conductivity on sulfonation degree in sPS-based membranes can be found in [24,26]. Both samples A and B show a higher proton conductivity at 80 °C compared to that exhibited at room temperature. As reported in the literature, the dependence of the proton conductivity on the temperature is Arrhenius-type [48,49], although deviations from this behavior can be observed at high temperatures and low hydration levels in some types of Nafion membranes [50,51]. However, at 80 °C, the sample B shows a much higher conductivity than the sample A, which seems to be an effect related to the incorporation of fullerenes into the membrane. An increase in conductivity of PEMs when fullerenes were added was observed in the case of Nafion 117. Conductivity of Nafion 117 and Nafion-C60 composite membrane as a function of RH and temperature is reported in [52]. The composite membranes performed better than the Nafion 117 over the temperature range from 20 °C to 80 °C and for different hydration levels between RH = 25% and 95%. Although the improvement in conductivity due to the addition of C60 (about 1 wt%, comparable to that in our sPS-based composite membranes) was moderate at high hydration level, under low humidity conditions, RH < 50%, the conductivity of the composite membranes was about three times higher than that of Nafion 117. On the other hand, the water uptake for the Nafion-C60 composite membranes shows only a little increase compared to Nafion 117 membranes. Although it is not clear why there is this tremendous increase in conductivity when fullerenes are added, possible morphological changes, which could not be observed at that low C60 loading, or the interfacial water between the C60 aggregates, which were revealed by the optical microscopy (OM) and scanning electron microscopy (SEM) observations, and the Nafion domains were suspected for this effect [52]. We can only speculate here that the same effects might be the reason for the observed much higher conductivity of the composite membrane at 80 °C compared to that of fullerenes-free membrane.

3.3. Microstructure Characterization

With SANS we investigated the evolution of the morphology of hydrated domains with variation of hydration level and temperature. As discussed in details in our previous publications [31,32], the cast films produce isotropic scattering patterns on the two-dimensional position sensitive detector, while uni-axially deformed films deliver on the detector clearly separated inter-lamellar peaks due to orientation of the lamellar stacks along the deformation axis. Thus, the two-dimensional scattering patterns from sample B are isotropic (Figure 3a), while those from sample A are anisotropic (Figure 3b). Therefore, on sample B a detailed qualitative analysis of scattering data at different RH and temperatures could be done, while a semi-quantitative analysis of data with structural models could be performed on sample A. The 010 crystalline peak should appear in the scattering patterns at very high Q if the sample preparation (stretching) and contrast would allow this [31]. In our study, the loading of deuterated polymer clathrates with d–toluene would however not enable this.

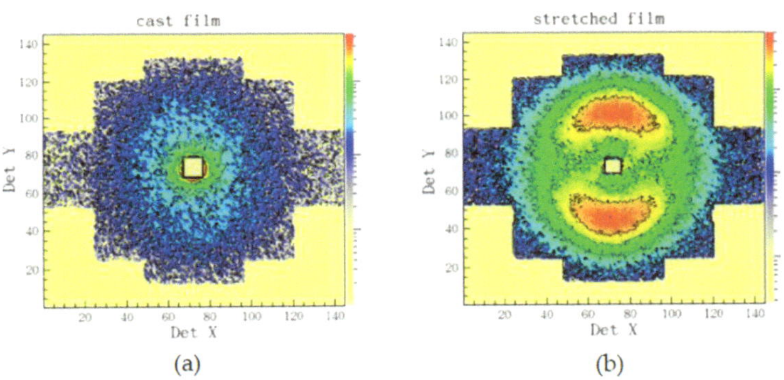

Figure 3. Example of typical two-dimensional neutron scattering patterns from sPS-based films as cast (**a**) or uni-axially deformed (**b**). The cast film delivers an isotropic scattering pattern on the two-dimensional position-sensitive neutron detector, while the stretched film, an anisotropic one, with the interlamellar correlation peaks from the oriented crystalline lamellar stacks appearing on meridian sectors (the stretching direction is vertical).

In Figure 4, the one-dimensional scattering patterns from the sample B at different hydration levels and temperatures are displayed. The SANS results were collected for the sample treatment in the beam starting at RH = 10% and temperature of 20 °C (the orange curve in Figure 4a) and continuing at RH = 30%, varying the temperature between 20 °C and 80 °C (the three upper curves in Figure 4a). Further, the hydration was raised to RH = 60% and the temperature was varied again between 20 °C and 80 °C, followed by drying the sample back to RH = 10% and the temperature of 20 °C (Figure 4b). Three scattering features could be observed in all scattering patterns: a) the power-law behavior in the small Q-regime, where the typical upturn behaving like Q^{-3} is due to the large-scale fractal character of the polymer film; b) an intermediate Q-regime between 0.01–0.1 $Å^{-1}$, where a broad feature corresponding to superposition of scattering signals from the inter-crystalline spacing (so-called "matrix knee") and sulfonated domains appears; and c) the high Q-regime (around 0.1–0.5 $Å^{-1}$), where the most characteristic feature is observed, namely the ionomer peak arising due to the correlation spacing between the dry or hydrated ionic clusters. After inspection of the evolution of scattering patterns and sample conditions, three main conclusions arise. (i) The scattering level in the intermediate Q-range, relevant for the length scale of the sulfonated and hydrated domains, increases with increasing the RH. This is due to absorption of more water by the film sample as the RH level increases, which leads to an increase in the neutron contrast between the hydrated (protonated) and the dry (deuterated) regions of the membrane. At the same time, the position of the ionomer peak shifts to lower Qs, as a consequence of the increase in the structural correlation length for the swelling ionic hydrophilic domains in the amorphous regions. (ii) The scattering level at constant RH decreases slightly with increasing the temperature. This may be due to a reduction in the absorbed water amount with increasing temperature that may be accompanied by morphological changes, hence changes in the contrast and size of the scattering objects (hydrated domains). (iii) The slight shift of the ionomer peak position towards higher Q values with increasing temperature at constant RH, accompanied by the slight decrease in intensity, as commented already in (ii).

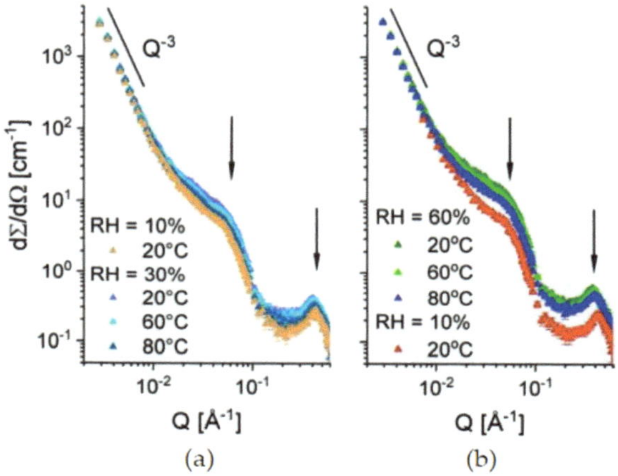

Figure 4. One-dimensional SANS patterns from sample B at different low (**a**) or intermediate (**b**) hydration levels and temperatures. The scattering features, which were observed in the scattering patterns and discussed in the text, are indicated by black arrows or the power-law scattering behavior.

This is an indication of a slight decrease in the correlation length between the ionic clusters with increasing temperature, which may be a consequence of weak morphological changes related to variation of the amount of the absorbed water. Also, morphological changes and changes in the micro-dynamics of the polymer matrix with increasing the temperature may induce such an effect. As reported earlier [32], the hydrated domain in sPS-based sulfonated membranes also includes segments of the sPS chain in addition to the sulfonic group attached to it. Therefore, it is worth studying in the future how the local micro-dynamics of the sPS at different temperatures may affect the conformation of the hydrated domains

After applying variation of RH and temperature on the membrane, the scattering pattern from the film sample B exposed again to RH = 10% and the temperature of 20 °C (red curve in Figure 4b) coincides with that in the initial state of the membrane. This can be very well observed in Figure 5a, where the evolution of the ionomer peak profile as the RH and temperature are varied and then set back to initial values of the sample treatment in beam. As expected, the peak position moves to lower Q values (orange arrow) when the RH is raised from 10% to 60%, while having the same temperature on the sample (20 °C), which is indicative for the swelling of the hydrated domains and the increase in the correlation length between the ionic clusters.

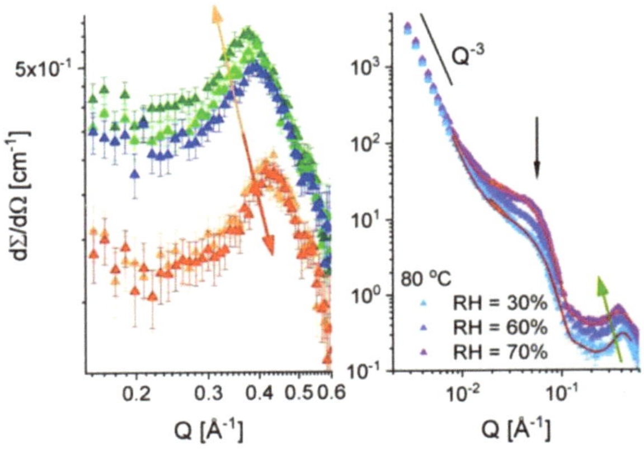

(a) (b)

Figure 5. The evolution of the ionomer peak in sample B (indicated by arrows) with increasing the RH and temperature followed by the return to the initial hydration and temperature state (**a**), and the one-dimensional scattering patterns from the same sample at 80 °C with varying the RH (**b**). The colors in Figure 5a are as in Figure 4. The red curves in the Figure 5b represent the model description of the experimental curves (see text) while the green arrow indicates the evolution of the ionomer peak with increasing the RH.

At RH = 60%, the peak position moves slightly to higher Q values as the temperature is increased. A slight shrinkage of the hydrated domain and a decrease in the correlation length between the ionic clusters seem to take place. Changing back the RH to 10% and the temperature to 20 °C makes the ionomer peak recover its initial profile (red arrow). Figure 5b, presents the scattering profile from the same membrane at 80 °C for different hydration levels. The evolution of the scattering features with increasing RH conforms to the discussion made above at the point (i). As mentioned in the beginning of this paragraph, the scattering at intermediate Q from cast films, which consist of functionalized semi-crystalline polymers such as sPS, arises as a superposition of scattering signals from the dry/hydrated amorphous domains and the crystalline regions (the inter-lamellar correlation). However, due to the random orientation of the crystalline lamellae in a cast film, the later signal is smeared out and has a weak contribution to the scattering at intermediate Q, which occurs mainly from the functionalized amorphous domains. When these domains are hydrated, their scattering contribution prevails at intermediate Q. Therefore, we tried to get a first insight on the morphology and volume fraction occupied by the hydrated domains (assumed as scattering particles) within the sample by interpreting the data in terms of the simple model of scattering from correlated spherical objects:

$$I(Q) = \varphi\Delta\rho^2 V_d P(Q)S(Q) + I_{ion} + I_{fract} + Bckgd \tag{3}$$

where P(Q) represents the particles form factor, which relates to the intra-particle correlations, and S(Q) the structure factor, which denotes the inter-particle correlation effects, and are described in details in [32]. The contrast $\Delta\rho = \rho_d - \rho_{env}$ is the difference between the SLD of the scattering hydrated domains ρ_d and their polymeric environment. Usually, the factor $I_0 = (\varphi\ \Delta\rho^2\ V_d)$ is called the "forward scattering" from the ensemble of scattering objects. The terms I_{ion} and I_{fract} represent the additional contribution at high Q, from the ionomer peak, which can be described by a Gaussian function, and at low Q, from the fractal behavior of the film, which can be described by a simple power-law term, P_3Q^{-3}, with P_3 the power-law constant [32]. A constant background, Bckgd, which arises mostly from the incoherent scattering contribution from the film sample, is added as a final term of the model. The red lines in Figure 5b, represent the model interpretation of the experimental data collected for

RH = 30% and 70%. The main parameters of the fitting procedure were the radius of the hydrated domains R_{sph}, assumed spherical, and the "forward scattering" from these domains, $(I_0)_{sph}$. Other free parameters were considered in the fitting procedure: the P_3; the area, width and position of the Gaussian that describes the ionomer peak; and the parameters defining the structure factor in Equation (3) [31]. Despite the multitude of free parameters, the reliability of the fitting procedure is high, because the three structures considered—the fractal behavior, the hydrated domains and the correlation between the ionic clusters, are very well separated by sizes and their contributions to the total scattering curve overlap only marginally over the wide Q-range covered in this experiment. This makes the parameters that describe them pretty well determined. Moreover, at RH = 30% it seems that the spherical water domains are still well separated from each other: The spherical form factor is well defined towards high Qs and no shoulder or peak-like feature is observed, as in the case of the data measured at RH = 70%, when the S(Q) contribution yields the additional scattering observable as a shoulder at around Q = 0.07 $Å^{-1}$ (indicated by the vertical arrow). The dimensions of the hydrated domains formed at RH = 30% and 70% are about R_{sph} = 48 Å and 54 Å, respectively, quite smaller to the values reported for the fullerene doped s–sPS membranes [32].

However, the C60–sPS composite membrane (sample B) investigated in this study was produced following a different procedure than the previously studied membranes. On the other hand, the volume fraction occupied by the hydrated domains in the amorphous region of the film sample B, which was estimated from the interpretation of the $(I_0)_{sph}$ by taking into account the crystallinity of the film, is about 2.35% and 8.52% at RH = 30% and 70%, respectively. The value obtained at RH = 70% seems to be larger than that obtained for the fullerene doped sPS membranes [32], which were characterized by a slightly higher sulfonation degree than the film sample B in this study. Although the values are not very different from each other, we should take into consideration that the current values were obtained for the sample heated at 80 °C, where, according to our qualitative SANS observations, the amount of the adsorbed water should be lower than that at room temperature. Apparently, the presence of fullerenes in the composite membrane and the tendency of fullerene to aggregate seem to have an effect on the morphology and content of the water domains in the composite membrane.

Figure 6 presents the scattering data from sample A at RH = 60% and two temperatures, 20 °C and 80 °C, as they were averaged over the equatorial and meridian sectors of the anisotropic two-dimensional scattering patterns that form the uni-axially deformed membrane (Figure 3b). The slight decrease of the scattering level with increasing the temperature is observed here too, as it was in the case of sample B. The scattering features from different microstructural levels occurring in the membrane as a consequence of its properties and treatment—the semi-crystalline character, the functionalization (sulfonation) and the hydration, are clearly revealed by the scattering patterns on different detection sectors due to alignment of some of these structures under stretching [31]. The low-Q power law behavior due to the fractal character of the membrane at a larger length scale, around 1000 Å, is visible in the scattering profiles collected on both the equatorial and meridian sectors. The ionomer peak, which is an isotropic scattering feature, is also visible in the high-Q data on both equatorial and meridian sectors. At intermediate Q, the isotropic scattering from the hydrated domains is well distinguished in the data on equatorial sectors, while the same feature on the meridian sectors is buried under the very strong scattering from the oriented lamellar stacks. Sample A was measured with the stretching direction positioned vertically in the neutron beam, therefore, the strong reflections due to the inter-lamellar correlation appear in the meridian direction (Figure 3b). The scattering data on meridian and equatorial sectors were fitted simultaneously for each temperature according to the model that is presented in detail in [32]. From the interpretation of the fitted parameters, the main geometrical and density information about the hydrated domains in the amorphous regions could be obtained in a similar way as described in [32]. The fitting procedure delivered the size of the hydrated domains and their "forward scattering". Supposing spherical hydrated domains, the volume fraction occupied by water in the amorphous region could be estimated from the evaluation of the forward scattering. The fitting procedure also delivered the SLD of the hydrated inter-lamellar amorphous regions, from which the volume fraction occupied by water in these regions could be estimated. The evaluated parameters are reported in Table 2. As it can be observed, the volume fraction in the membrane that is occupied by the water is lower at 80 °C than at 20 °C, which agrees with the observation made on sample B. The volume fractions occupied by water within the bulk and inter-lamellar amorphous regions are comparable at 20 °C and are quite similar to those evaluated for the fullerenes-doped sPS membranes for comparable sulfonation degree, which were reported in [32].

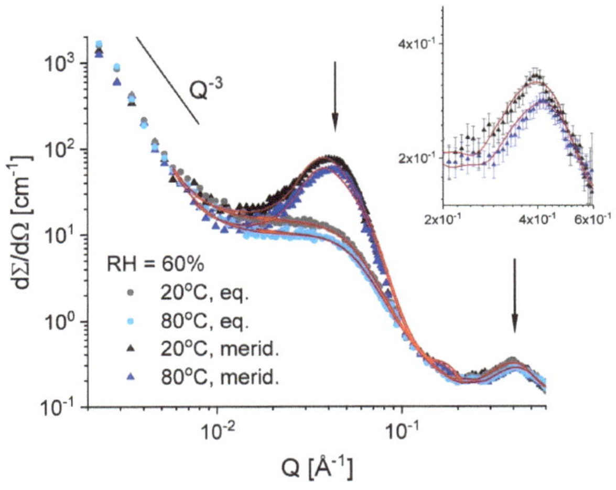

Figure 6. One-dimensional SANS patterns from the film sample A at RH = 60% and different temperatures. Experimental data (symbols) are averaged over the equatorial and meridian directions, with the lines corresponding to the model interpretation of the scattering profiles, as discussed in text. The main scattering features are indicated by the power-law and the vertical arrows. The inset presents in details the ionomer peak (the patterns over the meridian sectors, as long as this scattering feature appears isotropic in the two-dimensional scattering pattern).

Table 2. The structural parameters of the hydrated morphologies in samples A and B of this study, as delivered by the interpretation of the experimental data according to the models introduced in [32].

Parameter	Sample A, RH = 60%		Sample B, 80 °C	
	20 °C	80 °C	RH = 30%	RH = 70%
$(\varphi_{water})^{amorphous}$ (%)	3.24	2.75	2.35	8.52
R_{sph} (Å)	50.3	45.1	41.5	38.2
$(\varphi_{water})^{inter\text{-}lam}$ (%)	3.21	2.87	–	–
ξ_{ion} (Å)	16.10	14.95	16.75	15.06

On the other hand, at 80 °C it seems that the volume fraction of water within the inter-lamellar amorphous regions is slightly higher than that in the bulk amorphous domains. Apparently, at high temperature, the water desorbs easier from the bulk amorphous than from the interlamellar amorphous regions. The average size of the hydrated domains is also slightly smaller at 80 °C than at 20 °C (Table 2). The correlation length between the hydrated ionic clusters, $\xi_{ion} = 2\pi/Q_{ion}$, where Q_{ion} is the ionomer peak position, shows the same trend (Table 2).

Finally, the brief exposure of an old uni-axially deformed s–sPS film (sulfonation degree S = 19.5%, crystallinity of 22%, and doped with C60 fullerenes [32]) to the Fenton's test conditions [27] seems to have no effect on the crystalline phase of the membrane, as can be deduced from the inspection of the scattering patterns from the tested film, compared to those from the same sample before the application of the test. Figure 7 shows the scattering data averaged over the sectors parallel with the deformation axis of the membrane after it was kept for 1 h in Fenton's reagent at 60 °C (red symbols) in parallel with the SANS data from the same membrane at RH = 85% (blue symbols, data already discussed in [32]), before the Fenton's test was carried out. In both cases the inter-lamellar peak due to correlation effects between oriented lamellae in the crystalline phase of stretched film can be well observed at around Q * = 0.035 $Å^{-1}$, which is indicative of the preservation of membrane crystalline features regardless of its treatment. The profile of the ionomer peak at high Q values, however, is different, depending on the membrane treatment: it is clearly observed in the scattering pattern from the hydrated membrane at RH = 85%, while after immersing the polymer film into the Fenton's reagents, the peak turns into a very broad feature and shifts to lower Q values due to the extreme hydration of the membrane and the high incoherent scattering contribution of the absorbed water.

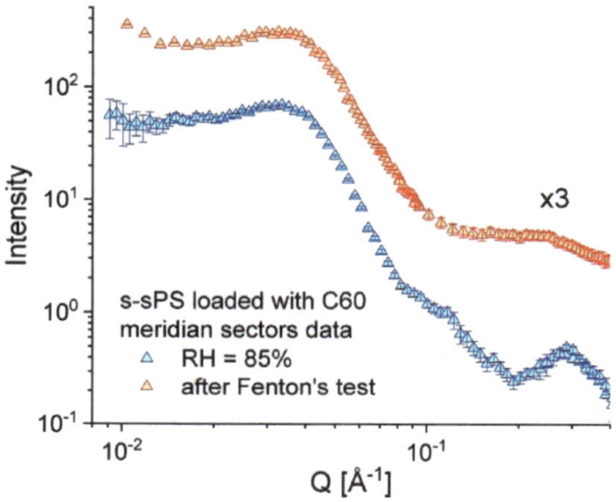

Figure 7. One-dimensional SANS patterns over the stretching direction from a uni-axially deformed s–sPS film incorporating C60 fullerenes, as measured in different treatment conditions: blue symbols—the sample at RH = 85% and room temperature (data reported [32]); red symbols (vertically shifted for clarity)—the sample exposed to the Fenton's test conditions.

This test offers a first hint about the resistance to oxidative degradation of the sulfonated membranes containing fullerenes that we have prepared based on sPS, and confirms the early observations made on similar systems, as reported in [27]. However, further systematic tests of film samples with different fullerene contents must be carried out over a longer time in different treatment conditions (higher temperature and involvement of higher concentration reagents), in order to clearly assess the oxidative stability and life time of the membranes of this type. On the other hand, it seems that the addition of fullerene does not significantly improve the mechanical properties of sPS-based membranes [27], which apparently are still poorer than those shown by Nafion, though thorough mechanical examination still needs to be performed for quantitative conclusion on this issue.

4. Conclusions

The structural results of this study enabled a characterization of the hydrated domains in the sPS-based membranes in different humidity and temperature conditions. Our complex structural study presented here completes the conclusions reported in the previous publications [31,32]. According to these, the water that is taken-up by the functionalized membranes accumulates around the agglomerations of sulfonic groups and gives rise to hydrated domains, mostly in the bulk amorphous region. The hydrated domains grow in size and number with increasing the hydration level while, on the other hand, they shrink slightly with increasing the temperature at constant hydration level, due to desorption of some water, mostly from the bulk amorphous regions. Despite these weak morphological changes at high temperatures, the conductivity of the membrane seems not to be affected, as observed in [30]. With a further increase of hydration level up to the complete membrane equilibration in liquid water, the water clusters grow and give rise to water channels.

According to our measurements, the sulfonated sPS–fullerenes composite membranes perform at high temperature much better than the fullerenes free membranes in terms of proton conductivity in liquid water. Apparently, this may be related to the formation of additional hydrated pathways in the composite membrane due to the interfacial water accumulated between the fullerene aggregates and the sulfonated polymer domains, like in the case of Nafion-fullerenes composite membranes [52], possibly in conjunction with changes in the dynamics of water and polymer at high temperatures. To better understand this observed effect in the sPS-based membranes and to verify these assumptions, further investigations of the micro-structure and micro-dynamics in such systems are needed, possibly involving also a much simpler system, such as the amorphous atactic polystyrene.

Author Contributions: Conceptualization, M.-M.S. and A.R.; methodology, M.-M.S. and A.R.; sample preparation, M.-M.S. and D.H.L.; investigation, M.-M.S. (FTIR, UV-Vis), Y.Z. (conductivity), F.Z. (XRD), Z.R. (PGAA), A.R. (FTIR, UV-Vis., SANS); data curation, A.R..; writing—original draft preparation, A.R.; writing—review and editing, M.-M.S., Y.Z., A.R. All authors have read and agreed to the published version of the manuscript.

Funding: This research received no external funding.

Conflicts of Interest: The authors declare no conflict of interest.

References

1. Kraytsberg, A.; Ein-Eli, Y. Review of Advanced Materials for Proton Exchange Membrane Fuel Cells. *Energy Fuels* **2014**, *28*, 7303–7330. [CrossRef]
2. Hayashi, A.; Nishihara, M.; Matsuda, J.; Sasaki, K. Polymer electrolyte Fuel Cells (PEFCs). In *Hydrogen Energy Engineering. Green Energy and Technology*; Sasaki, K., Li, H.-W., Hayashi, A., Yamabe, J., Ogura, T., Lyth, S.M., Eds.; Springer: Tokyo, Japan, 2016; pp. 301–311.
3. Kreuer, K.-D.; Paddison, S.J.; Spohr, E.; Schuster, M. Transport in Proton Conductors for Fuel-Cell Applications: Simulations, Elementary Reactions, and Phenomenology. *Chem. Rev.* **2004**, *104*, 4637–4678. [CrossRef] [PubMed]
4. Kusoglu, A.; Weber, A.Z. New Insights into Perfluorinated Sulfonic-Acid Ionomers. *Chem. Rev.* **2017**, *117*, 987–1104. [CrossRef] [PubMed]
5. Smitha, B.; Sridhar, A.; Khan, A.A. Solid Polymer Electrolyte Membranes for Fuel Cell Applications—A Review. *J. Membr. Sci.* **2005**, *259*, 10–26. [CrossRef]
6. Sahu, A.K.; Jalajakshi, A.; Pitchumani, S.; Shridhar, P.; Shukla, A.K. Endurance of Nafion-Composite Membranes in PEFCs Operating at Elevated Temperature under Low Relative-Humidity. *J. Chem. Sci.* **2012**, *124*, 529–536. [CrossRef]
7. Vinothkannan, M.; Hariprasad, R.; Ramakrishnan, S.; Kim, A.R.; Yoo, D.J. Potential Bifunctional Filler (CeO_2-ACNTs) for Nafion Matrix toward Extended Electrochemical Power Density and Durability in Proton-Exchange Membrane Fuel Cells Operating at reduced Relative Humidity. *ACS Sustain. Chem. Eng.* **2019**, *7*, 12847–12857. [CrossRef]
8. Alberti, G.; Narducci, R.; Di Vona, M.L.; Giancola, S. Annealing of Nafion 1100 in the Presence of an Annealing Agent: A Powerful Method for Increasing Ionomer Working Temperature in PEMFCs. *Fuel Cells* **2013**, *13*, 42–47. [CrossRef]
9. Okada, T. Theory for Water Management in Membranes for Polymer Electrolyte Fuel Cells—Part 1 The Effect of Impurity Ions at the Anode Side on the Membrane Performance. *J. Electroanal. Chem.* **1999**, *465*, 1–17. [CrossRef]
10. St. Pierre, J.; Wilkinson, D.P.; Knights, S.; Bos, M.L. Relationships between Water Management, Contamination and Lifetime Degradation in PEFC. *J. New Mater. Electrochem. Syst.* **2000**, *3*, 99–106.
11. Huguet, P.; Morin, A.; Gebel, G.; Deabate, S.; Sutor, A.K.; Peng, Z. In Situ Analysis of Water Management in Operating Fuell Cells by Confocal Raman Spectroscopy. *Electrochem. Commun.* **2011**, *13*, 418–422. [CrossRef]

12. Alberti, G.; Di Vona, M.L.; Narducci, R. New results on the visco-elastic behaviour of ionomer membranes and relations between TeRH plots and proton conductivity decay of Nafion_117 in the range 50–140 °C. *Int. J. Hydrog. Energy* **2012**, *37*, 6302–6307. [CrossRef]
13. Narducci, R.; Knauth, P.; Chailan, J.-F.; DiVona, M.L. How to Improve Nafion with Tailor Made Annealing. *RSC Adv.* **2018**, *8*, 27268–27274. [CrossRef]
14. Neelakandan, S.; Liu, D.; Wang, L.; Hu, M.; Wang, L. Highly Branched Poly (Arylene Ether)/Surface Functionalized Fullerene-based Composite Membrane Electrolyte for DMFC Applications. *Int. J. Energy Res.* **2019**, *43*, 3756–3767. [CrossRef]
15. Kim, A.R.; Gabunada, J.C.; Yoo, D.J. Amelioration in Physicochemical Properties and Single Cell Performance of Sulfonated Poly (ether ether ketone) Block Copolymer Composite Membrane using Sulfonated Carbon Nanotubes for Intermediate Humidity Fuel Cells. *Int. J. Energy Res.* **2019**, *43*, 2974–2989. [CrossRef]
16. Kim, A.R.; Vinothkannan, M.; Yoo, D.J. Sulfonated Fluorinated Multi-Block Copolymer Hybrid Containing Sulfonated (Poly Ether Ether Ketone) and Graphene Oxide: A Ternary Hybrid Membrane Architecture for Electrolyte Applications in Proton Exchange Membrane Fuel Cells. *J. Energy. Chem.* **2018**, *27*, 1247–1260. [CrossRef]
17. Kim, A.R.; Vinothkannan, M.; Park, C.J.; Yoo, D.J. Alleviating the Mechanical and Thermal Degradations of Highly Sulfonated Poly (Ether Ether Ketone) Blocks via Copolymerization with Hydrophobic Unit for Intermediate Humidity Fuel Cells. *Polymers* **2018**, *10*, 1346–1360. [CrossRef]
18. Hickner, M.A.; Ghassemi, H.; Kim, Y.S.; Einsla, B.R.; McGrath, J.E. Alternative Polymer System for Proton Exchange Membranes (PEMs). *Chem. Rev.* **2004**, *104*, 4587–4612. [CrossRef]
19. Hihashihara, T.; Matsumoto, K.; Ueda, M. Sulfonated Aromatic Hydrocarbon Polymers as Proton Exchange Membranes for Fuel Cells. *Polymer* **2009**, *50*, 5341–5357. [CrossRef]
20. Zhang, Y.; Li, J.; Ma, L.; Cai, W.; Cheng, H. Recent Developments on Alternative Proton Exchange Membranes: Strategies for Systematic Performance Improvement. *Energy Technol.* **2015**, *3*, 675–691. [CrossRef]
21. Zhao, Y.; Yoshida, M.; Oshima, T.; Koizumi, S.; Rikukawa, M.; Szekely, N.; Radulescu, A.; Richter, D. Elucidation of the Morphology of the Hydrocarbon Multiblock Copolymer Electrolyte Membranes for Proton Exchange Fuel Cells. *Polymer* **2016**, *86*, 157–167. [CrossRef]
22. Feng, S.; Kondo, S.; Kaseyama, T.; Nakazawa, T.; Kikuchi, T.; Selyanchyn, R.; Fujikawa, S.; Christiani, L.; Sasaki, K.; Nishihara, M. Development of Polymer-Polymer Type Charge-Transfer Blend Membranes for Fuel Cell Application. *J. Membr. Sci.* **2018**, *548*, 223–231. [CrossRef]
23. Escorihuela, J.; Narducci, R.; Compan, V.; Constantino, F. Proton Conductivity of Composite Polyelectrolyte Membranes with Metal-Organic Frameworks for Fuel Cell Applications. *Adv. Mater. Interfaces* **2019**, *6*, 1801146. [CrossRef]
24. Kim, J.-D.; Ohira, A.; Nakao, N. Chemically Crosslinked Sulfonated Polyphenylsulfone (CSPPSU) Membranes for PEM Fuel Cells. *Membranes* **2020**, *10*, 31. [CrossRef] [PubMed]
25. Serbanescu, O.S.; Voicu, S.I.; Thakur, V.K. Polysulfone Functionalized Membranes: Properties and Challenges. *Mater. Today Chem.* **2020**, *17*, 100302. [CrossRef]
26. Boriello, A.; Agoretti, P.; Ambrosio, L.; Fasano, G.; Pellegrino, M.; Venditto, V.; Guerra, G. Syndiotactic Polystyrene Films with Sulfonated Amorphous Phase and Nanoporous Crystalline Phase. *Chem. Mater.* **2009**, *21*, 3191–3196. [CrossRef]
27. Saga, S.; Matsumoto, H.; Saito, K.; Minagawa, M.; Tanioka, A. Polyelectrolyte Membranes Based on Hydrocarbon Polymer Containing Fullerenes. *J. Power Sources* **2008**, *176*, 16–22. [CrossRef]
28. Fasano, G.; Califano, R.; Pellegrino, M.; Venditto, V.; Guerra, G.; Boriello, A.; Ambrosio, L.; Sansone, L. Semicrystalline proton-conductive membranes with sulfonated amorphous phases. *Int. J. Hydrog. Energy* **2011**, *36*, 8038–8044.
29. Lavorgna, M.; Fusco, L.; Piscitelli, F.; Mensitieri, G.; Agoretti, P.; Borriello, A.; Mascia, L. Control of morphology of sulfonated syndio-polystyrene membranes through constraints imposed by siloxane netwroks. *Polym. Eng. Sci.* **2008**, *48*, 2389–2399. [CrossRef]
30. Borriello, A.; Napolitano, T. Proton conductivity and methanol permeability of sulfonated syndiotactic polystyrene membranes. *Soft Mater.* **2011**, *9*, 224–337. [CrossRef]

31. Schiavone, M.-M.; Tarallo, O.; Di Girolamo, R.; Caporaso, L.; Appavou, M.-S.; Revay, Z.; Radulescu, A. Structure and Morphology of Model Polymer Electrolyte Membranes based on Sulfonated Syndiotactic-Polystyrene in the Co-crystalline Phase Resolved by Small-Angle Neutron Scattering. *Solid State Ion.* **2018**, *320*, 392–406. [CrossRef]
32. Schiavone, M.-M.; Iwase, H.; Takata, S.; Radulescu, A. The multilevel structure of syndiotactic-polystyrene model polyelectrolyte membranes resolved by extended Q-range contrast variation SANS. *Membranes* **2019**, *9*, 136. [CrossRef] [PubMed]
33. Kaneko, F.; Radulescu, A.; Ute, K. Time-resolved SANS Studies on Guest Exchange Processes in Co-Crystals of Syndiotactic Polytyrene. *Polymer* **2013**, *54*, 3145–3149. [CrossRef]
34. Pandele, A.M.; Neacsu, P.; Cimpean, A.; Staras, A.I.; Miculescu, F.; Iordache, A.; Voicu, S.I.; Thakur, V.K.; Toader, O.D. Cellulose Acetate Membranes Functionalized with Resveratrol by Covalent Immobilization for Improved Osseointegration. *Appl. Surf. Sci.* **2018**, *438*, 2–13. [CrossRef]
35. Revay, Z.; Kudejova, P.; Kleszczm, K.; Sölradl, S.; Genreith, C. In-beam activation analysis at Heinz Maier-Leibnitz Zentrum, Garching. *Nucl. Instrum. Meth. A* **2015**, *799*, 114–123. [CrossRef]
36. Hermans, P.H.; Weidinger, A. Quantitative Investigation of the X-Ray Diffraction Picture of Some Typical Rayon Specimens Part, I. *Textil. Res.* **1961**, *31*, 558–571. [CrossRef]
37. Radulescu, A.; Szekely, N.K.; Appavou, M.-S.; Pipich, V.; Kohnke, T.; Ossovyi, V.; Staringer, S.; Schneider, G.J.; Amann, M.; Zhang-Haagen, B.; et al. Studying Soft-Matter and Biological Systems over a Wide Length-Scale from Nanometer and Micrometer Sizes at the Small-Angle Neutron Diffractometer KWS-2. *J. Vis. Exp.* **2016**, *118*, 54639. [CrossRef]
38. Moyses, S.; Spells, S.J. Conformationally Sensitive Infrared Vibrations of the Syndiotactic Polystyrene/Ethylbenzene Complex. *Macromolecules* **1999**, *32*, 2684–2689. [CrossRef]
39. Kaneko, F.; Seto, N.; Radulescu, A.; Schiavone, M.-M.; Allgaier, J.; Ute, K. Development of a Simultaneous SANS/FTIR Measuring System. *Chem. Lett.* **2015**, *44*, 497–499. [CrossRef]
40. Huang, T.; Jin, B.; Peng, R.F.; Chen, C.D.; Zheng, R.Z.; He, Y.; Chu, S.J. Synthesis and Characterization of [60]Fullerenes-Glycidyl Azide Polymer and Its Thermal Decomposition. *Polymers* **2015**, *7*, 896–908. [CrossRef]
41. Wang, C.; Ruan, L.; Chang, X.-L.; Zhang, X.; Yang, S.-T.; Guo, X.; Yuan, H.; Guo, C.; Shi, W.; Sun, B.; et al. The Isotopic Effect of ^{13}C-Labeled Large Carbon Cage (C_{70}) Fullerens and Their Formation Process. *RSC Adv.* **2015**, *5*, 76949–76956. [CrossRef]
42. Wu, Z.C.; Jelski, D.A.; George, T.F. Vibrational motions of Buckminsterfullerenes. *Chem. Phys. Lett.* **1987**, *137*, 291–294. [CrossRef]
43. Törpe, A.; Belton, D.J. Improved Spectrophotometric Analysis of Fullerenes C60 and C70 in High-solubility Organic Solvents. *Anal. Sci.* **2015**, *31*, 125–130. [PubMed]
44. Bensasson, R.V.; Bienvenue, E.; Dellinger, M.; Leach, S.; Seta, P. C60 in model Biological Systems. A Visible-UV Absorption Study of Solvent-Dependent Parameters and Solute Aggregation. *J. Phys. Chem.* **1994**, *98*, 3492–3500. [CrossRef]
45. Deguchi, S.; Alargova, R.G.; Tsujii, K. Stable dispersions of fullerenes, C60 and C70, in water. Preparation and Characterization. *Langmuir* **2001**, *17*, 6013–6017. [CrossRef]
46. De Rosa, C.; Guerra, G.; Petraccone, V.; Pirozzi, B. Crystal Structure of the Emptied Clathrate Form (e Form) of Syndiotactic Polystyrene. *Macromolecules* **1997**, *30*, 4147–4152. [CrossRef]
47. Chakraborty, C.; Malik, S.; Guenet, J.-M. Syndiotactic Polystyrene/Fullerene Composites: Elucidation of Structural Aspect. *Macromol. Symp.* **2011**, *303*, 56–62. [CrossRef]
48. Saito, M.; Hayamizu, K.; Okada, T. Temperature Dependence of Ion and Water Transport in Perfluorinated Ionomer Membranes for Fuel Cells. *J. Phys. Chem. B* **2005**, *109*, 3112–3119. [CrossRef]
49. Matos, B.R.; Goulart, C.A.; Santiago, E.I.; Muccillo, R.; Fonseca, F.C. Proton Conductivity of Perfluorosulfonate Ionomers at High Temperature and High Relative Humidity. *Appl. Phys. Lett.* **2014**, *104*, 091904. [CrossRef]
50. Rieke, P.R.; Vanderborgh, N.E. Temperature Dependence of Water Content and Proton Conductivity in Polyperfluorosulfonic Acid Membranes. *J. Membr. Sci.* **1987**, *32*, 313–328. [CrossRef]

51. Sumner, J.J.; Creager, S.E.; Ma, J.J.; DesMarteau, D.D. Proton Conductivity in Nafion® 117 and in a Novel BisE (perfluoroalkyl)sulfonylimide lonomer Membrane. *J. Electrochem. Soc.* **1998**, *145*, 107–110. [CrossRef]
52. Tasaki, K.; DeSousa, R.; Wang, H.; Gasa, J.; Venkatesan, A.; Pugazhendhi, P.; Loutfy, R.O. Fullerene Composite Proton Conducting Membranes for Polymer Electrolyte Fuel Cells Operating under Low Humidity Conditions. *J. Membr. Sci.* **2006**, *281*, 570–580. [CrossRef]

membranes

MDPI

Article

Structural, Morphological, Electrical and Electrochemical Properties of PVA: CS-Based Proton-Conducting Polymer Blend Electrolytes

Ayub Shahab Marf [1], Ranjdar M. Abdullah [1] and Shujahadeen B. Aziz [1,2,*]

1 Advanced Polymeric Materials Research Lab., Department of Physics, College of Science, University of Sulaimani, Qlyasan Street, Sulaimani 46001, Iraq; ayub.shahab@gmail.com (A.S.M.); ranjdar.abdullah@univsul.edu.iq (R.M.A.)

2 Department of Civil Engineering, College of Engineering, Komar University of Science and Technology, Sulaimani 46001, Iraq

* Correspondence: shujaadeen78@yahoo.com or shujahadeenaziz@gmail.com

Received: 11 March 2020; Accepted: 13 April 2020; Published: 15 April 2020

Abstract: Polymer blend electrolytes based on poly(vinyl alcohol):chitosan (PVA:CS) incorporated with various quantities of ammonium iodide were prepared and characterized using a range of electrochemical, structural and microscopic techniques. In the structural analysis, X-ray diffraction (XRD) was used to confirm the buildup of the amorphous phase. To reveal the effect of dopant addition on structural changes, field-emission scanning electron microscope (FESEM) was used. The protrusions of salt aggregates with large quantity were seen at the surface of the formed films at 50 wt.% of the added salt. The nature of the relationship between conductivity and dielectric properties was shown using electrochemical impedance spectroscopy (EIS). The EIS spectra were fitted with electrical equivalent circuits (EECs). It was observed that both dielectric constant and dielectric loss were high in the low-frequency region. For all samples, loss tangent and electric modulus plots were analyzed to become familiar with the relaxation behavior. Linear sweep voltammetry (LSV) and transference number measurement (TNM) were recorded. A relatively high cut-off potential for the polymer electrolyte was obtained at 1.33 V and both values of the transference number for ion (t_{ion}) and electronic (t_{elec}) showed the ion dominant as charge carrier species. The TNM and LSV measurements indicate the suitability of the samples for energy storage application if their conductivity can be more enhanced.

Keywords: PVA:CS polymer blend; NH_4I salt; XRD and FESEM; impedance; dielectric properties; TNM and LSV study

1. Introduction

New materials that follow green chemistry principles have been focused upon extensively due to the lower release of pollutants into the environment [1]. Two strong alternatives, namely batteries and supercapacitors as energy storage devices, appear to replace other non-sustainable energy sources, such as oil, nuclear fuel and other fossil fuels [2–4]. Solid polymer electrolytes (SPEs) have been extensively studied to examine their application to large-scale production, within capacitor systems [5]. Polymer electrolytes are membranes for the ion transport mechanism. For example, a poly(vinyl alcohol) (PVA) polymer host is as assumed to be one of the cornerstones for the preparation of various types of polymer electrolytes, because of its capability of dissolving a number of inorganic salts and also its biodegradability [6,7]. On the other hand, composites and natural polyphenols are crucial for membrane separation technology and the treatment of wastewater dyes [8,9]. PVA is mainly composed of vinyl alcohol groups (i.e., enriched with polar oxygen atoms) that can make complexes with the salt cations, thus forming polymer electrolyte complexes. Furthermore, PVA is characterized by several inherent properties, for instance, non-toxicity, affordability, a relatively high capacity of charge storage and exceptional mechanical properties [10,11]. The ability of PVA to form electrolytes is associated with its hydrophilic properties [12] and several hydroxyl groups attached to methane carbons (CH_2-CH) on the PVA backbone [13]. However, natural biopolymers have been considered as strong candidates in the preparation of polymer electrolytes for long-term usage, due to their biodegradability. In addition, natural polymers are relatively cost-effective and environmentally friendly, exhibit high solubility and are able to form films with desired shapes [1]. The most popular natural polymers are chitosan, starch and cellulose that have drawn the attention of many research groups [5]. In terms of structure, chitosan (CS) is a cationic polysaccharide within which β-1–4-linked 2-amino-2 deoxy-D-glucopyranose repeats in a sequence of a billion units. It is easily obtained from the alkaline N-acetylation reaction of chitin. This natural polymer is superior over polymers in terms of biodegradability, biocompatibility, low toxicity and affordability [14]. The presence of amino groups on the CS backbone makes this natural polymer into an exceptional biopolymer. This enables CS to create ion-conducting polymer electrolytes [15]. The amine groups within the CS structure act as electron-rich donors that interact with alkaline metal salts. Based on these facts, CS meets the requirements as a host polymer that encompasses ions using the salt solvation process [15,16]. Moreover, CS has shown membrane properties, such as low methanol permeability, the existence of functional groups within the backbone and hydrophilicity that makes CS practical in a relatively low humidity and high temperature environment [17].

The problem of PVA's low conductivity has motivated researchers to carry out studies to solve this issue. Kim et al. investigated PVA and lithium trifluoromethane sulfonate ($LiCF_3SO_3$, LiTf) salt in an attempt to improve the low ionic conductivity. It was found that the ionic conductivity of the PVA-based SPE could be increased with increasing salt concentration [18]. It has also been emphasized that the polymer blending approach is an appropriate way to increase ionic conductivity as a result of lowering the degree of crystallinity [19]. In this methodology, two or more polymers can be combined with or without forming primary chemical bonds, such as ionic and covalent bonds [17]. A modification of PVA's crystalline phase was achieved through chitosan polymer blending. The most common additives used in the preparation of proton (H^+)-conducting SPEs are strong inorganic acids (H_3PO_4 and H_2SO_4) and ammonium salts. It is important to mention that SPEs containing inorganic acids usually suffer from chemical degradation that makes them practically unusable [20,21]. Alternatively, proton-conducting SPEs having relatively high thermal stability and ionic conductivity can be achieved using ammonium salts. To know the degree of dissociation of ammonium salts, it is important to take into consideration the lattice energy. The lattice energy of a number of ammonium salts is given as follows: ammonium chloride (NH_4Cl), ammonium bromide (NH_4Br) and ammonium iodide (NH_4I) possess a lattice energy of 698 kJ/mol, 665.3 kJ/mol, and 634 kJ/mol, respectively [22,23]. Based on lattice energy, a biopolymer electrolyte can be enriched with ions using ammonium iodide (NH_4I) [24].

In condensed matter physics, ion conduction and dielectric relaxation in solid materials are two hot topics that are under intensive study. In particular, dealing with dielectric relaxation in SPEs has been found to be an appropriate way to obtain insight into the characteristics of cation–polymer interactions. This is because the dielectric constant is a measure of a polymer material's ability to dissolve inorganic salts [25,26].

The present work is aimed at studying PVA:CS systems incorporated with various quantities of NH_4I, examined through electrochemical impedance spectroscopy (EIS), X-ray diffraction (XRD) and scanning electron microscopy (SEM). The results of the present work are crucial to understanding the basic relationships between DC conductivity and dielectric properties. The study of tanδ and electric modulus is informative about the ion conduction mechanism in polymer electrolytes. On the other hand, the electrochemical investigations (TNM and LSV) indicate the suitability of the samples for electrochemical device application if their conductivity is enhanced using various approaches such as the addition of fillers or plasticizers. In future works, we will consider these approaches to improve the conductivity. From linear sweep voltammetry (LSV), the PVA:CS biopolymer electrolyte in this work was found to be electrochemically stable up to 1.33 V, which is important for use in electrical double-layer capacitor (EDLC) applications.

2. Experimental Details

2.1. Materials and Sample Preparation

All the chemicals were purchased from commercial suppliers (Sigma-Aldrich and Merck) and used as received. The chitosan (CS) with medium molecular mass (CAS 9012-76-4) was obtained from Sigma-Aldrich (Sigma-Aldrich, Warrington, PA, USA), and poly(vinyl alcohol) (PVA) and ammonium iodide (NH_4I) were purchased from Merck. All these materials were used as raw materials in the synthesis of the PVA:CS electrolyte systems. The blend polymer electrolyte synthesis procedure involves dissolving 0.5 g of CS in 30 mLof a 1 wt.% acetic acid solution under magnetic stirring continuously for several hours. Meanwhile, 0.5 g of PVA was dissolved in 20 mLof distilled water at 90 °C. This PVA solution was left to cool down to ambient temperature. Afterwards, the PVA solution and CS were mixed with a magnetic stirrer under continuous stirring until a homogenous solution was obtained. Consequently, different wt.% amounts of NH_4I were added to the final solution. The salt addition was varied from 10 to 50 wt.%, and this series of sample solutions was coded as PVCS1, PVCS2, PVCS3, PVCS4 and PVCS5. The coded solutions were incorporated with 50% CS and 50% PVA integrated with 10, 20, 30, 40 and 50 wt.% amounts of ammonium iodide (NH_4I), respectively. The solution mixtures were then poured into differently labeled Petri dishes and left to evaporate slowly at room temperature for a couple of weeks to form solvent-free films. Finally, for further drying, the films were put into desiccators that were filled with blue silica gel desiccant to ensure the removal of any trace amount of solvent or moisture. Table 1 summarizes the designation and composition of the various PVA:CS:NH_4I polymer blend electrolytes.

Table 1. Designation and composition of the various PVA:CS:NH_4I solid polymer electrolytes (SPEs).

Sample Designation	(PVA:CS)(0.5:0.5) (g)	NH_4Iwt.%
PVCS1	1	10
PVCS2	1	20
PVCS3	1	30
PVCS4	1	40
PVCS5	1	50

2.2. SEM and XRD Study

The surface morphology images of the prepared blend electrolyte films were acquired using SEM (FEI Quanta 200 FESEM). The structural texture study involved recording X-ray diffraction (XRD) patterns at room temperature using an X-ray diffractometer (Bruker AXS) with operating current and voltage of 40 mA and 40 kV, respectively.

2.3. Electrical Impedance Spectroscopy (EIS)

Impedance analyses of the prepared samples were performed using an LCR meter (HIOKI 3531 Z HiTester, Nagano, Japan), connected to a computer. Prior to taking measurements, the prepared films were cut into small discs of 2 cm diameter and sandwiched between two stainless steel electrodes with the aid of spring clips to ensure a good contact. The measurements were conducted within the frequency range of 50 Hz to 1 MHz at ambient temperature. The bulk resistance (R_b) was obtained from the intercept of the plot with the Z' axis. Both real and imaginary parts of the complex permittivity (ε^*)

and the complex modulus (M^*) as well as tanδwere extracted from the impedance data (i.e., Z' and Z'') using the following equations [27,28]:

$$\varepsilon' = \frac{Z''}{\omega C_0\left(Z'^2 + Z''^2\right)} \tag{1}$$

$$\varepsilon'' = \frac{Z'}{\omega C_0\left(Z'^2 + Z''^2\right)} \tag{2}$$

$$\tan\delta = \frac{Z'}{Z''} \tag{3}$$

$$M' = \frac{\varepsilon'}{\left(\varepsilon'^2 + \varepsilon''^2\right)} = \omega C_0 Z'' \tag{4}$$

$$M'' = \frac{\varepsilon''}{\left(\varepsilon'^2 + \varepsilon''^2\right)} = \omega C_0 Z' \tag{5}$$

where ω is the angular frequency of the applied field ($\omega = 2\pi f$) and ε' and ε'' are the dielectric constant and dielectric loss, respectively. The real and imaginary parts of the complex modulus M^* are represented by M' and M'', respectively. The C_o is the vacuum capacitance given by $\varepsilon_0 A/t$, where ε_o is a permittivity of free space, A is the electrode cross sectional area and t is the film thickness.

2.4. Transference Number Measurement (TNM)

In this work, both ionic (t_{ion}) and electronic (t_{el}) transference numbers were obtained from the cell polarization versus time at room temperature. Two stainless steel (SS) electrodes were used for the conducting sample of the SPE. The cell was polarized at a working voltage of 0.20 V. For this purpose, a V&A Instrument DP3003 digital DC power supply was used at room temperature.

2.5. Linear Sweep Voltammetry (LSV) Study

Linear sweep voltammetry (LSV) was recorded for SPEs to determine the electrochemical stability by measuring the oxidation decomposition voltage at room temperature. The LSV recording was carried out in the cell of two electrodes of stainless steel electrodes where the film sample was kept using a Digi-IVY DY2300 potentiostat (10 mV/s scan rate).

3. Results and Discussion

3.1. Structural Study

The XRD patterns of pure PVA and pure CS are presented in Figure 1a,b, respectively. In Figure 1a, a peak appears around 2θ = 18°, indicating the semi-crystalline behavior of pure PVA [11]. This is due to the existence of OH groups along the main chain of PVA, providing a relatively strong intermolecular and intramolecular hydrogen bonding. At the same time, a broad peak centered at 2θ = 40.7° refers to amorphous phases in the PVA structure.

From Figure 1b, a semi-crystalline feature of the CS polymer can be observed by two broad amorphous peaks centered at 2θ ranges from 33° to 45° and the pure CS exhibits several crystalline peaks at lower 2θ values [29]. Figure 2 provides insight into the PVA:CS blending system, indicating that there is no observable peak at 2θ =18.6°, which is the characteristic peak that appears for the pure PVA polymer. It is also observed that the hollow intensity decreases, and the broadening of line-width upon blending can be attributed to amorphous development [30].The XRD pattern for the PVA:CS blend electrolytes doped with various quantities of NH_4I is exhibited in Figure 3. Obviously, it can be seen that as NH_4I salt is increased up to 40 wt.% in the PVA:CS system, the hollow intensity substantially reduces and is accompanied by peak broadening. This indicates that the crystallinity of the blend polymer decreases at the expense of the amorphous region. At 50 wt.% of NH_4I addition, several crystalline peaks appear as a result of recrystallization of the NH_4I salt, and ion recombination could occur at high salt concentrations. It is worth mentioning that the crystalline regions in the polymer electrolytes decrease as a consequence of the complex formation. The complex formation results from the interaction between cations from the salt and the functional groups within the polar polymer backbone [31,32]. On the one hand, the periodic ordering of atom arrangements produces the lattice phases that are evidenced by scattering X-rays at certain directions [31]. On the other hand, the non-crystalline materials, such as the PVA:CS:NH_4I system, are characterized by a distorted structure within the atomic arrangement in the unit cells, as presented in Figure 3a. This distortion causes various X-ray scattering in all directions, resulting in the appearance of broadened peaks as well as relatively low peak intensity [31,33]. It is of great importance that broad peaks appear in polymer electrolytes, which indicates an improvement in the ion conductivity. In addition, the dominance of the amorphous region accelerates segmental motion within the polymer backbone that, in turn, increases the ion migration and ion conductivity as investigated by Fan et al. [13]. Similarly, Rangasamy et al. [8] believed that an increase in the amorphous content of the polymer film results in enhancing ion mobility as a consequence of providing free volume in the polymer network. It also causes an increase in the segmental motion of the polymer chains, leading to an increase in the flexibility of the polymer matrix. Consequently, the overall ionic conductivity of the polymer electrolyte can be improved.

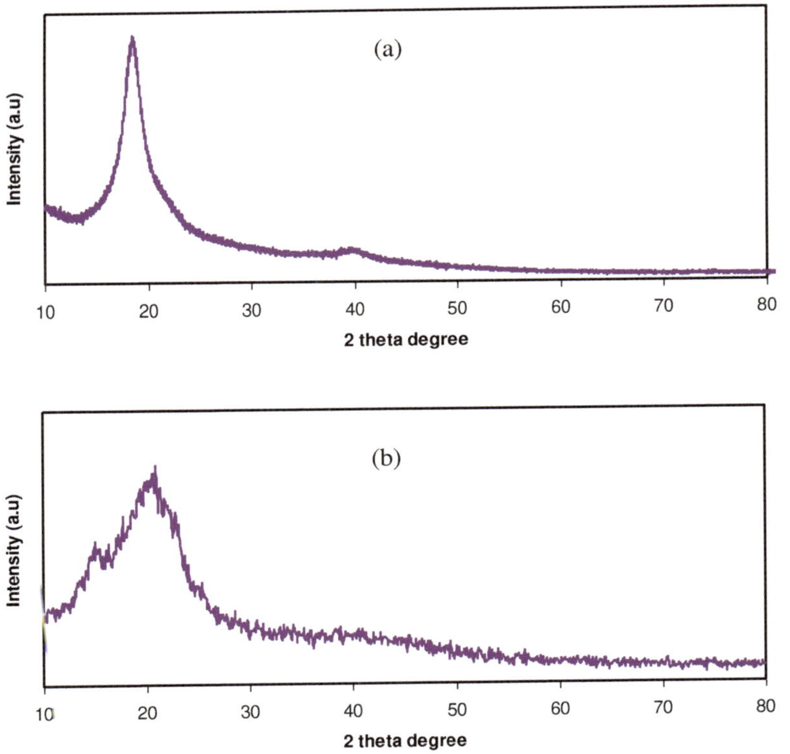

Figure 1. XRD pattern for (**a**) pure poly(vinyl alcohol) (PVA) and (**b**) pure chitosan (CS).

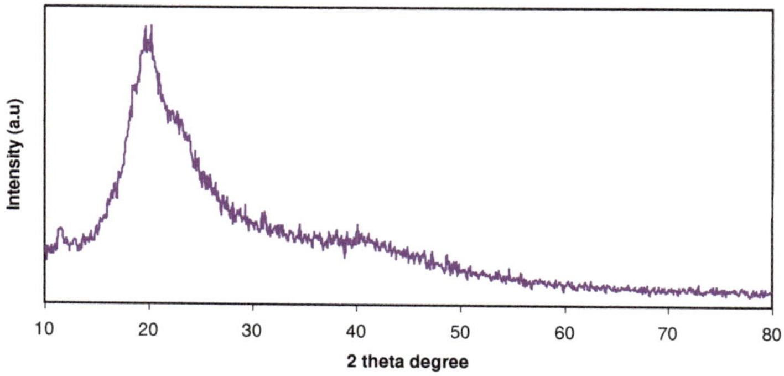

Figure 2. XRD pattern for PVA:CS [50:50] polymer blend films.

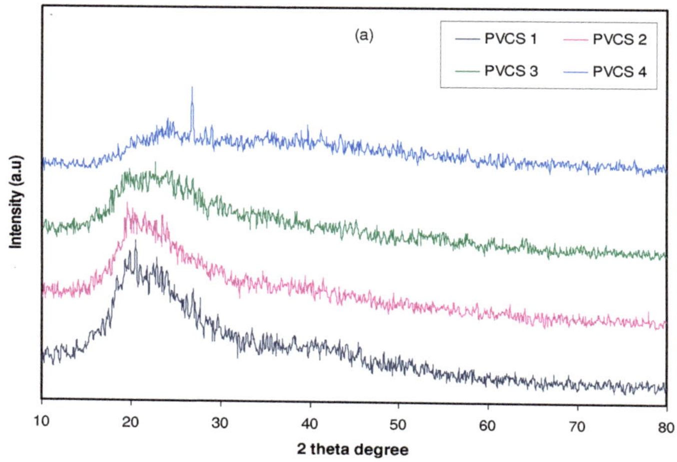

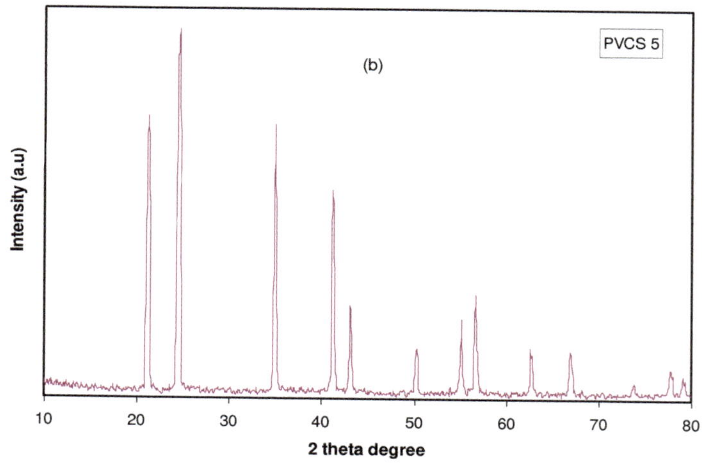

Figure 3. XRD pattern for PVA:CS [50:50] polymer blend electrolytes (**a**) from 10 to 40 wt.% of NH_4I and (**b**) for 50 wt.% of NH_4I.

3.2. Morphological Study

Figure 4a–e shows the SEM images of all SPE samples. It is clearly seen that the surfaces are relatively smooth to a large extent, in particular when the amount of NH_4I salt up to 40 wt.% was incorporated. Earlier studies have shown that, from the surface morphology, one can obtain information about polymer/salt complexation and reduction of transition metal salts, such as silver salts or copper salts in polymer electrolytes [34–37]. It is also well documented that uniform surface morphology could be an indication of the forming of a membrane without porosity [38]. Mobarak et al. [39] showed that smooth surface electrolytes indicate moving ions more freely and thus increasing DC conductivity. Thus, from surface analysis, it is possible to gain insight into the changes in structural and electrical properties of polymer composite systems [34]. A. K. Arof and coworkers [40] successfully established a correlation between the surface morphology of SPEs and DC conductivity. The study showed that a drop in DC conductivity at relatively high salt concentration was recorded for CS:PVA:xNH_4NO_3 electrolyte systems. This can be explained by the fact that ion aggregations lead to the formation of protrusions on the surface. In our earlier study, a large number of protrusions on the film surface of a CS:NaTf system were obviously seen at 50 wt.% NaTf and they even covered almost the whole electrolyte surface [41].

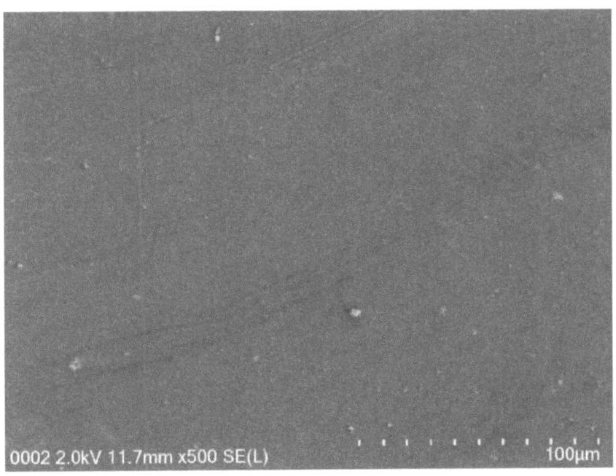

(a)

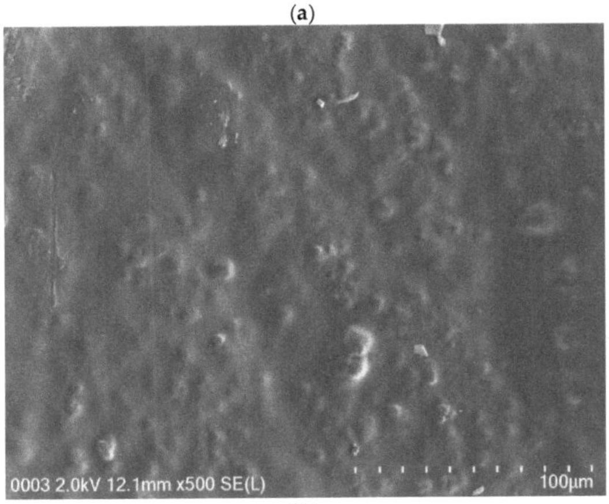

(b)

Figure 4. *Cont.*

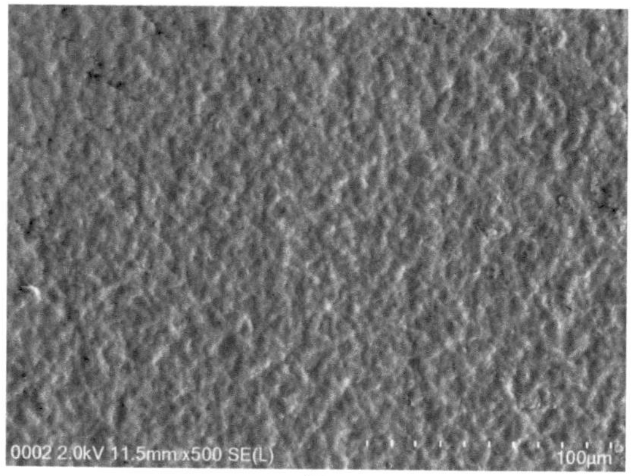

(**c**)

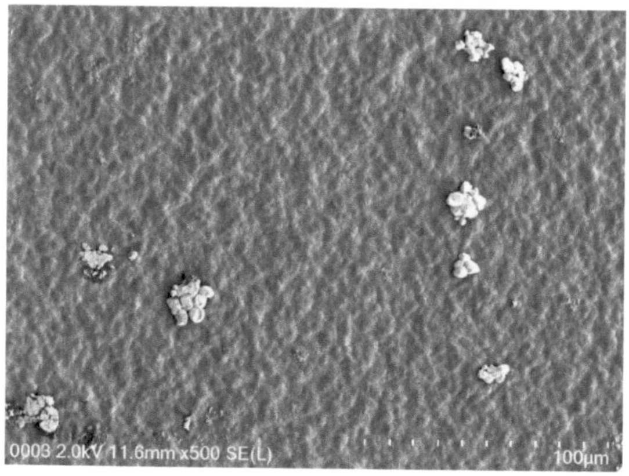

(**d**)

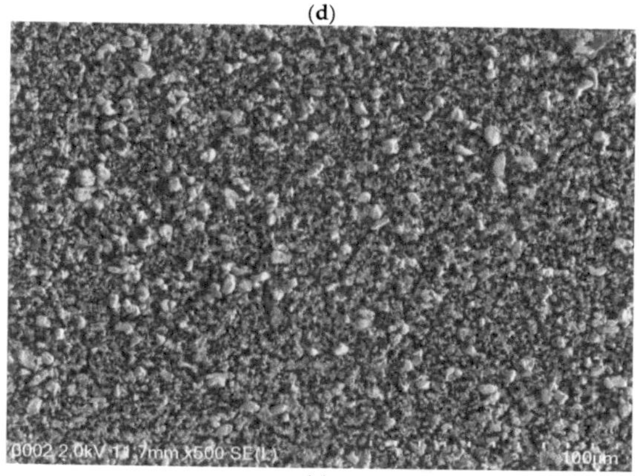

(**e**)

Figure 4. FESEM images for (**a**) PVCS 1, (**b**) PVCS 2, (**c**) PVCS 3, (**d**) PVCS 4 and (**e**) PVCS 5.

3.3. Impedance and AC Conductivity Study

The mechanism of ionic transport within polymer electrolyte systems is not fully understood, which becomes an obstacle to reaching the required ambient conductivity [25,42]. Nevertheless, impedance spectroscopy is a powerful technique to be used in the study of the ionic conductivity of polymer materials. Over the last decades, a new class of ion-conducting membrane materials has been the focus of researchers. Interest in these materials is returning to their wide use in solid-state electrochemical devices [43]. For this purpose, impedance spectroscopy was chosen to tackle the electrochemical properties of these materials, for instance, diffusion layer thickness, double-layer capacitance and charge transfer resistance [25]. From the impedance analysis, a plot for an ion-conducting polymer electrolyte is obtainable that consists of a small semicircle and a tail corresponding to high- and low-frequency regions, respectively. These responses provide insight into the bulk properties of the sample under study. It is observed that the depression of the semicircle size can be ascribed to the charge transfer at the interfacial region, as presented in Figure 5a–e. It is interesting to note that a capacitor-like material, a so-called pseudocapacitor, is formed at the sample/electrode interface resulting from double-layer growing [44]. A more interesting observation is the inclination caused by the blocking double-layer capacitance (i.e., electrode polarization) at the electrodes. The unparallel inclination of the straight line is other than the supposed value of 90° [45].

A direct relationship between the relatively high salt concentration and the bulk resistance can be established, as shown in Figure 5e. The impedance results are in strong agreement with the XRD results. The sharp peaks that correspond to the pure NH_4I salt obviously appear to be shifted as a consequence of the ion association that lowers conductivity [30]. All these suggest that it is straightforward to detect ion association within electrolyte polymers using EIS. From the data analysis, the bulk resistance (R_b) can be determined from the point where the real axis (Z_r) and the semicircle intersect. To calculate the DC conductivity of the samples, the following relationship can be used using sample dimensions and the R_b value [46]:

$$\sigma_{dc} = \left(\frac{1}{R_b}\right) \times \left(\frac{t}{A}\right) \tag{6}$$

where A and t are the surface area and thickness of the films, respectively. The tabulated results of the computed DC conductivities are presented in Table 2. It is clear that the highest DC conductivity is obtained at 40 wt.% of NH_4I and that the DC conductivity dropped as salt incorporation was introduced. The DC conductivity results are evidently in good agreement with the XRD results and surface morphology.

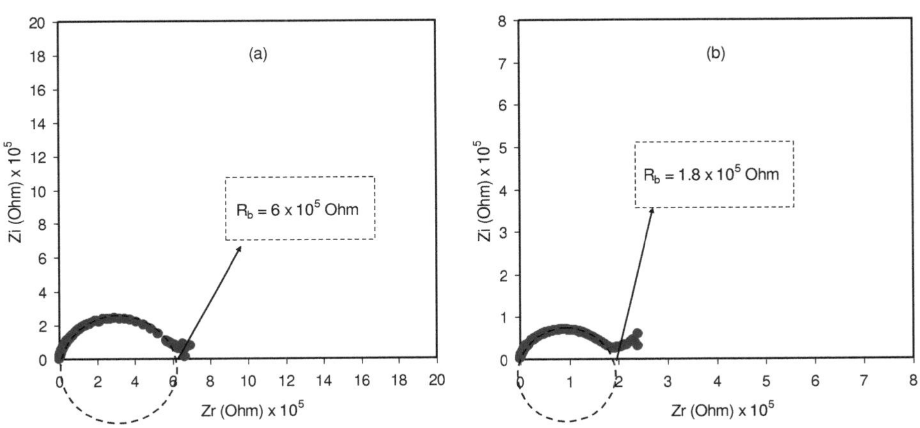

Figure 5. *Cont.*

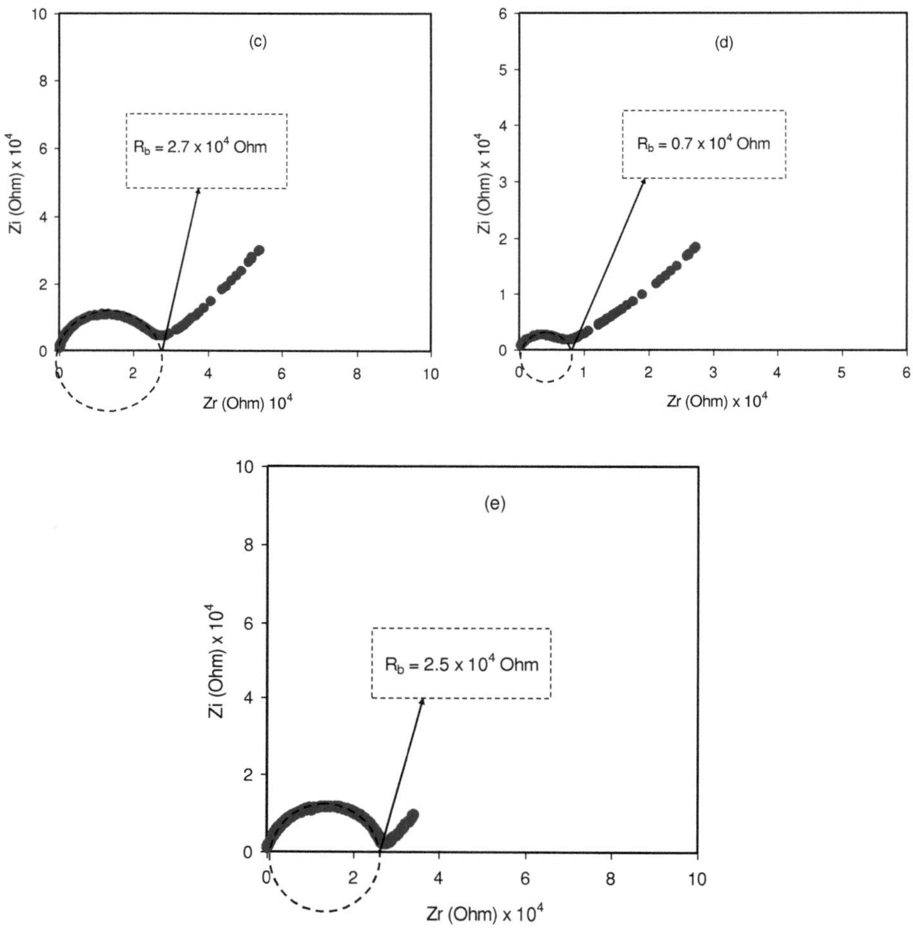

Figure 5. Complex impedance plots for (**a**) PVCS 1, (**b**) PVCS 2, (**c**) PVCS 3, (**d**) PVCS 4 and (**e**) PVCS 5.

Table 2. DC conductivity for PVA:CS:NH_4I blend electrolyte films from Equation (6) and AC conductivity spectra at room temperature.

Sample Designation	DC Conductivity (S/cm) (Using Equation (6))	DC Conductivity (S/cm) (Using AC Plot)
PVCS1	1.13×10^{-8}	1.18×10^{-8}
PVCS2	3.77×10^{-7}	3.8×10^{-7}
PVCS3	2.51×10^{-7}	2.3×10^{-7}
PVCS4	9.71×10^{-7}	7.6×10^{-7}
PVCS5	2.72×10^{-7}	2.0×10^{-7}

The electrical equivalent circuit (EEC) method is typically employed for examining EIS because the method is straightforward and rapid, as well as conveying a whole image of the PE system [3]. The Nyquist plot for the chosen PEs can be deduced with regard to the electrical equivalent circuit (EEC), comprising bulk resistance (R_b) for the species charge carriers in the PE films as well as two constant phase elements (CPEs) as revealed in the inserts in Figure 6. The high-frequency area reveals the parallel connection of R_b and CPE, while the low-frequency area displays CPE, specifically, the established EDLC between electrodes and electrolytes. The term CPE is more commonly used in EEC instead of ideal capacitor in the real system. The impedance of Z_{CPE} can be written as follows [3]:

$$Z_{CPE} = \frac{1}{C\omega^p}\left[\cos\left(\frac{\pi p}{2}\right) - i\sin\left(\frac{\pi p}{2}\right)\right] \quad (7)$$

where C denotes the CPE capacitance, ω stands for the angular frequency and P is linked to the divergence from the axis of the vertical EIS figures. Here, the real axis (Z_r) and the imaginary axis (Z_i) of the complex impedance (Z^*) correlated with the EEC (insert in Figure 6a) can be demonstrated as follows:

$$Z_r = \frac{R_b C_1 \omega^{p1}\cos\left(\frac{\pi p_1}{2}\right) + R_b}{2R_b C_1 \omega^{p1}\cos\left(\frac{\pi p_1}{2}\right) + R_b{}^2 C_1{}^2 \omega^{2p1} + 1} \quad (8)$$

$$Z_i = \frac{R_b C_1 \omega^{p1}\sin\left(\frac{\pi p_1}{2}\right)}{2R_b C_1 \omega^{p1}\cos\left(\frac{\pi p_1}{2}\right) + R_b{}^2 C_1{}^2 \omega^{2P1} + 1} \quad (9)$$

where C_1 stands for the CPE capacitance at the bulk. Furthermore, the real axis (Z_r) and the imaginary axis (Z_i) of the complex impedance (Z^*) correlated with the EEC (insert of Figure 6b–e) can be demonstrated as follows:

$$Z_r = \frac{R_b C_1 \omega^{p1}\cos\left(\frac{\pi p_1}{2}\right) + R_b}{2R_b C_1 \omega^{p1}\cos\left(\frac{\pi p_1}{2}\right) + R_b{}^2 C_1{}^2 \omega^{2p1} + 1} + \frac{\cos\left(\frac{\pi p_2}{2}\right)}{C_2\omega^{p2}} \quad (10)$$

$$Z_i = \frac{R_b C_1 \omega^{p1}\sin\left(\frac{\pi p_1}{2}\right)}{2R_b C_1 \omega^{p1}\cos\left(\frac{\pi p_1}{2}\right) + R_b{}^2 C_1{}^2 \omega^{2P1} + 1} + \frac{\sin\left(\frac{\pi p_2}{2}\right)}{C_2\omega^{p2}} \quad (11)$$

where C_1 stands for the CPE capacitance at the bulk and C_2 stands for the CPE capacitance at the electrode/electrolyte interface.

Table 3 presents the fitting parameters in the EEC. The R_b is moving away from the intersection of the semicircle or the line of the spike with the real part (see Figure 5). It is obvious from Figure 6 that, upon adding the salt, the semicircle at the high-frequency region becomes smaller up to 40 wt.% of the added salt.

Table 3. Fitting parameters of the EEC for electrolyte films at room temperature.

Sample	p1(rad)	p2(rad)	C_1(F)	C_2(F)
PVCS1	0.859	-	1.43×10^{-10}	-
PVCS2	0.71	0.8	3.33×10^{-10}	1.05×10^{-6}
PVCS3	0.8	0.5	2.00×10^{-9}	2.00×10^{-6}
PVCS4	0.78	0.42	5.00×10^{-9}	4.35×10^{-6}
PVCS5	0.89	0.52	1.00×10^{-9}	4.17×10^{-6}

Figure 7 shows the AC conductivity spectra at room temperature for all the samples. The AC conductivities are calculated using the following relation [27]:

$$\sigma'_{ac} = \left[\frac{Z'}{Z'^2 + Z''^2}\right] \times \left(\frac{t}{A}\right) \quad (12)$$

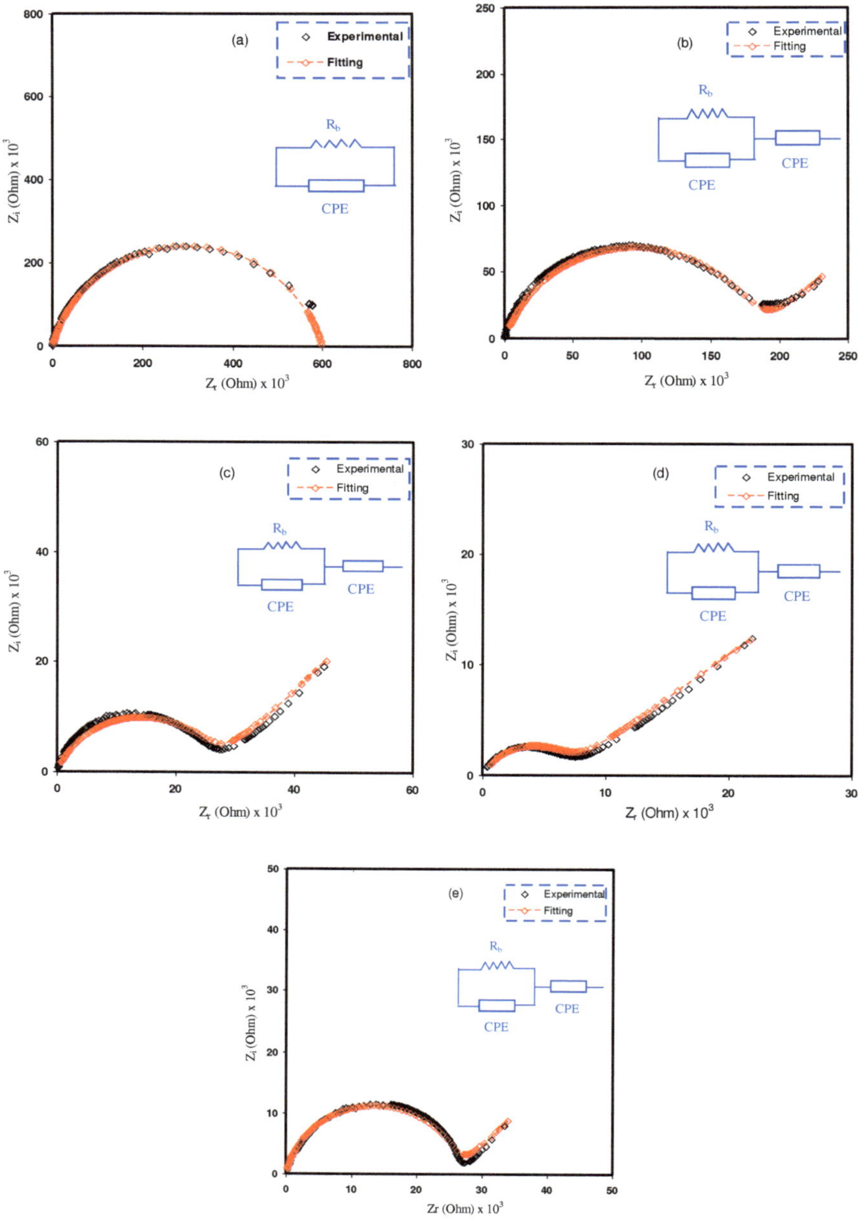

Figure 6. Experimental and EEC fitting plots for (**a**) PVCS 1, (**b**) PVCS 2, (**c**) PVCS 3, (**d**) PVCS 4 and (**e**) PVCS 5.

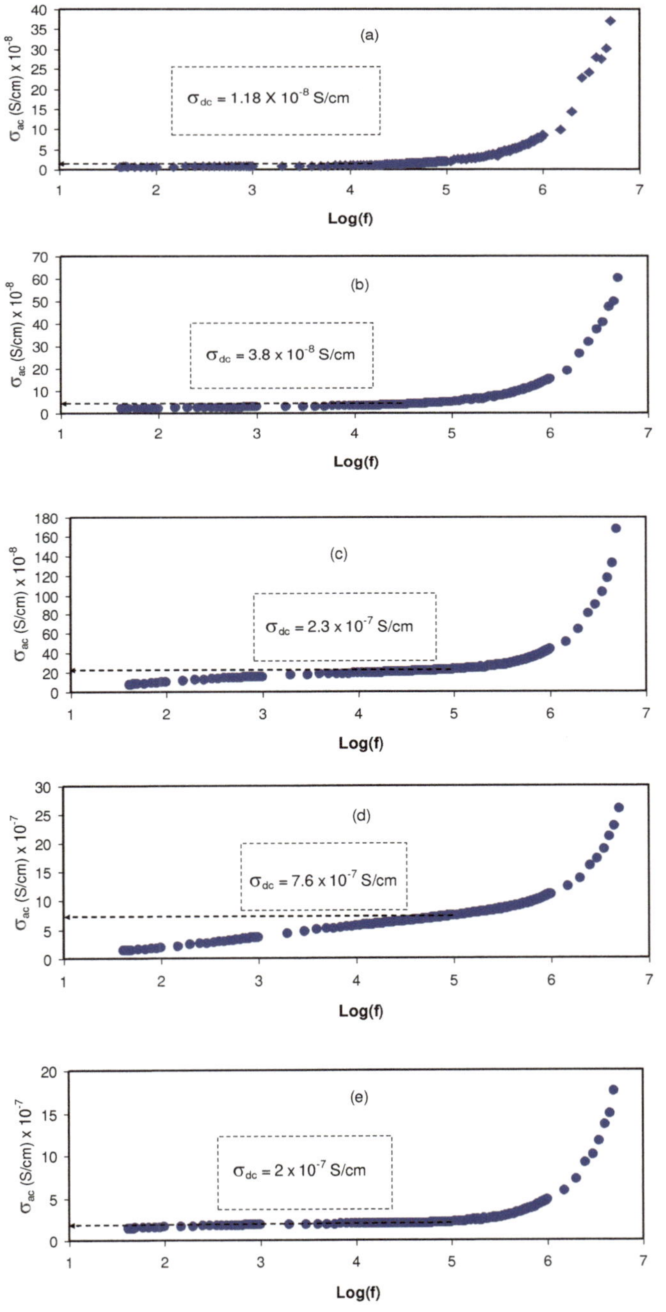

Figure 7. AC conductivity spectra for (**a**) PVCS 1, (**b**) PVCS 2, (**c**) PVCS 3, (**d**) PVCS 4 and (**e**) PVCS 5.

At a moderate salt concentration, three distinguished regions are seen, namely the low-frequency spike and the middle-frequency and high-frequency regions. At the low-frequency region, there is an electrode polarization that is directly proportional to the salt concentration. The middle-frequency region is seen almost in the form of a plateau that rises from the bulk DC conductivity. Interestingly, at the high-frequency region, the EIS response comes from conductivity relaxation that undergoes shifting as the salt concentration is increased [28]. An earlier study reported AC conductivity to the frequency of the electric signal correlation, which was then used in the analysis of electrical DC conductivity [35].

From this finding, it is possible to determine the DC conductivity by extrapolating the plateau region on the *y*-axis of the AC conductivity spectra. It is worth noting that the DC conductivity values (see the insets in Figure 7) are quite close to those obtained from the bulk resistance (R_b), as presented in Table 2. It is interesting to note that as the salt concentration is increased, the dispersion region becomes narrow. This is directly governed by electrode polarization (in the form of spikes). It is also seen that the AC conductivity increases as it moves toward the higher frequency region. From this increase in AC conductivity with frequency, one can suggest that the hopping conduction mechanism occurs where charge carrier hopping is enhanced between the localized states [47]. These observations in the study of AC conductivity have been confirmed in earlier work [27]. Analyzing AC conductivity spectra, it is easy to determine DC conductivity. It is also possible to compare the impedance spectra to show a strong relationship between the dispersion region of AC conductivity and the high-frequency semicircle (see the Nyquist plot in Figure 5).

The presence of a spike region is the response of adding a relatively high amount of salt, as shown in Figure 7. The phenomenon of AC conductivity results from charge carrier confinement throughout the whole body of the sample [27,47]. Based on the well-known Jonscher universal power low, specifying the nature of ion dynamics by calculating the frequency exponent (s) can be accurately obtained as follows [26–28]:

$$\sigma'_{ac}(\omega) = \sigma_{dc} + A\,\omega^{s} \qquad (0 < s < 1), \tag{13}$$

For most ion conductors, the second term of Equation (8) (i.e., $\sigma'_{ac}(\omega) = A\,\omega^{s}$) is followed by the dispersion region of AC conductivity where the frequency exponent (s) is less than unity [16]. It is also important to examine the first term, which is related to the plateau response of the AC spectra, and its extension to the *y*-axis can be used to estimate DC conductivity [26,28]. The DC conductivity values obtained from the AC spectra are presented in Table 2. It is noteworthy that it is of vital importance to show a good agreement between DC conductivity obtained from the AC spectra and the impedance results.

3.4. Study of Dielectric Properties

Dielectric constant analysis is an informative way to deal with the mechanism of ion transportation and the phase transitions within polymer electrolytes. It is well known that ion pairs, triplets and clusters cause a lowering of electrical conductivity. It is important to mention that ion pairs possess a higher permanent dipole moment compared to others. Importantly, the host polymer shows a relatively low dielectric constant despite the existence of ion pairs when the dielectric measurement is performed [25,48–52].Figures 8 and 9 explain the variation of dielectric constant and dielectric loss for all blended electrolyte samples at ambient temperature. It can also be seen that both dielectric parameter (ε', ε'') values are relatively high in the low-frequency region, indicating the electrode polarization phenomenon. This phenomenon results from local charge accumulation at the electrode/electrolyte interface region. More significantly, this phenomenon is the result of the difference in conductivity between two materials in contact with each other [46]. An increase in salt concentration results in an increase in dielectric constant and dielectric loss of the PVA:CS solid polymer electrolyte. This can be correlated to both the bond energy of the salt and the polymer and an increase in polarization [28,49]. In general, the stable structure of the polymer comes from two main forces, namely primary intra- and secondary intermolecular forces [50]. The primary forces comprise covalent bonding (2.2–8.6 eV) and ionic bonding (0.43–0.87 eV) that connect atoms in the polymer backbone chain, whereas the secondary forces are hydrogen bonding (0.13–0.30 eV), dipolar interactions (0.07–0.13 eV) and dispersion interactions (0.002–0.09 eV). In terms of dissociation energy, it is easy to break down the secondary forces compared to the primary ones. Based on this explanation, the secondary forces can easily be disrupted with salt addition that, in turn, impacts considerably on the polymer segmental motions within the polymer body. In other words, salt addition affects the dielectric behavior, charge transport and charge storage. As a consequent, a double-layer capacitance can develop from ion transport that accumulates between the sample and the electrodes. In addition, as the applied field frequency increases, the available drift time reduces. Accordingly, the dielectric constant decreases, and the required time for charge carrier drifting reduces; therefore, both real and imaginary parts of the dielectric properties decrease [51–53]. Principally, polarization results from charge orientation and ultimately disappears due to the inertia of the ions [49]. It is clear that a high dielectric constant is obtained for 40 wt.% of NH_4I addition and then it drops in the case of 50 wt.% because of electrode/electrolyte interface blocking. This is further confirmed by the DC conductivity values presented in Table 2. The physics of this DC conductivity and dielectric constant relationship can be qualitatively explained. In general, the famous expression for conductivity is formulated by the following:

$$\sigma = \sum n q \mu \tag{14}$$

where n is the charge carrier density, q is 1.6×10^{-19} C and μ is the mobility of the ions. Based on this formulation, it can be clearly seen that both ionic mobility and the number of charge carriers increase as the salt concentration is increased in the range 10 to 40 wt.%. However, further salt addition leads to a decrease in dielectric parameters, for example, at 50 wt.%. Moreover, the carrier density (i.e., number of the charge) is directly related to the bond dissociation energy U and dielectric constant ε', which can be understood via this relationship ($n = n_o \exp(-U/\varepsilon' KT)$). For example, the addition of salts into polymer matrices causes an increasing dielectric constant, as a consequence of increasing the charge carrier density [54]. Therefore, to determine the conductivity behavior of the solid polymer electrolyte, relationship between DC conductivity and dielectric constant must be established as well as to salt concentration at room temperature.

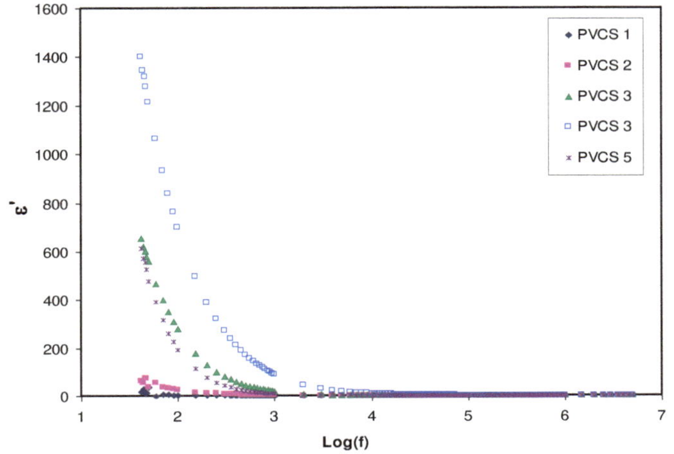

Figure 8. Dielectric constant versus log (f) for all polymer blend electrolytes.

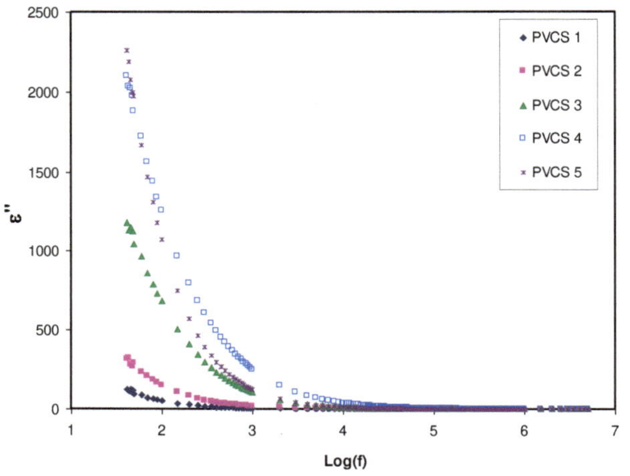

Figure 9. Dielectric loss versus log (f) for all polymer blend electrolytes.

The relaxation processes of polymer materials can be accurately examined via loss tangent peaks. The dipoles in the polymer electrolytes can certainly be interpreted on the basis of the dielectric relaxation. Figure 10 shows dielectric relaxation peaks of the loss tanδ versus frequency plot for all PVCS electrolyte samples at room temperature [52]. In the figure, there is a shift to a higher frequency region of the loss tangent peak, indicating the occurrence of dielectric relaxation. Furthermore, the shortest relaxation time of the PVCS4 electrolyte sample can be observed via the loss tangent plot. One finding of this study is that permanent or induced dipoles cause the electric conductivity and dielectric relaxation peaks. It has also been observed that induced diploes hidden the polarization relaxation of mobile charged species in the materials [55]. Based on Koops phenomenological model, the loss tangent shape can be interpreted [56]. Accordingly, the low-frequency dispersion curve possesses a negative slope, suggesting the loss of dominancy of conduction and that it is modeled via a parallel RC circuit in the adopting homogeneous system. The EECs presented in Figure 7 support precisely the Koops model. The loss tangent intensity increases to a maximum at a certain frequency value and then decreases as the frequency increases. This is related to the fact that active component (ohmic) of the current increases more rapidly than the reactive component (capacitive). More than one relaxation process causes the loss tangent peaks to be broad and to obey non-Debye type relaxation [57]. As previously shown, the tanδ plot is helpful in calculating transport parameters, namely the diffusion coefficient, carrier density and mobility [58].

Conductivity and relaxation dynamics are frequency dependent and are responsible for charged species motion and the induction of dipoles in the polymer electrolyte. The relaxation dynamics from dielectric relaxation can be investigated using electric modulus formalism [52,59]. To obtain bulk relaxation properties, the electric modulus is helpful. The electric polarization (EP), space charge injection phenomena and conduction effects can be understood via electric modulus formalism [27,60]. The evaluations of the real and imaginary parts of the electric modulus were carried out using Equations (4) and (5). Figures 11 and 12 exhibitthe real and imaginary parts of the electric modulus versus frequency for all PVCS electrolyte samples at room temperature. At low frequency, a long tail for either M' and M'' is seen and is related to the capacitance obtained from the double-layer charge building up at the interfacial region [46,59]. In comparison, the ε' and ε'' spectra and the M' and M'' behave in exactly the opposite manner. It is interesting that the dielectric constant's high value (see Figure 8) is seen at the low-frequency region. Based on the principles, the electric moduli (M' and M'') are made from the reciprocal of the complex dielectric constant, recording a minimum value at high frequency [61]. From the unusual response of M'', it is difficult to apply simple exponential Debye to interpret the relaxation process. It is extremely interesting to note the relaxation peak appearance in the M'' spectra (see Figure 12) and the disappearance in the dielectric loss spectra, as exhibited in Figure 9. The appearance of peaks in the imaginary part of the modulus spectra indicates the existence and contribution of huge amounts of electrode polarization in the dielectric loss parameter. This can be used in tackling the conduction process mechanism in polymer electrolytes, where it progresses via ion migration between coordinated sites within the polymer body and segmental relaxation. In other words, the appearance of peaks in the M'' spectra can be interpreted as a result of a couple of species motions, namely ionic and polymer segmental motions [62,63]. An interesting observation can be seen from the peak shifting of relaxation to the lower frequency side with an increase to 50 wt.% of NH_4I salt. This reveals a strong relationship between relaxation time and salt concentration, which is directly proportional. Basically, it has been documented that the relaxation time increases as the ionic mobility decreases [57]. Specifically, the relaxation time decreases as conductivity increases, as reported in previous work [52,58,59,64]. In the current study, it was confirmed that the relaxation time increases when conductivity decreases.

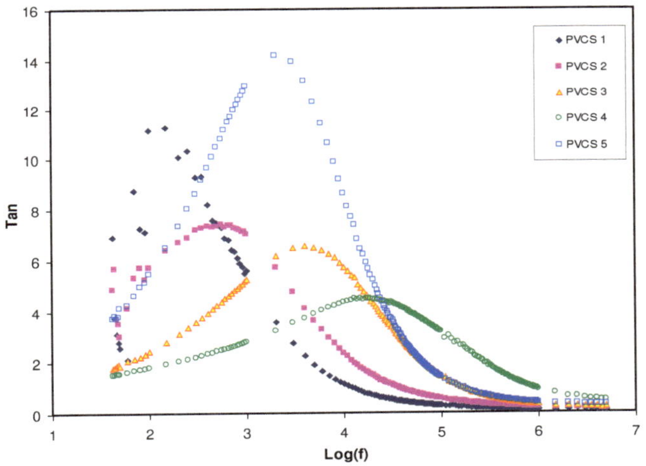

Figure 10. Tanδ versus log (f) for all polymer blend electrolytes.

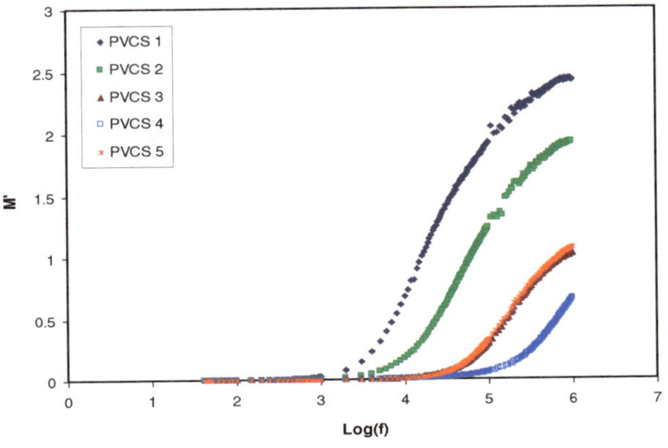

Figure 11. Real part of electric modulus versus log (f) for all polymer blend electrolytes.

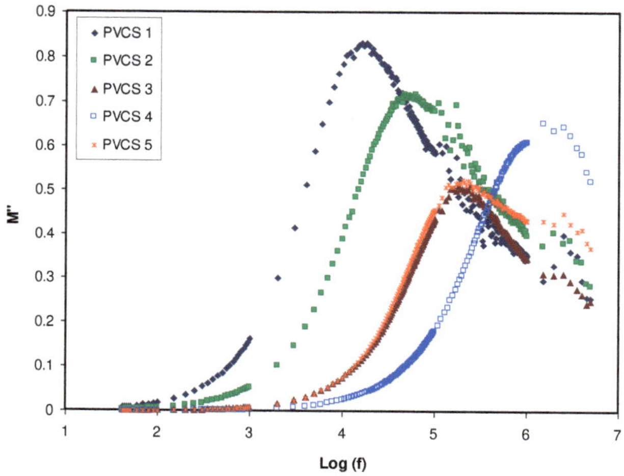

Figure 12. Imaginary part of electric modulus versus log (f) for all polymer blend electrolytes.

3.5. Electrochemical Characteristics

3.5.1. TNM Measurement

The heart of electrochemical devices is considered to be the ion-conducting electrolyte, for instance, in batteries and supercapacitors; thus, the study of electrolytes has priority [65]. The focus has been devoted to evaluating the contribution of both ions and electrons in the polymer electrolyte and metallic solid materials, respectively. It is clearly known that, at the interfacial region, charge transfer occurs between two different phases, i.e., the solid metallic electrode and the liquid (electrolyte) phases. It is of significant importance to determine the identity of species that are responsible for carrying a charge in the system under study. Herein, a 0.8 V of the DC polarization method was applied to the working electrode, and transfer number measurement (TNM) was performed. The information obtained about the TNM plots for PVCS4 and PVCS5 electrolyte samples from the polarization responses at room temperature versus time can be seen in Figure 13a,b. The nature of ion transport was evaluated via Wagner's polarization method in an attempt to calculate the total ionic transference number (t_{ion}) from the current–time plot. From the plot, there is a sudden drop in current within the system at the beginning, indicating the extent of contribution to conductivity by ions versus electrons [12]. The values of t_{ion} and t_{el} were estimated using the following equations [66]:

$$t_{ion} = \frac{I_i - I_{ss}}{I_i} \quad (15)$$

$$t_{el} = 1 - t_{ion} \quad (16)$$

where I_i and I_{ss} are the initial current and the steady-state current, respectively. From the TNM plots, as the potential is swept, at the beginning stage, there is a huge current rise of 0.89 μA and 0.41 μA for PVCS4 and PVCS5, respectively. This results from the contribution of both electrons and ions in the conduction. The value of ion transport t_{ion} was extracted from the initial and steady-state currents of the PVCS4 and PVCS5 samples, and found to be 0.88 and 0.75, respectively. It can be seen that the value of t_{ion} of PVCS4 is higher than PVCS5. The general response profile consists of three distinct regions, namely the initial current rise, current decay and steady-state current. At the initial state, the current flows across the cell at the blocking electrode under the impact of an applied voltage are due to electrode charging. In other words, at the initial stage of the electrochemical course, the DC potential results in the current creation that is proportional to both the ion migration and electron moving. The electric field encourages mobile species to be pushed to move in the electrolyte as a result of the migration of ions. Subsequently, the current drops within a short time period, where ion drifting is equivalent to ion diffusion. At steady state, the diffusion layer develops from a concentration gradient induced by electrode polarization at the interface region where polarization occurs for a long period of time. Additionally, the DC polarization at the interfacial region faces a substantial resistance from the passive layer formation by ions. This means the current comes solely from electrons, i.e., the entire current is from electrons with no contribution by ions. Based on the ion transfer number values of 0.88 and 0.75 of the two samples, the main contributor of the carrying charge are ions rather than electrons [67]. Therefore, from the TNM measurement it can be concluded that PVCS4 is a preferable system for electrochemical device applications, in particular if the conductivity of the system is maximized to 10^{-3}–10^{-4} S/cm. The TNM measurement setup is shown in Figure 14. The initial current is very important, and as the cell is subjected to 0.2 V, the first current response in the multimeter during the switch-on was taken as initial current. The initial current is high due to the contribution of both ions and electrons.

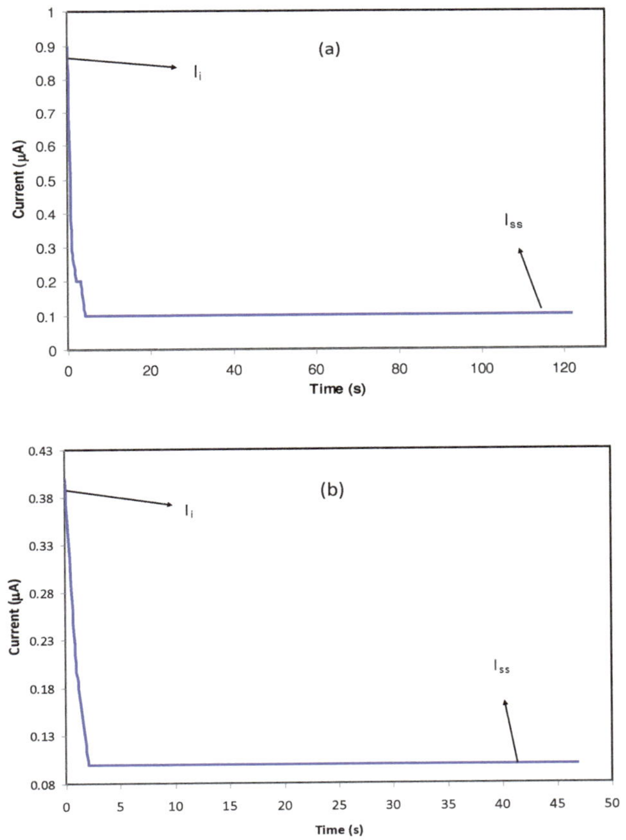

Figure 13. Polarization current versus time for (**a**) PVCS4 and (**b**) PVCS5 electrolyte films.

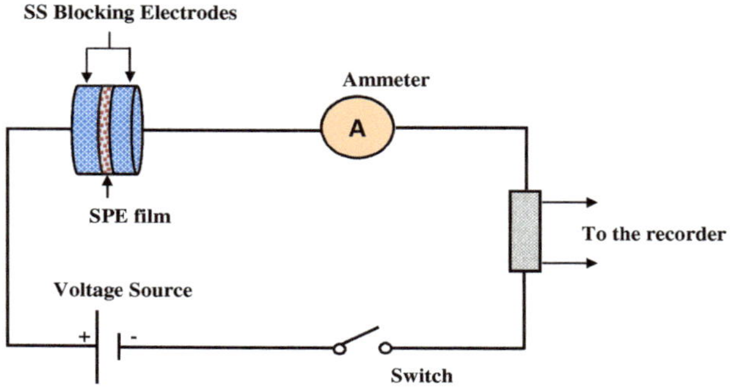

Figure 14. TNM measurement setup.

3.5.2. Linear Sweep Voltammetry (LSV) Studies

In order to determine the electrochemical stability of the polymer electrolyte under study, linear sweep voltammetry (LSV) is a straightforward and powerful technique to use. LSV was performed for the highest DC conductivity and TNM value sample.The LSV response of the polymer electrolyte PVA:CS:NH_4I (PVCS4) at 10 mV/s sweep rate with room temperature is shown in Figure 15. It can be seen that the current rose considerably and sharply at 1.33 V, which indicates the electrolyte decomposition. This potential cut-off from the LSV response suggests the eligibility of the electrolyte for use in proton-based energy devices. For comparison, this result is quite close to that documented for a polymer electrolyte based on ammonium salt found in the literature. Ng and Mohamad reported the cut-off potential of 1.8 V for a chitosan-based membrane made of ammonium nitrate salt and ethylene carbonate plasticizer at room temperature [68]. In another study conducted by Kadir et al., the potential window of 1.7 V was reported for CS:poly(vinyl alcohol) (PVA) [69]. Comparably, in our earlier report on PVA:dextran:NH_4I system, 1.3 V was obtained [70]. There is harmony between the potential window that is lower than 1 V and undesired consequences, such as solvent evaporation and leakage in supercapacitor systems [10]. Consequently, these results help to decide on the eligibility of PVA:CS:NH_4I as the electrolyte of choice in EDLC applications.

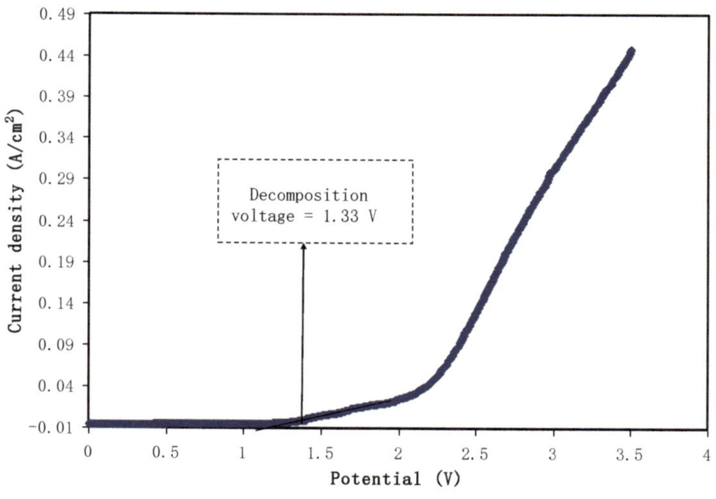

Figure 15. Linear sweep voltammetry (LSV) plot for the highest conducting (PVCS4) blend electrolyte film.

4. Conclusions

In conclusion, the preparation and characterization of polymer blend electrolytes based on poly(vinyl alcohol):chitosan (PVA:CS) polymers were studied. The electrical characterizations (ε′, ε″, tanδ, M′ and M″) were carried out in order to understand that ion transport occurs through the coupling between ion and polymer segmental motion. The electrochemical investigations (TNM and LS) were performed in order to examine the suitability of the samples for energy storage device applications. The structural analysis from X-ray diffraction (XRD) revealed the structural change upon the addition of NH_4I salt. As the salt concentration increased to 50 wt.% NH_4I, several crystalline sharp peaks were observed in the XRD spectra due to the salt's protruded appearance. The protruded appearance on the samples' surface was evidently shown at high salt concentrations in the field-emission scanning electron microscope (FESEM) images. The XRD and FESEM results support each other. The optimum salt content is 40 wt.% of NH_4I to reach the maximum DC conductivity (9.1×10^{-7} S/cm). The most amorphous system exhibits the highest DC conductivity. To obtain further insight into the electrical characteristics of the ion-conducting films, the EIS spectra were fitted with electrical equivalent circuits (EECs). The conductivity measurements of the samples were correlated with their dielectric properties. At the low-frequency region, high values of dielectric constant and dielectric loss were observed due to electrode polarization. From loss tangent and electric modulus plots, the broad nature of the peaks appeared in the tanδ and imaginary parts of the electric modulus, indicating the distribution of relaxation times. From the TNM measurements, both ionic (t_{ion}) and electronic (t_{ele}) transference numbers were evaluated. It was found that the t_{ion} values for PVCS4 and PVCS5 samples were 0.88 and 0.75, respectively. This indicates that the system incorporated with 40 wt.% of NH_4I salt exhibits a high ion transference number. A potential cut-off of 1.33 V was recorded for the electrolyte system as decomposition voltage. This allowed us to conclude that the PVA:CS biopolymer electrolyte in this work is electrochemically stable up to above 1 V, which is important for use in electrical double-layer capacitor (EDLC) applications.

Author Contributions: Conceptualization, R.M.A. and S.B.A.; Formal analysis, A.S.M.; Investigation, A.S.M.; Methodology, A.S.M. and S.B.A.; Project administration, S.B.A.; Supervision, R.M.A. and S.B.A.; Validation, R.M.A. and S.B.A.; Writing—original draft, A.S.M.; Writing—review and editing, R.M.A. and S.B.A. All authors have read and agreed to the published version of the manuscript.

Funding: This research was funded by the Ministry of Higher Education and Scientific Research-Kurdish National Research Council (grant number: KNRC), Kurdistan Regional Government/Iraq.

Acknowledgments: The authors gratefully acknowledge the financial support for this study received from the Ministry of Higher Education and Scientific Research-Kurdish National Research Council (KNRC), Kurdistan Regional Government/Iraq.

Conflicts of Interest: The authors declare no conflict of interest.

References

1. Shukur, M.F.; Yusof, Y.M.; Zawawi, S.M.M.; Illias, H.A.; Kadir, M.F.Z. Conductivity and transport studies of plasticized chitosan-based proton conducting biopolymer electrolytes. *Phys. Scr.* **2013**, *157*, 014050. [CrossRef]
2. Aziz, S.B.; Woo, T.J.; Kadir, M.F.Z.; Ahmed, H.M. A conceptual review on polymer electrolytes and ion transport models. *J. Sci. Adv. Mater. Devices* **2018**, *3*, 1–17. [CrossRef]
3. Aziz, S.B.; Hamsan, M.H.; Ranjdar, M.; Abdullah, M.F.Z.; Kadir, A. Promising polymer blend electrolytes based on chitosan: Methyl cellulose for EDLC application with high specific capacitance and energy density. *Molecules* **2019**, *24*, 2503. [CrossRef]
4. Hamsan, M.H.; Aziz, S.B.; Azha, M.A.S.; Azli, A.A.; Shukur, M.F. YusnizaMohdYusof, Saifful Kamaluddin Muzakir, Ninie SA Manan, MohdFakhrulZamaniKadir, Solid-state double layer capacitors and protonic cell fabricated with dextran from Leuconostocmesenteroides based green polymer electrolyte. *Mater. Chem. Phys.* **2020**, *241*, 122290. [CrossRef]

5. Aziz, N.A.; Majid, S.R.; Arof, A.K. Synthesis and characterizations of phthaloyl chitosan-based polymer electrolytes. *J. Non Cryst. Solids* **2012**, *358*, 1581–1590. [CrossRef]
6. Alipoori, S.; Mazinani, S.; Aboutalebi, S.H.; Sharif, F. Review of PVA-based gel polymer electrolytes in flexible solid-statesupercapacitors: Opportunities and challenges. *J. Energy Storage* **2020**, *27*, 101072. [CrossRef]
7. Sunitha, V.R.; Kabbur, S.K.M.; Pavan, G.S.; Sandesh, N.; Suhas, M.R.; Lalithnarayan, C.; Laxman, N.; Radhakrishnan, S. Lithium ion conduction in PVA-based polymer electrolyte system modified with combination of nanofillers. *Ionics* **2019**, *26*, 823–829. [CrossRef]
8. Guo, D.; Xiao, Y.; Li, T.; Zhou, Q.; Shen, L.; Li, R.; Xu, Y.; Lin, H. Fabrication of high-performance composite nanofiltration membranes for dye wastewater treatment: Mussel-inspired layer-by-layer self-assembly. *J. Colloid Interface Sci.* **2020**, *560*, 2732–2783. [CrossRef]
9. Zhang, Y.; Cheng, X.; Jiang, X.; Urban, J.J.; Lau, C.H.; Liu, S.; Shao, L. Robust natural nanocomposites realizing unprecedented ultrafast precise molecular separations. *Mater. Today* **2020**. [CrossRef]
10. Wang, J.; Zhao, Z.; Song, S.; Ma, Q.; Liu, R. High Performance Poly (vinyl alcohol)-Based Li-IonConducting Gel Polymer Electrolyte Films for Electric Double-Layer Capacitors. *Polymers* **2018**, *10*, 1179. [CrossRef]
11. Aziz, S.B. Modifying poly (vinyl alcohol) (PVA) from insulator to small-bandgap polymer: A novel approach for organic solar cells and optoelectronic devices. *J. Electron. Mater.* **2016**, *45*, 736–745. [CrossRef]
12. Yan, L.; Yang, X.; Long, J.; Cheng, X.; Pan, D.; Huang, Y.; Shao, L. Universal unilateral electro-spinning/spraying strategy to construct water-unidirectional Janus membranes with well-tuned hierarchical micro/nanostructures. *Chem. Commun.* **2020**, *56*, 478–481. [CrossRef]
13. Ramaswamy, M.; Malayandi, T.; Subramanian, S.; Srinivasalu, J.; Rangaswamy, M.; Soundararajan, V. Development and Study of Solid Polymer Electrolyte Based on Polyvinyl Alcohol: Mg(ClO4)2. *Polym. Plast. Technol. Eng.* **2017**, *56*, 992–1002. [CrossRef]
14. Shujahadeen, B.A.; Hamsan, M.H.; Kadir, M.F.Z.; Woo, H.J. Design of polymer blends based on chitosan: POZ with improved dielectric constant for application in polymer electrolytes and flexible electronics. *Adv. Polym. Technol.* **2020**, *2020*, 8586136. [CrossRef]
15. Aziz, S.B. Role of dielectric constant on ion transport: Reformulated Arrhenius equation. *Adv. Mater. Sci. Eng.* **2016**, *2016*. [CrossRef]
16. Aziz, S.B.; Abidin, Z.H.Z. Electrical conduction mechanism in solid polymer electrolytes: New concepts to arrhenius equation. *J. Soft Matter* **2013**, *2013*, 8. [CrossRef]
17. Ikram, S.; Ahmed, S.; Ali, S.W.; Agarwal, H. Chitosan-Based Polymer Electrolyte Membranes for Fuel Cell Applications. In *Composite Polymer Electrolyte Membranes Organic-Inorganic*; Inamuddin, D., Mohammad, A., Asiri, A., Eds.; Springer: Cham, Switzerland, 2017. [CrossRef]
18. Gunawan, I.; Sugeng, B. Synthesis and characterization of PVA blended LiClO4 as electrolyte material for battery Li-ion. In Proceedings of the IOP Conference Series: Materials Science and Engineering, Busan, Korea, 25–27 August 2017; Volume 223, p. 012039. [CrossRef]
19. Aziz, S.B.; Brza, M.A.; Mishra, K.; Hamsan, M.H.; Karim, W.O.; Abdullah, R.M.; Abdulwahid, R.T. Fabrication of high performance energy storage EDLC device from proton conducting methylcellulose: Dextran polymer blend electrolytes. *J. Mater. Res. Technol.* **2019**, *2020*, 1137–1150. [CrossRef]
20. Prajapati, G.K.; Roshan, R.; Gupta, P.N. Effect of plasticizer on ionic transport and dielectric properties of PVA–H3PO4 proton conducting polymeric electrolytes. *J. Phys. Chem. Solids* **2010**, *71*, 1717–1723. [CrossRef]
21. Gao, H.; Lian, K. Proton-conducting polymer electrolytes and their applications in solid supercapacitors: A review. *R. Soc. Chem.* **2014**, *4*, 33091–33113. [CrossRef]
22. Jonson, D.A. *Some Thermodynamic Aspects of inorganic Chemistry*, 2nd ed.; Cambridge University Press: Cambridge, UK, 1982; ISBN 0521242045.
23. Aziz, S.B.; Abidin, Z.H.Z.; Arof, A.K. Effect of silver nanoparticles on the DC conductivity in chitosan–silver triflate polymer electrolyte. *Phys. B Condens. Matter* **2010**, *405*, 4429–4433. [CrossRef]
24. Aziz, S.B.; Hamsan, M.H.; Abdullah, R.M.; Abdulwahid, R.T.; Brza, M.A.; Marif, A.S.; Kadir, M.F.Z. Protonic EDLC Cell based on chitosan (CS): Methylcellulose (MC) Solid Polymer Blend Electrolytes. *Ionics* **2020**. [CrossRef]
25. Aziz, S.B. Li+ ion conduction mechanism in poly (ε-caprolactone)-based polymer electrolyte. *Iran. Polym. J.* **2013**, *22*, 877. [CrossRef]
26. Aziz, S.B.; Abidin, Z.H.Z. Ion-transport study in nanocomposite solid polymer electrolytes based on chitosan: Electrical and dielectric analysis. *J. Appl. Polym. Sci.* **2015**, *132*, 41774. [CrossRef]

27. Aziz, S.B. Occurrence of electrical percolation threshold and observation of phase transition in Chitosan(1-x): AgIx ($0.05 \leq x \leq 0.2$)-based ion-conducting solid polymer composites. *Appl. Phys. A* **2016**, *122*, 706. [CrossRef]
28. Aziz, S.B.; Abidin, Z.H.Z. Electrical and morphological analysis of chitosan: AgTf solid electrolyte. *Mater. Chem. Phys.* **2014**, *144*, 2802–2886. [CrossRef]
29. Abidin, Z.H.Z.; Kadir, M.F.Z. Innovative method to avoid the reduction of silver ions to silver nanoparticles in silver ion conducting based polymer electrolytes. *Phys. Scr.* **2015**, *90*, 3.
30. Hamsan, M.H.; Shukur, M.F.; Aziz, S.B.; Kadir, M.F.Z. Dextran from leuconostocmesenteroides doped ammonium salt based green polymer electrolyte. *Bull. Mater. Sci.* **2019**, *42*, 57. [CrossRef]
31. Aziz, S.B.; Abdullah, O.G.; Hussein, S.A. Role of Silver Salts Lattice Energy on Conductivity Drops in Chitosan Based Solid Electrolyte: Structural, Morphological and Electrical Characteristics. *J. Electron. Mater.* **2018**, *47*, 3800–3808. [CrossRef]
32. Aziz, S.B.; Kadir, M.F.Z.; Abidin, Z.H.Z. Structural, morphological and Electrochemical Impedance Study of CS:LiTf based Solid Polymer Electrolyte:Reformulated Arrhenius Equation for Ion Transport Study. *Int. J. Electrochem. Sci.* **2016**, *11*, 9228–9244. [CrossRef]
33. Yusuf, S.N.F.; Azzahari, A.D.; Yahya, R.; Majid, S.R.; Careem, M.A.; Arof, A.K. From crab shell to solar cell: A gel polymer electrolyte based on N-phthaloylchitosan and its application in dye-sensitized solar cells. *RSC Adv.* **2016**, *6*, 27714. [CrossRef]
34. Shujahadeen, B.A.; Wrya, O.K.; Hewa, O.G. The deficiency of chitosan: AgNO3 polymer electrolyte incorporated with titanium dioxide filler for device fabrication and membrane separation technology. *J. Mater. Res. Technol.* 2020. [CrossRef]
35. Aziz, S.B.; Abdullah, R.M.; Rasheed, M.A.; Ahmed, H.M. Role of ion dissociation on DC conductivity and silver nanoparticle formation in PVA:AgNt based polymer electrolytes: Deep insights to ion transport mechanism. *Polymers* **2017**, *9*, 338. [CrossRef]
36. Aziz, S.B.; Rasheed, M.A.; Abidin, Z.H. Optical and electrical characteristics of silver ion conducting nanocomposite solid polymer electrolytes based on chitosan. *J. Electron. Mater.* **2017**, *46*, 6119. [CrossRef]
37. Aziz, S.B. Morphological and optical characteristics of chitosan ($1-x$): Cuox ($4 \leq x \leq 12$) based polymer nano-composites: Optical dielectric loss as an alternative method for tauc's model. *Nanomaterials* **2017**, *7*, 444. [CrossRef]
38. Monisha, S.; Mathavan, T.; Selvasekarapandian, S.; Benial, A.M.F.; Aristatil, G.; Mani, N.; Premalatha, M. Investigation of bio polymer electrolyte based on cellulose acetate-ammonium nitrate for potential use in electrochemical devices. *Carbohyd. Polym.* **2017**, *157*, 38–47. [CrossRef] [PubMed]
39. Mobarak, N.N.; Ahmad, A.; Abdullah, M.P.; Ramli, N.; Rahman, M.Y.A. Conductivity enhancement via chemical modification of chitosan based green polymer electrolyte. *Electrochim. Acta* **2013**, *92*, 161–167. [CrossRef]
40. Kadir, M.F.Z.; Majid, S.R.; Arof, A.K. Plasticized chitosan–PVA blend polymer electrolyte based proton battery. *Electrochim. Acta* **2010**, *55*, 1475–1482. [CrossRef]
41. Aziz, S.B.; Abdullah, O.G.; Rasheed, M.A.; Ahmed, H.M. Effect of high salt concentration (HSC) on structural, morphological, and electrical characteristics of chitosan based solid polymer electrolytes. *Polymers* **2017**, *9*, 187. [CrossRef]
42. Natesan, B.; Karan, N.K.; Katiyar, R.S. Ion relaxation dynamics and nearly constant loss behavior in polymer electrolyte. *Phys. Rev. E* **2006**, *74*, 042801. [CrossRef]
43. Nasef, M.M.; Saidi, H.; Dahlan, K.Z.M. Preparation of composite polymer electrolytes by electron beam-induced grafting: Proton and lithium ion-conducting membranes. *Nucl. Instrum. Methods Phys. Res. Sect. B Beam Interact. Mater. Atoms.* **2007**, *265*, 168–172. [CrossRef]
44. Dillip, K.; Pradhan, R.N.P.; Choudhary, B.K. Samantaray. Studies of structural, thermal and electrical behavior of polymer nanocomposite electrolytes. *Express Polym. Lett.* **2008**, *2*, 630–638.
45. Fonseca, C.P.; Cavalcante, F.C., Jr.; Amaral, F.A.; Souza, C.A.Z.; Neves, S. Thermal and Conduction Properties of a PCL-biodegradable Gel Polymer Electrolyte with $LiClO_4$, LiF3CSO3, and LiBF4 Salts. *Int. J. Electrochem. Sci.* **2007**, *2*, 52–63.
46. Al-Muntaser, A.A.; Abdelghany, A.M.; Abdelrazek, E.M.; Elshahawy, A.G. Elshahawy, Enhancement of optical and electrical properties of PVC/PMMA blend films doped with Li4Ti5O12 nanoparticles. *J. Mater. Res. Technol.* **2020**, *9*, 789–797. [CrossRef]

47. Hassib, H.; Razik, A.A. Dielectric properties and AC conduction mechanism for 5,7-dihydroxy-6-formyl-2-methylbenzo-pyran-4-one bis-schiff base. *Solid State Commun.* **2008**, *147*, 345–349. [CrossRef]
48. Eliasson, H.; Albinsson, I.; Mellander, B.E. Conductivity and dielectric properties of AgCF3SO3-PPG. *Mater. Res. Bull.* **2000**, *35*, 1053–1065. [CrossRef]
49. Salem, S.M. Effect of iron on the electrical properties of lead–bismuth glasses. *J. Mater. Sci.* **2009**, *44*, 5760–5767. [CrossRef]
50. Das-Gupta, D.K. Molecular processes in polymer electrets. *J. Electrost.* **2001**, *515*, 159–166. [CrossRef]
51. Shujahadeen, B.A.; Kadir, M.F.Z.; Hamsan, M.H.; Woo, H.J.; Brza, M.A. Development of Polymer Blends Based on PVA: POZ with Low Dielectric Constant for Microelectronic Applications. *Sci. Rep.* **2019**, *9*, 13163. [CrossRef]
52. Aziz, S.B.; Abdullah, R.M. Crystalline and amorphous phase identification from the tanδ relaxation peaks and impedance plots in polymer blend electrolytes based on [CS: AgNt] x: PEO (x-1)(10≤ x ≤ 50). *Electrochim. Acta* **2018**, *285*, 30–46. [CrossRef]
53. Kunanuruksapong, R.; Sirivat, A. Electrical properties and electrochemical responses of acrylic elastomers and styrene copolymers: Effects of temperature. *Appl. Phys. A Mater. Sci. Process.* **2008**, *92*, 313–320. [CrossRef]
54. Ramya, C.S.; Selvasekarapandian, S.; Hirankumar, G.; Savitha, T.; Angelo, P.C. Investigation on dielectric relaxations of PVP–NH4SCN polymer electrolyte. *J. Non-Cryst Solids* **2008**, *354*, 1494–1502. [CrossRef]
55. Jayathilaka, P.A.R.D.; Dissanayake, M.A.K.L.; Albinsson, I.; Mellander, B.E. Dielectric relaxation, ionic conductivity and thermal studies of the gel polymer electrolyte system PAN/EC/PC/LITFSI. *Solid State Ion.* **2003**, *156*, 179–195. [CrossRef]
56. Khatri, P.; Behera, B.; Srinivas, V.; Choudhary, R.P.N. Structural and dielectric properties of Ba3V2O8 ceramics. *Curr. Appl. Phys.* **2009**, *9*, 515–519. [CrossRef]
57. Aziz, S.B.; Al-Zangana, S.; Brza, M.A.; Saeed, S.R.; Abdulwahid, R.T.; Kadir, M.F.Z. Study of Dielectric Properties and Ion Transport Parameters in Chitosan-Barium Nitrate Based Solid Polymer Electrolytes. *Int. J. Electrochem. Sci.* **2019**, *14*, 11580–11595. [CrossRef]
58. Aziz, S.B.; B Marif, R.; Brza, M.A.; Hamsan, M.H.; Kadir, M.F.Z. Employing of Trukhan Model to Estimate Ion Transport Parameters in PVA Based Solid Polymer Electrolyte. *Polymers* **2019**, *11*, 1694. [CrossRef]
59. Aziz, S.B. The mixed contribution of ionic and electronic carriers to conductivity in chitosan based solid electrolytes mediated by CuNt salt. *J. Inorg. Organomet. Polym.* **2018**, *28*, 1942–1952. [CrossRef]
60. Smaoui, H.; Mir, L.E.; Guermazi, H.; Agnel, S.; Toureille, A. Study of dielectric relaxations in zinc oxide-epoxy resin nanocomposites. *J. Alloys Compd.* **2009**, *477*, 316–321. [CrossRef]
61. Belattar, J.; Graça MP, F.; Costa, L.C.; Achour, M.E.; Brosseau, C. Electric modulus-based analysis of the dielectric relaxation in carbon black loaded polymer composites. *J. Appl. Phys.* **2010**, *107*, 124111. [CrossRef]
62. Pradhan, D.K.; Choudhary, R.N.P.; Samantaray, B.K. Studies of Dielectric Relaxation and AC Conductivity Behavior of Plasticized Polymer Nanocomposite Electrolytes. *Int. J. Electrochem. Sci.* **2008**, *3*, 597–608.
63. Sengwa, R.J.; Choudhary, S.; Sankhla, S. Low frequency dielectric relaxation processes and ionic conductivity of montmorillonite clay nanoparticles colloidal suspension in poly(vinyl pyrrolidone)-ethylene glycol blends. *Express Polym. Lett.* **2008**, *2*, 800–809. [CrossRef]
64. Aziz, S.B.; Karim, W.O.; Brza, M.A.; Abdulwahid, R.T.; Saeed, S.R.; Al-Zangana, S.; Kadir, M.F.Z. Ion Transport Study in CS: POZ Based Polymer Membrane Electrolytes Using Trukhan Model. *Int. J. Mol. Sci.* **2019**, *20*, 5265. [CrossRef]
65. Aziz, S.B.; Abdullah, R.M.; Kadir, M.F.Z.; Ahmed, H.M. Non suitability of silver ion conducting polymer electrolytes based on chitosan mediated by barium titanate (BaTiO3) for electrochemical device applications. *Electrochim. Acta* **2019**, *296*, 494–507. [CrossRef]
66. Aziz, S.B.; Hamsan, M.H.; Brza, M.A.; Kadir, M.F.Z.; Abdulwahid, R.T.; Ghareeb, H.O.; Woo, H.J. Fabrication of energy storage EDLC device based on CS: PEO polymer blend electrolytes with high Li+ ion transference number. *Results Phys.* **2019**, *15*, 102584. [CrossRef]
67. Basha, S.; Rao, M.C. Spectroscopic and Electrochemical Properties of [PVA/PVP] : [MgCl26H2O] Blend Polymer Electrolyte Films. *Int. J. Polym. Sci.* **2018**, *2018*. [CrossRef]
68. Ng, L.S.; Mohamad, A.A. Effect of temperature on the performance of proton batteries based on chitosan–NH4NO3–EC membrane. *J. Membr. Sci.* **2008**, *325*, 653–657. [CrossRef]

69. Kadir, M.F.Z.; Arof, A.K. Application of PVA–chitosan blends polymer electrolyte membrane in electrical double layer capacitor. *Mater. Res. Innov.* **2013**, *15*, 217–220. [CrossRef]
70. Aziz, S.B.; Brza, M.A.; Hamsan, M.H.; Kadir, M.F.Z.; Muzakir, S.K.; Abdulwahid, R.T. Effect of ohmic-drop on electrochemical performance of EDLC fabricated from PVA: Dextran: NH4I based polymer blend electrolytes. *J. Mater. Res. Technol.* **2020**. [CrossRef]

membranes

Review

New Perspectives on Fuel Cell Technology: A Brief Review

Norazlianie Sazali [1,*], Wan Norharyati Wan Salleh [2], Ahmad Shahir Jamaludin [3] and Mohd Nizar Mhd Razali [3]

1 Faculty of Mechanical & Automotive Engineering Technology, Universiti Malaysia Pahang, Pekan 26600, Pahang, Malaysia
2 Advanced Membrane Technology Research Centre (AMTEC), School of Chemical and Energy, Faculty of Engineering, Universiti Teknologi Malaysia, Skudai 81310, Johor Darul Takzim, Malaysia; hayati@petroleum.utm.my
3 Faculty of Manufacturing & Mechatronic Engineering Technology, Universiti Malaysia Pahang, Pekan 26600, Pahang, Malaysia; shahir@ump.edu.my (A.S.J.); mnizar@ump.edu.my (M.N.M.R.)
* Correspondence: azlianie@ump.edu.my

Received: 16 April 2020; Accepted: 9 May 2020; Published: 13 May 2020

Abstract: Energy storage and conversion is a very important link between the steps of energy production and energy consumption. Traditional fossil fuels are a natural and unsustainable energy storage medium with limited reserves and notorious pollution problems, therefore demanding a better choice to store and utilize the green and renewable energies in the future. Energy and environmental problems require a clean and efficient way of using the fuels. Fuel cell functions to efficiently convert oxidant and chemical energy accumulated in the fuel directly into DC electric, with the by-products of heat and water. Fuel cells, which are known as effective electrochemical converters, and electricity generation technology has gained attention due to the need for clean energy, the limitation of fossil fuel resources and the capability of a fuel cell to generate electricity without involving any moving mechanical part. The fuel cell technologies that received high interest for commercialization are polymer electrolyte membrane fuel cells (PEMFCs), solid oxide fuel cells (SOFCs), and direct methanol fuel cells (DMFCs). The optimum efficiency for the fuel cell is not bound by the principle of Carnot cycle compared to other traditional power machines that are generally based on thermal cycles such as gas turbines, steam turbines and internal combustion engines. However, the fuel cell applications have been restrained by the high cost needed to commercialize them. Researchers currently focus on the discovery of different materials and manufacturing methods to enhance fuel cell performance and simplify components of fuel cells. Fuel cell systems' designs are utilized to reduce the costs of the membrane and improve cell efficiency, durability and reliability, allowing them to compete with the traditional combustion engine. In this review, we primarily analyze recent developments in fuel cells technologies and up-to-date modeling for PEMFCs, SOFCs and DMFCs.

Keywords: fuel cell technology; energy; polymer electrolyte membrane fuel cells (PEMFCs); solid oxide fuel cells (SOFCs); direct methanol fuel cells (DMFCs)

1. Introduction

Energy is required in our everyday lives. Rapid increment in total population and stable personal income growth are a few factors that cause a rising demand for energy. It is estimated that by the year 2035, global population will exceed 8.7 billion, meaning that an additional of 1.6 billion people will need energy [1]. The main problem faced is the rising energy demand and decreasing fossil fuel supply, along with issues concerning the implementation of traditional fossil fuels on human health. There is an immediate need to use green alternative and sustainable energy to replace existing non-renewable fossil fuels. It is noted that there is an increase in renewable energy generation produced globally. Based on literature review, renewable power capacity of approximately 1560 GW was utilized at the end of 2013, nearly double the 895 GW recorded at the beginning of 2004 [2]. Nevertheless, renewable power plants were reported to have many disadvantages. One of the disadvantages is that renewable power plants are typically located far from the demand site, which causes difficulty in transporting renewable energy. With current centralized power generation and distribution networks, increasing distributed renewable power plants, such as photovoltaic arrays and wind farms, results in a major effect on grid stability. Hence, the curtailment method was applied to resolve these expensive problems and further escalating issues. Other than the storing energy technique, fuel cell technology is one of the recent technologies that provides a fast solution to the above-mentioned problems.

Fuel cells have potential in various applications, such as portable power, stationary electricity generation, vehicle propulsion and in large electrical plants [3,4]. The category of fuel cells is dependent on many elements, for example, conditions during operation (pressure, humidity, temperature), fuel cell structure (application system and scale), and the complexion of the fuel cell's polymer electrolyte [5]. DuPont Company produced a cation-exchange membrane, also known as Nafion® in the middle of 1960s with a backbone of polytetrafluoroethylene, perfluorinated vinyl ether suspended side chains eliminated by ionic sulfonate groups [6]. Its properties of excellent chemical and thermal strengths, as well as its high-proton-causing Nafion®, are now being used commercially. The structure of the Nafion membrane consists the cluster channel that is labeled as the first unit for its component. The 4-nm structure of the Nafion is linked together with the water structure that having the diameter of 1 nm that are equally discrete within the hydrophobic backbones is imagined in Figure 1 [7].

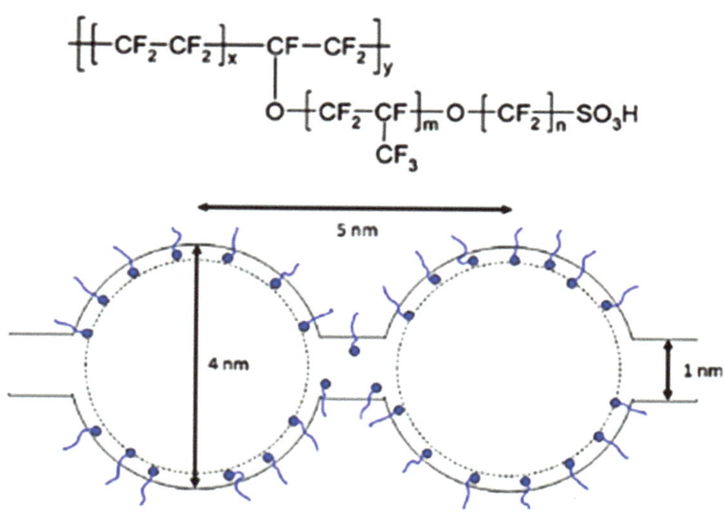

Figure 1. Structure of the Nafion with the presented cluster [7].

Researchers attempts to obtain a robust polymer electrolyte membrane with properties of the high conductivity of protons, little water or fuel crossover, high chemical and thermal stability, and excellent mechanical characteristics [8]. Therefore, to overcome the disadvantages of Nafion® and to create brand-new membrane materials of better or similar quality for the application of fuel cells, scientists are manufacturing feasible PEMs via the polymeric materials functionalization [7]. Previous studies reported the sulfonated poly (arylene ether sulfone) (SPAES) fabrication and alteration via functionalization in modifying membrane morphology to enhance the features of fuel cells, such as the conductivity of protons, the permeability of methanol and water absorption [9]. Moreover, the data acquired from SCOPUS®, peer-reviewed literature's citation database and the largest abstract show that there are currently increasing interests in SPAES for fuel cells used [10].

In a hydrogen fuel cell engine, water and heat are the only components of the electrochemical reactions. Carbon dioxide emission can be reduced using the superior energy efficiency of fuel cell engines if hydrogen is generated from hydrocarbons reforming or from electrolyzers powered by fossil-based electricity [11,12]. Emissions can be reduced to zero if hydrogen is generated from renewable sources like wind, solar thermal and nuclear power. For portable devices powered by batteries, fuel cells can be used effectively, from portable power tools needing a few hundred watts to cell phones needing a few watts of power. Hossain and groups mentioned that fuel cells are focused on the studies regarding energy conversion. Meanwhile, the battery, such as a lithium ion battery, refers to the energy storage. Both of these have managed to captive lots of attention [13]. Fuel cells are found to be more cost-effective compared to batteries. This statement has been proven by Haghi et al. through the site analysis by using fuel-cell-powered and battery-powered forklifts for reducing greenhouse gas (GHG) emissions in the province of Ontario, Canada [14]. The comparison of the usage for both of the fuel cell and battery power has found that battery-powered forklifts are more cost-effective compared to fuel cell-powered forklifts when lower levels of discounted power are available. However, with an increase in social cost of carbon (SCC) and discounted power available, fuel-cell-powered forklifts become more cost-effective. The benefit of fuel-cell-powered over battery-powered in higher levels of available discounted power is due to the lower operation and maintenance cost of fuel cells compared to batteries and the lower seasonal storage cost of hydrogen compared to batteries.

The power generation separation and energy storage units of a system are the significant benefits of fuel cells compared to batteries. For certain application, a module with a larger cell area and more fuel storage will be used if more power and energy is needed. Fuel cells have the main benefits of their ability to work without disruption or recharging for a more extended period compared to rechargeable batteries [15]. Unlike batteries, fast refueling with liquid methanol or hydrogen helps fuel cells to extend their operation. Recently, the application of fuel cells shows an increment in the sectors of grid connection, domestic usage and most of it in automotive fields. The fuel cell systems functioned as the supplier in terms of electric power in order to accommodate the electrical consumption for the local loads [16–18]. The standard blueprint of the grid-connected fuel cell is visualized in Figure 2. There are a few main components of the grid connected fuel-cell-based system, which are stacks, a DC-AC converter, a step-up transformer, a filter and an AC grid.

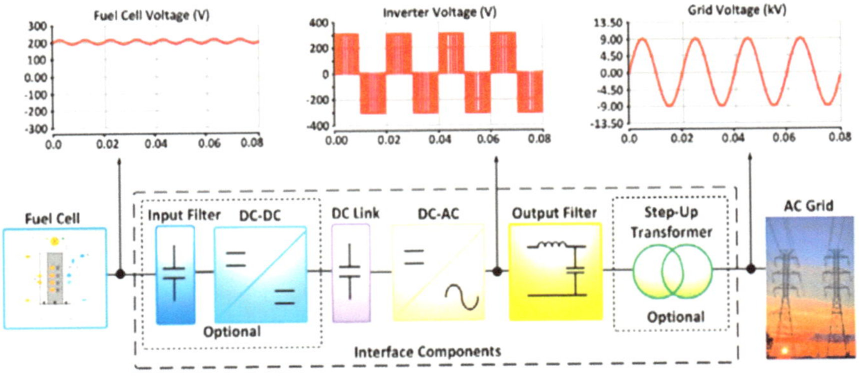

Figure 2. Standard design of a grid-connected fuel cell system [16].

The fuel cell system usually starts with electrons being released from the anode fuel oxidation, protons (ions) move across an electrolyte layer, and electrons are needed to reduce the cathode oxidant.

The optimal output is the possible most massive electrons flow over the highest electrical potential [19]. Even though oxidants like halogens have shown high-efficiency performance, oxygen is preferable due to their availability. Besides, hydrogen from pure ammonia, hydrocarbon fuels (methanol, methane) or carbon monoxide is typically used by fuel cells. In the grid connection of fuel cells systems, there are a few design and different concepts of the systems that are sources from the basic principles of the fuel cells. Conferring to these features, there are six main kinds of fuel cell that are used to initiate electrical power which are proton exchange membrane fuel cells (PEMFCs), solid-oxide fuel cells (SOFCs), alkaline fuel cells (AFCs), direct methanol fuel cells (DMFCs), phosphoric acid fuel cells (PAFCs) and molten carbonate fuel cells (MCFCs). Different classifications of the fuel cell together with the power rating and the benefit for each types of the fuel cell is displayed in Figure 3.

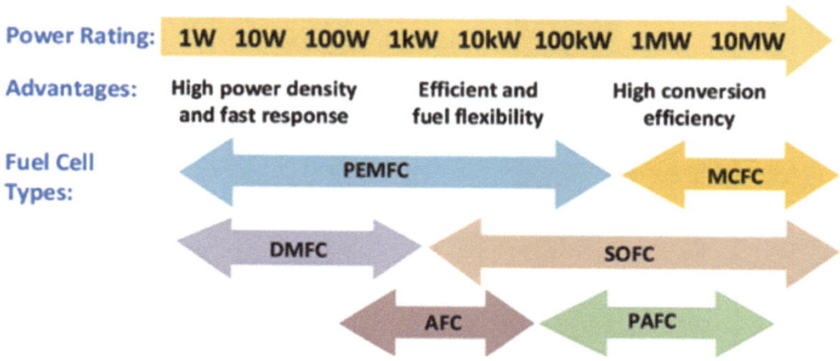

Figure 3. Different groups of fuel cells based on their power ratings and advantages.

The term used to describe the fuel cell type depend on the type of conductor utilized for protons (ions) or electrolyte, except for DMFCs in which its nature is determined by the fuel employed [20–23]. Usually, the electrolyte employed in DMFCs is a similar type of membrane utilized in a PEMFC, which is known as a fuel cell using hydrogen-rich gas or hydrogen gas (hydrocarbon reformer production) as a fuel [24–26]. The first row is the electrolyte, while the second column is the chosen parameters in operating procedure. Alkaline fuel cells need pure hydrogen, while hydrogen-rich gas from a hydrocarbon reformer can be tolerated by the phosphoric acid fuel cell (PAFC) [16,27]. Different types of fuel cell and their optimized process temperature have been widely studied by various researches. Molten carbonate fuel cells (MCFCs) operates at approximately 650 LC or above, which means that it needs to be heated to nearly 650 LC before undergo the operating procedure [28,29]. Alkaline fuel cells are capable of operating over a more comprehensive temperature range and do not generally require the heating process prior to operation. PEMFCs are currently operating at smaller than 100 LC, which is constrained by the Nafion-based polymer electrolyte membrane operating temperature range [30]. Higher temperature operation gives the advantages of decreasing the electrocatalyst's sensitivity to CO in the anode stream and promoting water recovery and thermal management issues.

2. Materials and Methods

2.1. Fuel Cell Varieties and Development

The fuel cell is an energy conversion device that is functioned to convert chemical energy to electrical energy as well as heat. The common fuel cell system consists of a few mains part (the anode, cathode, electrolyte and external circuit called the load). The operation of the fuel cell system is quite simplel, regardless of the intricate layout. The anode will continuously be supplied with hydrogen fuel, meanwhile the cathode is nourished with the oxidant in the air. In the anode, the supplied hydrogen is diverted into two types, the hydrogen positive ion, H^+, and the negative ion, H^-. Conceptually, the pathway between the anode and cathode is separated by electrolyte. The presence of the electrolyte only allowed the movement of the H^+ ions from the anode to cathode and inhibits the travel of the H^- ions by functioning as an insulator. In the fuel cell system, there are three main reactions steps that occur at the anode and cathode, which are represented in the equations below [16]:

$$\text{Anode: } H_2 \rightarrow 2H^+ + 2e^- \tag{1}$$

$$\text{Cathode: } \frac{1}{2}O_2 + 2H^+ + 2e^- \rightarrow H_2O \tag{2}$$

Modeling fuel cell system is worthy as it is a convenient tool to gain a better understanding of the internal operating process to enhance the design of the fuel cell. Running a modeled fuel cell is faster and cheaper than running a real-scale system and this helps to speed up the design process. The design of robust computational fuel cell models has been dedicated to extensive research efforts over the past ten years [31–35]. This study involved heat transfer modelling, numerical analysis and simulation, material matters, species flow/mass transfer, electrochemical kinetics, system integration and water management. Almost all fuel cell generates high-efficiency electrical energy in the range between 40% and 60% depending on the fuel's lower heating value (LHV) [36]. The efficiency of fuel conversion is higher compared to internal-combustion-engine-driven generators. At smaller scales, performance advantage is more important as the efficiency of fuel cells is almost constant with volume. However, fuel cells of high temperatures can be paired with gas turbines, thus surpassing the efficiency of massive combined power plants while emitting lower levels of SO_x, NO_x and CO_x [37].

2.2. Direct Methanol Fuel Cell (DMFCs)

Intensive progress in polymer electrolyte membranes for DMFCs designs has been made in recent years in the aspect of cost reduction and its practicality along with other related technological advances. An overview of the DMFC technology development indicates that some DMFC materials currently being developed met the Department of Environment (DOE) specifications [38,39]. Technological differences between the DOE specifications and the current technology are: (i) cheap and robust membranes, for example, polyfuel-produced hydrocarbon membranes (5000 h lifetime in passive DMFCs); (ii) low platinum anode catalysts or high-performance non-platinum (<0.2 mg cm^{-2}); (iii) high-performance non-platinum cathode catalysts with low metal load (0.2–0.5 mg cm^{-2}), such as palladium alloys; (iv) more oxidation-resistant non-carbon cathode supports, for example, porous titanium. Currently, direct methanol fuel cell (DMFC) technologies are under an evolvement process and are considered to be employed to replace or complement the Li-ion batteries in a variety of applications, for example, military uses, portable electronics, also small power range automotives such as forklifts, materials handling vehicles (MHVs) and scooters [23,40].

Comprehensive research and development efforts were made to decrease primary losses in DMFCs by identifying durable and active catalysts that are capable of lessening kinetic losses, through material selections, manufacturing and engineering aspects to reduce ohmic losses, also by choosing appropriate operating conditions to mitigate mass transport losses. Due to increasing expertise in numerous interests, DMFCs' initial quality has risen to a level that is suitable for practical applications, even though there are still issues related to durability and cost. Recently, the literatures reported on DMFCs long-term activity is increasing [23]. A range of diagnostic tools is used to classify the mechanisms and routes of DMFCs performance degradation. Various aspects of research and development work have focused on DMFCs' durability and performance, such as polymer electrolyte membranes and catalyst materials. Moreover, mass transport phenomena have been summarized in some review articles [41–44]. However, available studies only addressed individual aspects of DMFC quality and durability without providing a detailed image of mechanisms for degradation. It shows the need for a detailed report addressing entire DMFC deterioration problems in durability operations in accordance with the different performance restoration methods used to rejuvenate performance losses [45,46]. This overview paper briefly provides a review about recent studies from both industry and academia on DMFCs' lifetime operations, as well as a detailed analysis on the significant routes of performance degradation, followed by proposed methods to restore performance losses. With the aim of gaining insight into degradation mechanisms, durability studies of DMFCs were done at different periods. Specific in-situ electrochemical techniques as well as ex-situ analytical methods were also used to classify membrane electrode assembly (MEAs) for life tests or failure to have a deep understanding of the MEA status and mechanisms for DMFC degradation [47].

Kulikovsky and co-workers successfully demonstrated a two-dimensional mathematical modeling for DMFCs [48]. The model was based on the equations of mass and energy conservation. The liquid velocity is controlled by the membrane phase potential, which is the electroosmotic effect and pressure gradients. Based on the findings, methanol is regulated by the pressure gradient near the fuel channel, and diffusion transport dominates in the membrane and active layers. Shaded zones were created in front of the current collectors in which methanol is lacking. An observation made by a previous researcher concluded that pulsed methanol feeding can result in a notable and sustained increase in the time-averaged cell voltage combined with a significant reduction in the DMFC system's overall methanol consumption [49]. Their model has proven to be able to describe the DMFC's stationary behaviors quantitatively. In addition, even dynamic behavior can be described qualitatively due to the changes in the concentration of methanol feed. Jeng and Chen have introduced the DMFC anode with a mathematical model [50]. This type of model takes consideration on the ohmic and kinetic resistance effects through the catalyst surface, especially the mass transport in the entire proton exchange membrane and the anode compartment. It investigates the effect of key parameters on the performance of anode and methanol crossover. Methanol crossover causes an extensive volume of wasted methanol being fed into the fuel cell for a DMFC operating under high a concentration of methanol feed also low current density condition, resulting in low fuel efficiency.

Kulikovsky produced a DMFC anode-side analytical model [51]. The model considers the non-Tafel kinetics of methanol oxidation's electrochemical reaction, methanol crossover, and methanol transmission across the backing layer. The model provides an ideal resolution to the performance issue of a DMFC's anode catalyst layer. A semi-analytical DMFC model was developed by previous study [52]. This model can be quickly solved and able to be included in DMFC simulations at the real-time system level. This model deems the kinetics of the anode's multi-step methanol oxidation reaction and the cathode's mixed oxygen potential because of methanol crossover. Argyropoulos and groups analyzed a DMFC model for the estimation of cell voltage against a liquid feed DMFC's current density response [53]. The model is formed according to a semi-empirical method where methanol oxidation and kinetics of oxygen reduction are incorporated for the fuel cell electrodes with effective mass transport coefficients. In the mathematical modeling of a DMFC by Chen and co-workers, efforts toward furthering model heat transfer have not been made; they focused on proving that the experimental data supports the expected impact of operating temperature on diffusion coefficients [54]. The conclusion obtained is the higher operating temperature results in higher power density. This supports the dependence of power density on temperature as the study is lacking a heat transfer model.

Membrane Electrode Assembly (MEA)

Membrane electrode assembly (MEA) includes a multi-layer structure. MEA is considered as the DMFC core component system, functioning to host the main oxidant and fuel electrochemical reactions to produce electricity [47,55,56]. A typical configuration of MEA consists of a polymer electrolyte membrane (PEM), cathode and anode catalyst layers (CLs), gas diffusion layer (GDLs), and microporous layer (MPLs) that are also known as backing layers. MEA structure is delicately built with porosity in micro/nano-scale due to its ability to control many transportation processes in DMFC's electrochemical reactions. There are several methods available for producing MEAs using various procedures and materials. MEAs' durability and performance depend on the manufacturing process under certain conditions.

MEA's working environment is very harsh in DMFC [57]. Both catalyst layers and membrane must resist the intense oxidizing and reducing conditions, presence or formation of liquid water, the evolution of CO_2 gas, the temperature at 80 °C and higher, high ionomer and acidic environment, and high electrical current passage. Electrodes' delamination from the membrane and changes in morphology, for instance, cracked and altered pore structure, resulting in increasing kinetic and mass transport losses, are generally the most typical degradation phenomena occurred in the MEA system through long-term operations. Jiang et al. performed a DMFC durability procedure on MEAs for 5000 h using Nafion® bonded electrodes and Nafion® 117 membrane [58]. After 2000 h, an interfacial delamination was discovered between membrane and anode, that degraded the performance of the cell leads to an increase in interfacial resistance. Electrodes were physically separated from the membrane. Liu et al. as well as other researchers stated that long-term testing on DMFCs causes electrodes interfacial delamination [59].

2.3. Polymer Electrolyte Membrane Fuel Cell (PEMFCs)

Standard PEMFCs, as visualized in Figure 4, used the fuel of hydrogen gas, and are a competitor of DMFC in a remote or portable power generator [60]. Conceptually, the typical PEMFCs consist of a few important units, including MEA that is located in between of flow fields plates (FFPs) of the cathode and anode, into which flow canal are fluted. However, PEMFC has a problem in term of fuel delivery processes as pure hydrogen needs high-costs fuel transmission infrastructure; Moreover, on-site fuel processors employing liquid fuels require a long start-up time, as well as expensive and bulky [26,61].

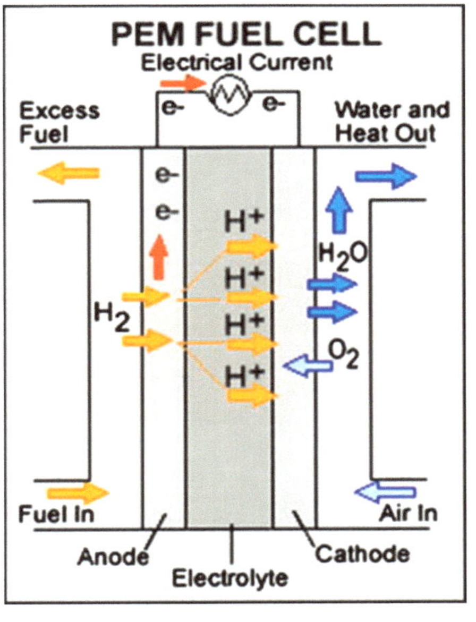

Figure 4. Polymer electrolyte membrane fuel cell overview [60].

As expected, there are various technical challenges in the improvement of fuel cell technology. The maximum theoretical voltage where a fuel cell can work is influenced by operating temperature. Higher temperatures are associated with lower theoretical efficiency and lower theoretical maximum voltages. Higher temperature operation often improves waste heat efficiency [62,63]. It is important to highlight that there is a medium temperature range that works well and is reliable for a certain type of fuel cell. Therefore, in fuel cell systems, the aim of thermal management is to make sure that stack operation within the specific range of temperature.

In PEMFCs, the generation of heat appeared due to the entropic heat reaction and the presence of irredeemable that is connected to the hydrogen. Aside from that, the stimulation of the electrochemical reaction and ohmic resistances in contradiction of pathway of proton and electron flow as well as the heat transport of hydrogen to anode also affect the heat present in PEMFCs systems as in Figure 5 [16,64]. The total heat produced in the system can be measured by equating the voltage of single cell with output voltage of 100% effectual PEMFCs. Generally, the produced heat in the PEMFCs is about 60% of the reacted hydrogen energy. Half of the reacted hydrogen is separated from the system by extra reactant and also the latent heat resulted from the vaporized water. The remainder of the generated heat is excluded from the systems via natural convection process. The flow of the hydrogen energy in the PEMFCs with the thermal insulation protection is presented in the Figure 6.

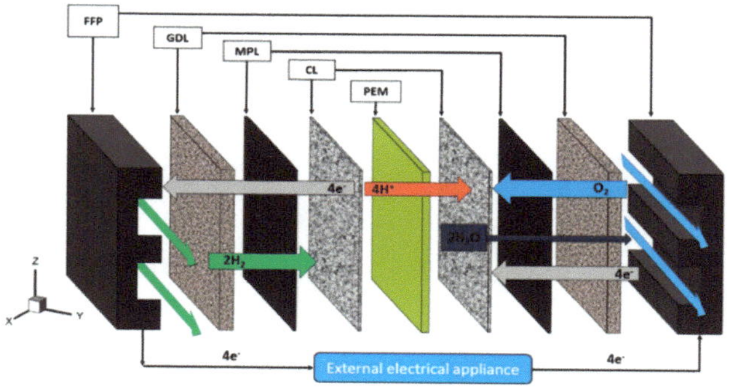

Figure 5. Main components in polymer electrolyte membrane fuel cells (PEMFCs) [64].

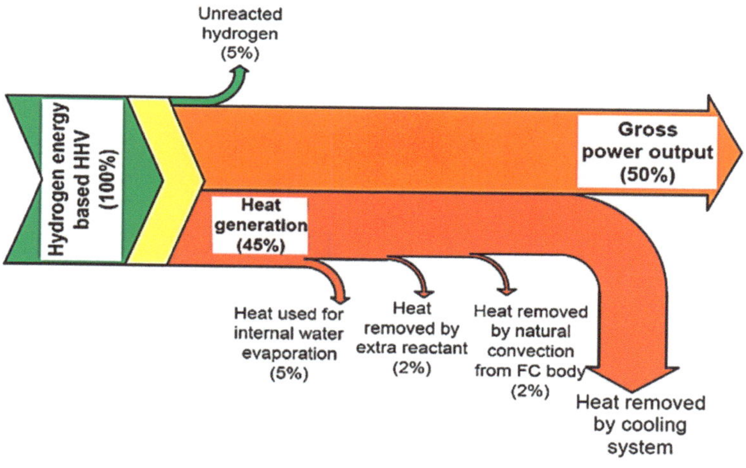

Figure 6. Hydrogen flow in PEMFCs [36].

A methanol-fed fuel cell system is designed to demonstrate the number of processes present in a fuel cell that involves heat and mass transfer [40]. A stack of PEM fuel cells used in the process is fueled from a methanol-reformer by hydrogen-rich air. The limitation of stack waste heat is caused by the low operating temperature of the PEM fuel cell [65,66]. With the purpose of effectively recovering the low-temperature heat, as stated later, a new cooling system was integrated into the fuel cell system. The methanol is pumped into a mixing chamber in the methanol tank and allowed to be mixed at a reasonable ratio with liquid water pumped from a water tank. Upon passing across an expansion valve, the mixture pressure is significantly decreased. The mixture then reached an evaporator or heat exchanger and vaporized during the absorption of a heat fuel cell stack cooler. High proton conductivity (0.1 S cm^{-1} at 120 °C), excellent thermal and chemical stability, high mechanical strength, fair durability, and compatibility with other fuel cell components are the crucial features required for PEMs. Currently, one degree of sulfonation has successfully improved the long-term durability of the sulfonated polymer [67]. The alteration of functionalized polymers with hydrophilic polymers has been documented to improve the thermal characteristics and organic phase interaction for thermal-specific applications [68]. Withal, polymer alteration with inorganic materials, for instance, silica, metal oxides, clays, carbon nanotubes, and others shows higher improvisation toward the fuel cell characteristics of PEM [69].

The fundamentals theory and the practical operation of a PEMFC involve various mathematical models presented in this section. Peng and groups had constructed the equivalent modeling of membrane hydration dynamic inside PEMFC in order to minimize the membrane micro-flooding. From the results, it was found that the implementation of the studied model able to improve the maximum net power boost can be estimated as being up to 3.74%, which is essential for the optimal operation of the integrated PEMFC system to achieve a higher system efficiency [70]. In another study by Salimi et al., the neural network modeling is found able to increase the power output of the PEMFC systems [71].Through the designated model named an artificial neural network (ANN), the operating performance increased up to 28.9%. A comprehensive stack model is developed based on the integration of a 1 + 1 dimensional multiphase stack sub-model and a flow distribution sub-model has been developed [72]. The purpose of the constructed model is to study the flow distributions as well as reactions, phase changes, and transport processes inside the PEMFC. From the obtained data analysis, the uniform flow assumption not only overestimates the stack output performance but also underestimates the fuel cell voltage variations. Besides, neglecting the non-uniform flow distribution may lead to higher predictions of the overall stack temperature and lower predictions of the temperature variations among different fuel cells. In other approaches by Chugh and colleagues, the low temperature of PEMFC performance is deeply studied via the mathematical modeling, which is MATLAB. The model predicts an increase in PEMFC performance with an increase in operating temperature, pressure and reactant humidity. An increase in stack operating temperature from 50 to 80 °C was seen to increase the stack voltage by 25%, because of lowering the activation potential and ohmic resistance. However, a further increase in operating temperature results in membrane dehydration. Similarly, a 30% increase in stack output power was observed upon increasing the operating pressure from 0 to 100 kPag. The further increase in pressure to 200 kPag showed a 60% increase in the output power [73]. In PEMFC, the water transport behavior in the gas diffusion layer (GDL) and bipolar plate (BPP) affected by the nonuniform compression on the GDL. Thus, Xu et al. studied these effects via the constructed model to obtain the relationship between the GDL deformation and assembly clamping force based on the energy method [74]. From the proposed model, the results show that drainage pressure increases monotonically with the assembly clamping force. In addition, thin GDL is conducive to improving drainage capacity. However, due to the combined effect of through-plane and in-plane mass transport in GDL, the maximum pressure first decreases and then increases with the thickness of GDL. GDL with a thickness of 0.2 mm is regarded as the best size to guarantee good water transport for the baseline case.

The introduction of combination between inorganic materials and PEMs, resulting in the development of nanocomposite membranes, in which the nanostructures lead to the improvement of mechanical and thermal stability of the membranes [75]. Proton conduction is the dominant aspect in the membrane's analysis for fuel cell potential applications in which high conductivity is important. The two main mechanisms can elucidate the protons' transfer in a hydrated polymer membrane, which is vehicular and Grotthus [76,77]. The Grotthus mechanism performs through the migration of protons across polymer matrices from one hydrolyzed ion to another [78]. The protons generated at the anode by hydrogen oxidation are added to water molecules to produce hydronium ions. The result of the Grotthus mechanism is the conductivity of a perfluorinated sulfonic acid membrane, for instance,

Nafion®. The value of the ion exchange capacity (IEC) affects the transfer of Grotthus type due to the recommended loading quantity of ionizable groups in the membrane of fuel cells [79]. Hydronium ions pass through either the aqueous medium with one or more methanol or water molecules via electro-osmotic drag in the membrane for the vehicle mechanism. Consequently, the molecules of methanol or water function as vehicles for the diffusion of protons in the polymeric membrane. Cationic complexes are created after joining protons with molecules of water or methanol. An integral feature in the vehicular mechanism is the free volume existence in polymeric chains of proton exchange membranes [80]. This method can be employed to choose inorganic additives to enhance polymeric membrane proton conductivity at low RH conditions and high temperatures [81].

2.4. Solid Oxide Fuel Cell (SOFCs)

The capability of the SOFCs to act as the sustainable energy supply has been explored and scientifically reported since 1990s. The pros and cons of the SOFCs is tabulated in Table 1 [82,83]. SOFCs are high-temperature fuel cells that have recently attracted the most attention for applications in cooling, heating and power generation systems. There are two types of electrolytes in SOFCs, which are oxygen ion-conducting (SOFC–O^{2-}) and proton-conducting (SOFC–H^+). Both the modeling for SOFC–O^{2-} and SOFC–H^+ is illustrated in Figure 7. As can be seen in Figure 7, in SOFC–O^2, the oxygen molecules freely pass through the electrolyte and react with hydrogen gas at anode side. By reacting the ion of oxygen with the proton, the steam forms at anode side. Meanwhile, the steam is produced and exits the cathode side as the hydrogen molecules from the anode reacts SOFC–H^+ electrolyte. Li et al. cited that SOFC–H^+ offers a low working temperature to prolong the lifetime of the cell [84]. The authors mentioned that the poor chemical stability of $BaCeO_3$-based SOFC–H^+ limits the practical applications. Thus, $BaZrO_3$-based proton-conducting oxides are intensively studied because they are chemically stable while offering high bulk conductivity. Xu et al. mentioned that the first-generation SOFC cathodes, including La1-xSrxMO3 (M = Mn and Fe), show good chemical stability as well as excellent chemical and thermal compatibility with electrolyte materials. However, these cathodes are not fully practical in certain applications due to their low performance. Thus, $La_{0.5}Sr_{0.5}FeO_{3-\delta}$ with Pr-doping were successfully fabricated in order to minimize the existing limitations [85]. The complex oxide of $BaCe_{0.7-x}Zr_{0.2}Y_{0.1}Fe_x0_{3-\delta}$ was successfully designed by Tarutina et al. for SOFC–H^+ [86]. Based on the findings, the Fe-doping has a positive effect on the densification of the materials which leading to improve grain growth at reduced sintering temperature. Working from a different perspective, for Mojaver and colleagues, the energy efficiency of an SOFC–O^2-based system, is higher, which is 60.20% compared to SOFC–H^+ with 54.06%. The sum of the unit cost of the product (SUPC) of SOFC–O^{2-} is lower (48.24 $/GJ compared to 48.75 $/GJ) rather than SOFC–$H^+$. In addition, the power produced by SOFC–O^{2-} is 18 kW greater than SOFC–H^+. Directly, this led to improving the system power from 147.9 kW for SOFC–H^+ and 156.4 kW in the case of the SOFC–O^{2-} [87].

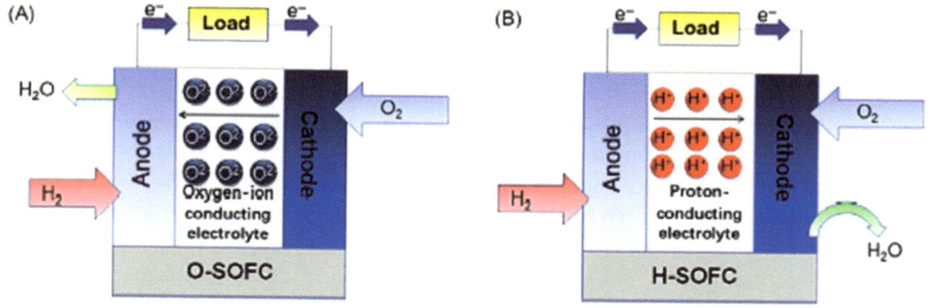

Figure 7. Modeling of (**A**) Oxygen ion-conducting SOFC–O^{2-} and (**B**) Proton-conducting SOFC–H^+ [88].

This subtopic discusses the SOFC's models and operations, considering the thermal management requirements and material-based restrictions. The fundamental theory and the practical operation of an SOFC involve various mathematical models presented by researchers. The basic interest of models is the ideal efficiency of combined cycle plants from SOFC. Chan et al. studied to construct a thermodynamic model for simple hydrogen and methane fed SOFC power systems in which heat recovered was used to pre-heat air and fuel [89]. Winkler and Lorenz hypothesized that the simply combined efficiency of the SOFC and the gas turbine cycle ranges from 60% to 70% [90]. Besides, they proposed a cycle of RH–SOFC–GT–ST which stands for ReHeat–SOFC–Gas Turbine–Steam Turbine that was proven to have more than 80 percent efficiency and supports the theoretical thermodynamic model's predictions. Jurado developed a dynamic model to compute low-order linear system models of SOFCs from the time domain to study the potential effects of fuel cells on future distribution systems [91,92].

Table 1. Benefits and the limitations of solid oxide fuel cells (SOFCs) [82,83].

Benefits	The consistency of the size and air flow in SOFC stack size is maintained. Pressure value is maintained along with the pressured existing SOFC stacks. Turbine inlet temperature values close to stack discharge conditions. Available air temperature values near to SOFC cathode inlet. Promising electrical integration at continuous current level.
Limitations	Commercial microturbines not specially premeditated for SOFC. Substantial impact of ambient temperature value. Plant exhaust flow temperature unable to decrease less than 200 to 250 °C. The controllability of dynamic issues.

This model applied the Box–Jenkins algorithm for calculating a linear system's transfer function from input and output samples, which is able to modulate reactive and real power regarding of changes in frequency and voltage on the grid. The analysis of energy balance was carried out by Van Herle et al. on a biogas-fed SOFC combined with heat and a small gas engine system [93]. A numerical model for SOFC was developed by Petruzzi et al. [94]. This is built for the convenience of luxury cars as an auxiliary power unit (APU). The model functions to simulate the thermal-electrochemical behaviour during operation in all possible conditions. A simulation model of an SOFC power plant was developed by Padullés et al. to be used in common commercial power system simulation package [95]. Many researchers have studied chemical equilibrium issues for an SOFC using internal reforming also shifting reactions in situations where there is the usage of natural gas, methane or biogas as fuel. Pre-reformer fuel gas consists of H_2, CO_2, CO, CH_4 and H_2O (vapour). In the cell, a combination of shifting and reforming reactions occurs. A fuel plug-flow model (natural gas surrogate) at the anode channels was studied by Walters et al. [96]. The model developed takes into account the basic gas-phase chemical kinetics of oxidation and pyrolysis of oil, also the limiting case of local chemical balance.

Guo et al. carried out a study on the efficiency of methane oxidative coupling affected by the different operating parameters [97]. Two mathematical models focusing on plug flow and well-mixed flow was implemented to explain SOFCs behavior. Intensive studies have been done related to the potential electrical losses in an SOFC's operation, involving ohmic loss, activation polarization and losses because of mass transportation resistance. SOFC systems operate between 900 and 1000 °C, higher than any other type of fuel cell system [98]. Virkar et al. analyzed the overall SOFC stack resistance dependency quantitatively as a function of transport characteristics, cathode thickness, interconnecting contact area, anode thickness, electrolyte thickness and interconnecting contact spacing associated with interfaces and each region (resistance of charge transfer) [99]. A ladder network approach was used to study the closed-form analytical expressions. From a mathematical model and experimental research, Fukunaga et al. studied the relation between the three-phase boundary (TPB) distance and the over-potential [100]. They found that values less than 20 lm is the effective thickness of (TPB) length. Based on an analytical model, the optimization of cermet SOFC electrodes and limiting behavior have been explained. The absence of a liquid phase causes the modeling of heat transfer in an SOFC to become more tractable.

Iwata et al. presented their SOFC model, which discussed the relationship between the profile of high temperature and the current density [101]. The studies revealed that temperature depends on power density. Numerical methods were used to analyze the coupled flow processes, mass/heat transfer, electrochemistry and chemical reaction. A tubular SOFC thermal transport model was developed by Haynes and Wepfer, and they stated that the primary heat transfer mechanism between the cell and the air supply tube was radiation [102]. Larrain et al. implemented a parameter estimation approach to investigate parameters for a simple kinetic and thermal model used for small SOFC (20 cm^2 of anode-supported electrolyte with an active area of 1 cm^2) [103]. Khaleel et al. incorporated MARC, a commercial finite element analysis code, with an electrochemical (EC) module formed in-house to simulate SOFCs of the planar type [104]. This EC module measures the distribution of heat generation, current density as well as oxidant and fuel concentration, including MARC's temperature profile. MARC conducts thermal and flow evaluation according to the boundary and initial of flow and thermal conditions, as well as the heat generation measured by the EC module. The operating conditions of a rectangular planar SOFC with an integrated air preheater were examined by Costamagna and co-workers [105].

The fuel cell system is built with the purpose of reducing the high-cost preheater for external air using lower airflow rates along with lower inlet temperature. An analytical form for the gas-flow distribution in a planar SOFC stack, assuming that the stack is seen as a hydraulic resistance network, has been reported by Boersma and Sammes [106,107]. Research has directed towards minimizing its thickness and searching new ion conductors to lower the electrolyte's resistivity. Dotelli et al. employed a digital simulation technique based on images to simulate the composite electrolyte's electrical behavior [108]. Voronoi tessellation is used to convert the two-phase polyhedral microstructures into a random electrical network [109]. The actual and imaginary part of the electrical network impedance was calculated by the method of the transfer matrix. Scott et al. formed a mathematical model explaining the distribution along with the electrolyte thickness of the electron holes and electrons concentration as well as the potential [110].

Interest in research on modeling SOFCs has also been growing over the past couple of years [111–113]. Modeling works in a fuel cell usually focused on an area or factor. There are also other conditions tabulated in the table. Spatial dimensions, for example, could be in the range of simple zero dimensions to complex three dimensions. The model's state is known to be a steady or transient state. At elevated operating temperatures, only gas remains in SOFCs. In order to study the integration of fuel cells with other energy storage components and power, modeling is required to be performed at the macrosystem level.

3. Conclusions

The fuel cell system design can be seen as a decision-making process that comprises of identifying potential design alternatives and selecting the most appropriate one. It can be classified as good design if it meets the design specifications as well as a trade-off between the various design goals. The specifications and goals for a fuel cell system consist of performance, dimension, which is weight and size, emissions, output power, rapid start-up and rapid response to changes in load, lifetime, and operability in intense environments and noise, which will be important in certain applications. Considerable attention is being paid to the utilization of computer-based and modeling optimization in fuel cell systems design. One advantage of this method is the positive effect on high cost and design cycle time savings, as well as its improved operation and design. The performance of optimum development depends primarily on the method by which the prototype is developed. It is crucial to identify the important factors and those that can be compromised without having an adverse effect on the design. Modeling is carried out to capture the designer's interest aspects of the fuel cell system. A mathematical model that represents particular fuel cell system aspects and estimates its characteristics can be in a form of algebraic equations, differential equations, or a process or subroutine based on a computer. The model can involve various alternatives to the design that can be achieved by changing parameters, variables, constraints or conditions. The principle explained in the preceding step contributes to the basis for comparing the various alternatives to design. Then, the prototype can be combined with a numerical optimization algorithm to produce improved designs iteratively. This can lead to one or more optimal solutions. Modeling and optimization will assist the designer in further consideration of shortlisting the designs. However, optimization will not always produce a better design appropriate for manufacturing. In this situation, the iteration of the preceding points is required to confirm that suitable fuel cell phenomena are captured in the model and exact governing equations are employed, the assumptions' validity employed in modeling is analyzed, as well as to confirm that the design specifications and goals are modified and altered if necessary. The final design will result in either a final prototype or an improvement of an existing design to be developed in the future.

Funding: This research was funded by Internal Research Grant Scheme [grant number; RDU1803133].

Acknowledgments: The authors would like to thank the Ministry of Higher Education and University Malaysia Pahang for the financial support provided under the Internal Research Grant Scheme (Project Number: RDU1803133) in completing this work.

Conflicts of Interest: The authors declare no conflict of interest.

References

1. Senjyu, T.; Howlader, A.M. Operational aspects of distribution systems with massive DER penetrations. *Integr. Distrib. Energy Resour. Power Syst.* **2016**, 51–76. [CrossRef]
2. Bilgili, M.; Özbek, A.; Sahin, B.; Kahraman, A. An overview of renewable electric power capacity and progress in new technologies in the world. *Renew. Sustain. Energy Rev.* **2015**, *49*, 323–334. [CrossRef]
3. Ajanovic, A.; Haas, R. Economic prospects and policy framework for hydrogen as fuel in the transport sector. *Energy Policy* **2018**, *123*, 280–288. [CrossRef]
4. Hames, Y.; Kaya, K.; Baltacioğlu, E.; Turksoy, A. Analysis of the control strategies for fuel saving in the hydrogen fuel cell vehicles. *Int. J. Hydrog. Energy* **2018**, *43*, 10810–10821. [CrossRef]
5. Salleh, M.T.; Jaafar, J.; Mohamed, M.A.; Norddin, M.; Ismail, A.F.; Othman, M.; Rahman, M.A.; Yusof, N.; Aziz, F.; Salleh, W.N.W. Stability of SPEEK/Cloisite®/TAP nanocomposite membrane under Fenton reagent condition for direct methanol fuel cell application. *Polym. Degrad. Stab.* **2017**, *137*, 83–99. [CrossRef]
6. Kim, D.J.; Jo, M.J.; Nam, S.Y. A review of polymer–nanocomposite electrolyte membranes for fuel cell application. *J. Ind. Eng. Chem.* **2015**, *21*, 36–52. [CrossRef]
7. Karimi, M.B.; Mohammadi, F.; Hooshyari, K. Recent approaches to improve Nafion performance for fuel cell applications: A review. *Int. J. Hydrog. Energy* **2019**, *44*, 28919–28938. [CrossRef]
8. Li, H.; Zhang, G.; Wu, J.; Zhao, C.; Jia, Q.; Lew, C.M.; Zhang, L.; Zhang, Y.; Han, M.; Zhu, J.; et al. A facile approach to prepare self-cross-linkable sulfonated poly(ether ether ketone) membranes for direct methanol fuel cells. *J. Power Sources* **2010**, *195*, 8061–8066. [CrossRef]
9. Kim, K.; Heo, P.; Han, J.; Kim, J.; Lee, J.-C. End-group cross-linked sulfonated poly(arylene ether sulfone) via thiol-ene click reaction for high-performance proton exchange membrane. *J. Power Sources* **2018**, *401*, 20–28. [CrossRef]
10. Pan, L.; Ott, S.; Dionigi, F.; Strasser, P. Current challenges related to the deployment of shape-controlled Pt alloy oxygen reduction reaction nanocatalysts into low Pt-loaded cathode layers of proton exchange membrane fuel cells. *Curr. Opin. Electrochem.* **2019**, *18*, 61–71. [CrossRef]
11. Liu, X.; Reddi, K.; Elgowainy, A.; Lohse-Busch, H.; Wang, M.; Rustagi, N. Comparison of well-to-wheels energy use and emissions of a hydrogen fuel cell electric vehicle relative to a conventional gasoline-powered internal combustion engine vehicle. *Int. J. Hydrog. Energy* **2020**, *45*, 972–983. [CrossRef]
12. Sun, T.; Zhang, X.; Chen, B.; Liu, X. Coordination control strategy for the air management of heavy vehicle fuel cell engine. *Int. J. Hydrog. Energy* **2019**. [CrossRef]

13. Hossain, S.; Abdalla, A.M.; Suhaili, S.B.H.; Kamal, I.; Shaikh, S.P.S.; Dawood, M.K.; Azad, A.K. Nanostructured graphene materials utilization in fuel cells and batteries: A review. *J. Energy Storage* **2020**, *29*, 101386. [CrossRef]
14. Haghi, E.; Shamsi, H.; Dimitrov, S.; Fowler, M.; Raahemifar, K. Assessing the potential of fuel cell-powered and battery-powered forklifts for reducing GHG emissions using clean surplus power; a game theory approach. *Int. J. Hydrog. Energy* **2020**. [CrossRef]
15. Cano, Z.P.; Banham, D.; Ye, S.; Hintennach, A.; Lu, J.; Fowler, M.; Chen, Z. Batteries and fuel cells for emerging electric vehicle markets. *Nat. Energy* **2018**, *3*, 279–289. [CrossRef]
16. Inci, M.; Turksoy, O. Review of fuel cells to grid interface: Configurations, technical challenges and trends. *J. Clean. Prod.* **2019**, *213*, 1353–1370. [CrossRef]
17. Widera, B. Renewable hydrogen implementations for combined energy storage, transportation and stationary applications. *Therm. Sci. Eng. Prog.* **2020**, *16*, 100460. [CrossRef]
18. Wang, Y.; Diaz, D.F.R.; Chen, K.S.; Wang, Z.; Adroher, X.C. Materials, technological status, and fundamentals of PEM fuel cells—A review. *Mater. Today* **2020**, *32*, 178–203. [CrossRef]
19. Siwal, S.S.; Thakur, S.; Zhang, Q.; Thakur, V. Electrocatalysts for electrooxidation of direct alcohol fuel cell: Chemistry and applications. *Mater. Today Chem.* **2019**, *14*, 100182. [CrossRef]
20. Mallick, R.; Thombre, S.B.; Shrivastava, N.K. Vapor feed direct methanol fuel cells (DMFCs): A review. *Renew. Sustain. Energy Rev.* **2016**, *56*, 51–74. [CrossRef]
21. Jaafar, J.; Ismail, A.F.; Matsuura, T.; Nagai, K. Performance of SPEEK based polymer–nanoclay inorganic membrane for DMFC. *J. Membr. Sci.* **2011**, *382*, 202–211. [CrossRef]
22. Kamarudin, S.K.; Achmad, F.; Daud, W.R.W. Overview on the application of direct methanol fuel cell (DMFC) for portable electronic devices. *Int. J. Hydrog. Energy* **2009**, *34*, 6902–6916. [CrossRef]
23. Kamarudin, S.K.; Daud, W.R.W.; Ho, S.L.; Hasran, U. Overview on the challenges and developments of micro-direct methanol fuel cells (DMFC). *J. Power Sources* **2007**, *163*, 743–754. [CrossRef]
24. Asensio, F.J.; Martin, J.I.S.; Zamora, I.; Saldaña, G.; Oñederra, O. Analysis of electrochemical and thermal models and modeling techniques for polymer electrolyte membrane fuel cells. *Renew. Sustain. Energy Rev.* **2019**, *113*, 109283. [CrossRef]
25. Moreno, N.G.; Molina, M.C.; Gervasio, D.; Robles, J.F.P. Approaches to polymer electrolyte membrane fuel cells (PEMFCs) and their cost. *Renew. Sustain. Energy Rev.* **2015**, *52*, 897–906. [CrossRef]
26. Chandan, A.; Hattenberger, M.; El-Kharouf, A.; Du, S.; Dhir, A.; Self, V.; Pollet, B.G.; Ingram, A.; Bujalski, W. High temperature (HT) polymer electrolyte membrane fuel cells (PEMFC)—A review. *J. Power Sources* **2013**, *231*, 264–278. [CrossRef]
27. Alcaide, F.; Cabot, P.-L.; Brillas, E. Fuel cells for chemicals and energy cogeneration. *J. Power Sources* **2006**, *153*, 47–60. [CrossRef]
28. Arshad, A.; Ali, H.M.; Habib, A.; Bashir, M.A.; Jabbal, M.; Yan, Y. Energy and exergy analysis of fuel cells: A review. *Therm. Sci. Eng. Prog.* **2019**, *9*, 308–321. [CrossRef]
29. Wee, J.-H. Carbon dioxide emission reduction using molten carbonate fuel cell systems. *Renew. Sustain. Energy Rev.* **2014**, *32*, 178–191. [CrossRef]
30. Rahman, S.; Masdar, M.S.; Rosli, M.; Majlan, E.; Husaini, T.; Kamarudin, S.; Daud, W. Overview biohydrogen technologies and application in fuel cell technology. *Renew. Sustain. Energy Rev.* **2016**, *66*, 137–162. [CrossRef]
31. Jeon, D.H. Computational fluid dynamics simulation of anode-supported solid oxide fuel cells with implementing complete overpotential model. *Energy* **2019**, *188*, 116050. [CrossRef]
32. Wang, H.-N.; Zhu, X.; Zhang, B.; Ye, D.; Chen, R.; Liao, Q.; Sui, P.-C.; Djilali, N. Two-phase computational modelling of a membraneless microfluidic fuel cell with a flow-through porous anode. *J. Power Sources* **2019**, *420*, 88–98. [CrossRef]
33. Ly, H.; Birgersson, E.; Vynnycky, M. Computationally efficient multi-phase models for a proton exchange membrane fuel cell: Asymptotic reduction and thermal decoupling. *Int. J. Hydrog. Energy* **2011**, *36*, 14573–14589. [CrossRef]
34. Vorobev, A.; Zikanov, O.; Shamim, T. A computational model of a PEM fuel cell with finite vapor absorption rate. *J. Power Sources* **2007**, *166*, 92–103. [CrossRef]
35. Sharma, A.K.; Birgersson, E.; Khor, S. Computationally-efficient hybrid strategy for mechanistic modeling of fuel cell stacks. *J. Power Sources* **2014**, *247*, 481–488. [CrossRef]

36. Nguyen, H.Q.; Shabani, B. Proton exchange membrane fuel cells heat recovery opportunities for combined heating/cooling and power applications. *Energy Convers. Manag.* **2020**, *204*, 112328. [CrossRef]
37. Amoatey, P.; Omidvarborna, H.; Baawain, M.S.; Al-Mamun, A. Emissions and exposure assessments of SOX, NOX, PM10/2.5 and trace metals from oil industries: A review study (2000–2018). *Process. Saf. Environ. Prot.* **2019**, *123*, 215–228. [CrossRef]
38. Kamarudin, S.; Hashim, N. Materials, morphologies and structures of MEAs in DMFCs. *Renew. Sustain. Energy Rev.* **2012**, *16*, 2494–2515. [CrossRef]
39. Sharma, S.; Pollet, B.G. Support materials for PEMFC and DMFC electrocatalysts—A review. *J. Power Sources* **2012**, *208*, 96–119. [CrossRef]
40. Akbari, E.; Buntat, Z.; Nikoukar, A.; Kheirandish, A.; Khaledian, M.; Afroozeh, A.; Khaledian, M. Sensor application in Direct Methanol Fuel Cells (DMFCs). *Renew. Sustain. Energy Rev.* **2016**, *60*, 1125–1139. [CrossRef]
41. Deng, R.; Xia, Z.; Sun, R.; Wang, S.; Sun, G. Nanostructured ultrathin catalyst layer with ordered platinum nanotube arrays for polymer electrolyte membrane fuel cells. *J. Energy Chem.* **2020**, *43*, 33–39. [CrossRef]
42. Zhao, J.; Shahgaldi, S.; Ozden, A.; Alaefour, I.E.; Li, X.; Hamdullahpur, F. Effect of catalyst deposition on electrode structure, mass transport and performance of polymer electrolyte membrane fuel cells. *Appl. Energy* **2019**, *255*, 113802. [CrossRef]
43. Star, A.G.; Wang, G.; Medina, S.; Pylypenko, S.; Neyerlin, K. Mass transport characterization of platinum group metal-free polymer electrolyte fuel cell electrodes using a differential cell with an integrated electrochemical sensor. *J. Power Sources* **2020**, *450*, 227655. [CrossRef]
44. Lee, C.; Lee, J.; Zhao, B.; Fahy, K.; Lamanna, J.; Baltic, E.; Hussey, D.; Jacobson, D.; Schulz, V.; Bazylak, A. Temperature-dependent gas accumulation in polymer electrolyte membrane electrolyzer porous transport layers. *J. Power Sources* **2020**, *446*, 227312. [CrossRef]
45. Tiwari, J.N.; Tiwari, R.N.; Singh, G.; Kim, K.S. Recent progress in the development of anode and cathode catalysts for direct methanol fuel cells. *Nano Energy* **2013**, *2*, 553–578. [CrossRef]
46. Wang, J. System integration, durability and reliability of fuel cells: Challenges and solutions. *Appl. Energy* **2017**, *189*, 460–479. [CrossRef]
47. Brouzgou, A.; Song, S.; Tsiakarasbcd, P. Low and non-platinum electrocatalysts for PEMFCs: Current status, challenges and prospects. *Appl. Catal. B Environ.* **2012**, *127*, 371–388. [CrossRef]
48. Kulikovsky, A. Direct methanol–hydrogen fuel cell: The mechanism of functioning. *Electrochem. Commun.* **2008**, *10*, 1415–1418. [CrossRef]
49. Pramanik, H.; Basu, S. Modeling and experimental validation of overpotentials of a direct ethanol fuel cell. *Chem. Eng. Process. Process. Intensif.* **2010**, *49*, 635–642. [CrossRef]
50. Jeng, K.; Chen, C. Modeling and simulation of a direct methanol fuel cell anode. *J. Power Sources* **2002**, *112*, 367–375. [CrossRef]
51. Kulikovsky, A. Analytical model of the anode side of DMFC: The effect of non-Tafel kinetics on cell performance. *Electrochem. Commun.* **2003**, *5*, 530–538. [CrossRef]
52. Rosenthal, N.S.; Vilekar, S.A.; Datta, R. A comprehensive yet comprehensible analytical model for the direct methanol fuel cell. *J. Power Sources* **2012**, *206*, 129–143. [CrossRef]
53. Kulikovsky, A. Comment on "A one dimensional model of a methanol fuel cell anode" [K. Scott, P. Argyropoulos, J. Power Sources 137 (2004) 228]. *J. Power Sources* **2005**, *148*, 54. [CrossRef]
54. Chen, R.; Zhao, T. Mathematical modeling of a passive-feed DMFC with heat transfer effect. *J. Power Sources* **2005**, *152*, 122–130. [CrossRef]
55. Theodosiou, P.; Greenman, J.; Ieropoulos, I. Towards monolithically printed Mfcs: Development of a 3d-printable membrane electrode assembly (mea). *Int. J. Hydrog. Energy* **2019**, *44*, 4450–4462. [CrossRef]
56. Sun, W.; Zhang, W.; Su, H.; Leung, P.; Xing, L.; Xu, L.; Yang, C.; Xu, Q. Improving cell performance and alleviating performance degradation by constructing a novel structure of membrane electrode assembly (MEA) of DMFCs. *Int. J. Hydrog. Energy* **2019**, *44*, 32231–32239. [CrossRef]
57. Wang, Z.; Shao, Y.; Zuo, P.-J.; Wang, X.-P.; Yin, G.-P. Durability studies of unsupported Pt cathodic catalyst with working time of direct methanol fuel cells. *J. Power Sources* **2008**, *185*, 1066–1072. [CrossRef]
58. Jiang, R.; Rong, C.; Chu, D. Fuel Crossover and Energy Conversion in Lifetime Operation of Direct Methanol Fuel Cells. *J. Electrochem. Soc.* **2007**, *154*, B13. [CrossRef]

59. Liu, J.; Zhou, Z.; Zhao, X.; Xin, Q.; Sun, G.; Yi, B. Studies on performance degradation of a direct methanol fuel cell (DMFC) in life test. *Phys. Chem. Chem. Phys.* **2004**, *6*, 134. [CrossRef]
60. Wang, Y.; Chen, K.S.; Mishler, J.; Cho, S.C.; Adroher, X.C. A review of polymer electrolyte membrane fuel cells: Technology, applications, and needs on fundamental research. *Appl. Energy* **2011**, *88*, 981–1007. [CrossRef]
61. Iulianelli, A.; Basile, A. Sulfonated PEEK-based polymers in PEMFC and DMFC applications: A review. *Int. J. Hydrog. Energy* **2012**, *37*, 15241–15255. [CrossRef]
62. Slater, J.; Chronopoulos, T.; Panesar, R.; Fitzgerald, F.; Garcia, M. Review and techno-economic assessment of fuel cell technologies with CO_2 capture. *Int. J. Greenh. Gas Control* **2019**, *91*, 102818. [CrossRef]
63. Valente, A.; Iribarren, D.; Dufour, J. End of life of fuel cells and hydrogen products: From technologies to strategies. *Int. J. Hydrog. Energy* **2019**, *44*, 20965–20977. [CrossRef]
64. Xing, L.; Shi, W.; Su, H.; Xu, Q.; Das, P.; Mao, B.; Scott, K. Membrane electrode assemblies for PEM fuel cells: A review of functional graded design and optimization. *Energy* **2019**, *177*, 445–464. [CrossRef]
65. Abdel-Rehim, A.A. The influence of electromagnetic field on the performance and operation of a PEM fuel cell stack subjected to a relatively low electromagnetic field intensity. *Energy Convers. Manag.* **2019**, *198*, 111906. [CrossRef]
66. Niroumand, A.M.; Homayouni, H.; Goransson, G.; Olfert, M.; Eikerling, M. In-situ diagnostic tools for hydrogen transfer leak characterization in PEM fuel cell stacks part III: Manufacturing applications. *J. Power Sources* **2020**, *448*, 227359. [CrossRef]
67. Haragirimana, A.; Ingabire, P.B.; Zhu, Y.; Lu, Y.; Li, N.; Hu, Z.; Chen, S. Four-polymer blend proton exchange membranes derived from sulfonated poly(aryl ether sulfone)s with various sulfonation degrees for application in fuel cells. *J. Membr. Sci.* **2019**, *583*, 209–219. [CrossRef]
68. Pasini, D.; Nitti, A. Free radical cyclopolymerization: A tool towards sequence control in functional polymers. *Eur. Polym. J.* **2020**, *122*, 109378. [CrossRef]
69. Simya, O.K.; Radhakrishnan, P.; Ashok, A. Engineered Nanomaterials for Energy Applications. In *Handbook of Nanomaterials for Industrial Applications*; Elsevier BV: Amsterdam, The Netherlands, 2018.
70. Peng, F.; Ren, L.; Zhao, Y.; Li, L. Hybrid dynamic modeling-based membrane hydration analysis for the commercial high-power integrated PEMFC systems considering water transport equivalent. *Energy Convers. Manag.* **2020**, *205*, 112385. [CrossRef]
71. Nanadegani, F.S.; Lay, E.N.; Iranzo, A.; Salva, J.A.; Sundén, B. On neural network modeling to maximize the power output of PEMFCs. *Electrochim. Acta* **2020**, *348*, 136345. [CrossRef]
72. Yang, Z.; Jiao, K.; Liu, Z.; Yin, Y.; Du, Q. Investigation of performance heterogeneity of PEMFC stack based on 1+1D and flow distribution models. *Energy Convers. Manag.* **2020**, *207*, 112502. [CrossRef]
73. Chugh, S.; Chaudhari, C.; Sonkar, K.; Sharma, A.; Kapur, G.; Ramakumar, S. Experimental and modelling studies of low temperature PEMFC performance. *Int. J. Hydrog. Energy* **2020**, *45*, 8866–8874. [CrossRef]
74. Xu, Y.; Qiu, D.; Yi, P.; Lan, S.; Peng, L. An integrated model of the water transport in nonuniform compressed gas diffusion layers for PEMFC. *Int. J. Hydrog. Energy* **2019**, *44*, 13777–13785. [CrossRef]
75. Wu, J.; Yuan, X.Z.; Martin, J.J.; Wang, H.; Zhang, J.; Shen, J.; Wu, S.; Mérida, W. A review of PEM fuel cell durability: Degradation mechanisms and mitigation strategies. *J. Power Sources* **2008**, *184*, 104–119. [CrossRef]
76. Zhang, J.; Liu, Y.; Lv, Z.; Zhao, T.; Li, P.; Sun, Y.; Wang, J. Sulfonated Ti_3C_2Tx to construct proton transfer pathways in polymer electrolyte membrane for enhanced conduction. *Solid State Ion.* **2017**, *310*, 100–111. [CrossRef]
77. Yu, L.; Yue, B.; Yan, L.; Zhao, H.; Zhang, J. Proton conducting composite membranes based on sulfonated polysulfone and polysulfone-g-(phosphonated polystyrene) via controlled atom-transfer radical polymerization for fuel cell applications. *Solid State Ion.* **2019**, *338*, 103–112. [CrossRef]
78. Delemotte, L.; Van Keulen, S.; Roethlisberger, U.; Gianti, E.; Carnevale, V.; Klein, M.L. Does Proton Conduction in the Voltage-Gated Proton Channel hH V 1 Involve Grotthus Hopping via Acidic Residues? *Biophys. J.* **2017**, *112*, 163a–164a. [CrossRef]
79. Vijayakumar, V.; Son, T.Y.; Kim, H.J.; Nam, S.Y. A facile approach to fabricate poly(2,6-dimethyl-1,4-phenylene oxide) based anion exchange membranes with extended alkaline stability and ion conductivity for fuel cell applications. *J. Membr. Sci.* **2019**, *591*, 117314. [CrossRef]

80. Yadav, V.; Rajput, A.; Sharma, P.P.; Jha, P.K.; Kulshrestha, V. Polyetherimide based anion exchange membranes for alkaline fuel cell: Better ion transport properties and stability. *Colloids Surf. A Physicochem. Eng. Asp.* **2020**, *588*, 124348. [CrossRef]
81. Wang, Y.; Wang, S.; Liu, S.; Li, H.; Zhu, K. Optimization of reactants relative humidity for high performance of polymer electrolyte membrane fuel cells with co-flow and counter-flow configurations. *Energy Convers. Manag.* **2020**, *205*, 112369. [CrossRef]
82. Damo, U.M.; Ferrari, M.; Turan, A.; Massardo, A. Solid oxide fuel cell hybrid system: A detailed review of an environmentally clean and efficient source of energy. *Energy* **2019**, *168*, 235–246. [CrossRef]
83. Raza, R.; Zhu, B.; Rafique, A.; Naqvi, M.R.; Lund, P. Functional ceria-based nanocomposites for advanced low-temperature (300–600 °C) solid oxide fuel cell: A comprehensive review. *Mater. Today Energy* **2020**, *15*, 100373. [CrossRef]
84. Li, J.; Wang, C.; Wang, X.; Bi, L. Sintering aids for proton-conducting oxides—A double-edged sword? A mini review. *Electrochem. Commun.* **2020**, *112*, 106672. [CrossRef]
85. Xu, X.; Wang, H.; Ma, J.; Liu, W.; Wang, X.; Fronzi, M.; Bi, L. Impressive performance of proton-conducting solid oxide fuel cells using a first-generation cathode with tailored cations. *J. Mater. Chem. A* **2019**, *7*, 18792–18798. [CrossRef]
86. Tarutina, L.R.; Vdovin, G.K.; Lyagaeva, J.G.; Medvedev, D.A. $BaCe_{0.7}$–$xZr_{0.2}Y_{0.1}FexO_3$–δ derived from proton-conducting electrolytes: A way of designing chemically compatible cathodes for solid oxide fuel cells. *J. Alloys Compd.* **2020**, *831*, 154895. [CrossRef]
87. Mojaver, P.; Chitsaz, A.; Sadeghi, M.; Khalilarya, S. Comprehensive comparison of SOFCs with proton-conducting electrolyte and oxygen ion-conducting electrolyte: Thermoeconomic analysis and multi-objective optimization. *Energy Convers. Manag.* **2020**, *205*, 112455. [CrossRef]
88. Xu, X.; Bi, L. Proton-conducting electrolyte materials. *Intermed. Temp. Solid Oxide Fuel Cells* **2020**, 81–111. [CrossRef]
89. Mehrpooya, M.; Sadeghzadeh, M.; Rahimi, A.; Pouriman, H. Technical performance analysis of a combined cooling heating and power (CCHP) system based on solid oxide fuel cell (SOFC) technology—A building application. *Energy Convers. Manag.* **2019**, *198*, 111767. [CrossRef]
90. Winkler, W.; Lorenz, H. Design studies of mobile applications with SOFC–heat engine modules. *J. Power Sources* **2002**, *106*, 338–343. [CrossRef]
91. Jurado, F. Modeling SOFC plants on the distribution system using identification algorithms. *J. Power Sources* **2004**, *129*, 205–215. [CrossRef]
92. Jurado, F.; Valverde, M.; Cano, A. Effect of a SOFC plant on distribution system stability. *J. Power Sources* **2004**, *129*, 170–179. [CrossRef]
93. Van Herle, J.; Wang, L.; Leuenberger, S.; Favrat, D. Energy balance model of a SOFC cogenerator operated with biogas. *J. Power Sources* **2003**, *118*, 375–383. [CrossRef]
94. Petruzzi, L.; Cocchi, S.; Fineschi, F. A global thermo-electrochemical model for SOFC systems design and engineering. *J. Power Sources* **2003**, *118*, 96–107. [CrossRef]
95. Padulles, J.; Ault, G.; McDonald, J. An integrated SOFC plant dynamic model for power systems simulation. *J. Power Sources* **2000**, *86*, 495–500. [CrossRef]
96. Walters, K.M.; Dean, A.; Zhu, H.; Kee, R.J. Homogeneous kinetics and equilibrium predictions of coking propensity in the anode channels of direct oxidation solid-oxide fuel cells using dry natural gas. *J. Power Sources* **2003**, *123*, 182–189. [CrossRef]
97. Guo, Z.; Liu, W.; Yang, C.; Gao, L.; Thangavel, S.; Wang, L.; He, Z.; Cai, W.; Wang, A. Computational and experimental analysis of organic degradation positively regulated by bioelectrochemistry in an anaerobic bioreactor system. *Water Res.* **2017**, *125*, 170–179. [CrossRef]
98. Sammes, N.; Galloway, K.; Serincan, M.; Suzuki, T.; Yamaguchi, T.; Awano, M.; Colella, W. Solid Oxide Fuel Cells. In *Handbook of Climate Change Mitigation*; Springer Science and Business Media LLC: New York, NY, USA, 2012; pp. 1703–1727.
99. Virkar, A.V.; Chen, J.; Tanner, C.W.; Kim, J.-W. The role of electrode microstructure on activation and concentration polarizations in solid oxide fuel cells. *Solid State Ion.* **2000**, *131*, 189–198. [CrossRef]
100. Fukunaga, H.; Ihara, M.; Sakaki, K.; Yamada, K. The relationship between overpotential and the three phase boundary length. *Solid State Ion.* **1996**, *86*, 1179–1185. [CrossRef]

101. Iwata, M.; Hikosaka, T.; Morita, M.; Iwanari, T.; Ito, K.; Onda, K.; Esaki, Y.; Sakaki, Y.; Nagata, S. Performance analysis of planar-type unit SOFC considering current and temperature distributions. *Solid State Ion.* **2000**, *132*, 297–308. [CrossRef]
102. Haynes, C.; Wepfer, W.J. 'Design for power' of a commercial grade tubular solid oxide fuel cell. *Energy Convers. Manag.* **2000**, *41*, 1123–1139. [CrossRef]
103. Larrain, D.; Van Herle, J.; Wang, L.; Favrat, D. Generalized model of planar SOFC repeat element for design optimization. *J. Power Sources* **2004**, *131*, 304–312. [CrossRef]
104. Khaleel, M.; Lin, Z.; Singh, P.; Surdoval, W.; Collin, D. A finite element analysis modeling tool for solid oxide fuel cell development: Coupled electrochemistry, thermal and flow analysis in MARC®. *J. Power Sources* **2004**, *130*, 136–148. [CrossRef]
105. Costamagna, P.; De Giorgi, A.; Moser, G.; Pellaco, L.; Trucco, A. Data-driven fault diagnosis in SOFC-based power plants under off-design operating conditions. *Int. J. Hydrog. Energy* **2019**, *44*, 29002–29006. [CrossRef]
106. Boersma, R.; Sammes, N. Computational analysis of the gas-flow distribution in solid oxide fuel cell stacks. *J. Power Sources* **1996**, *63*, 215–219. [CrossRef]
107. Boersma, R.; Sammes, N. Distribution of gas flow in internally manifolded solid oxide fuel-cell stacks. *J. Power Sources* **1997**, *66*, 41–45. [CrossRef]
108. Dotelli, G.; Sora, I.N.; Schmid, C.; Mari, C. Composite materials as electrolytes for solid oxide fuel cells: Simulation of microstructure and electrical properties. *Solid State Ion.* **2002**, *152*, 509–515. [CrossRef]
109. Rong, G.; Jin, M.; Shuai, L.; Guo, X. Centroidal Voronoi tessellation in universal covering space of manifold surfaces. *Comput. Aided Geom. Des.* **2011**, *28*, 475–496. [CrossRef]
110. Scott, K.; Taama, W.; Cruickshank, J. Performance and modelling of a direct methanol solid polymer electrolyte fuel cell. *J. Power Sources* **1997**, *65*, 159–171. [CrossRef]
111. Siegel, C. Review of computational heat and mass transfer modeling in polymer-electrolyte-membrane (PEM) fuel cells. *Energy* **2008**, *33*, 1331–1352. [CrossRef]
112. Radenahmad, N.; Azad, A.T.; Saghir, M.; Taweekun, J.; Abu Bakar, M.S.; Reza, S.; Azad, A.K. A review on biomass derived syngas for SOFC based combined heat and power application. *Renew. Sustain. Energy Rev.* **2020**, *119*, 109560. [CrossRef]
113. Saied, M.; Ahmed, K.; Nemat-Alla, M.; Ahmed, M.; El-Sebaie, M. Performance study of solid oxide fuel cell with various flow field designs: Numerical study. *Int. J. Hydrog. Energy* **2018**, *43*, 20931–20946. [CrossRef]

membranes

MDPI

Review

Modelling the Proton-Conductive Membrane in Practical Polymer Electrolyte Membrane Fuel Cell (PEMFC) Simulation: A Review

Edmund J. F. Dickinson * and Graham Smith

National Physical Laboratory, Hampton Road, Teddington TW11 0LW, UK; graham.smith@npl.co.uk
* Correspondence: edmund.dickinson@npl.co.uk

Received: 1 October 2020; Accepted: 22 October 2020; Published: 28 October 2020

check for updates

Abstract: Theoretical models used to describe the proton-conductive membrane in polymer electrolyte membrane fuel cells (PEMFCs) are reviewed, within the specific context of practical, physicochemical simulations of PEMFC device-scale performance and macroscopically observable behaviour. Reported models and their parameterisation (especially for Nafion 1100 materials) are compiled into a single source with consistent notation. Detailed attention is given to the Springer–Zawodzinski–Gottesfeld, Weber–Newman, and "binary friction model" methods of coupling proton transport with water uptake and diffusive water transport; alongside, data are compiled for the corresponding parameterisation of proton conductivity, water sorption isotherm, water diffusion coefficient, and electroosmotic drag coefficient. Subsequent sections address the formulation and parameterisation of models incorporating interfacial transport resistances, hydraulic transport of water, swelling and mechanical properties, transient and non-isothermal phenomena, and transport of dilute gases and other contaminants. Lastly, a section is dedicated to the formulation of models predicting the rate of membrane degradation and its influence on PEMFC behaviour.

Keywords: PEM; PEFC; PEMFC; ionomer; polymer electrolyte membrane; polymer electrolyte membrane fuel cell; proton exchange membrane; proton exchange membrane fuel cell

1. Introduction

In this review article, we summarise and evaluate the diversity of methods applied in the literature to describe theoretically the transport phenomena within a proton-conductive polymer electrolyte membrane (PEM, hereafter generally abbreviated to "membrane"), as applied in practical simulation methods for low-temperature polymer electrolyte membrane fuel cell (PEMFC) applications. Within the context of such applications, we identify the specific equations most often used, and sources of empirical experimental data for quantifying parameters for specific materials, especially membranes based on perfluorosulfonic acid (PFSA) ionomers, such as Nafion™.

We specifically place our focus on the bulk proton-conductive membrane which acts as a barrier to gas crossover in PEMFC devices. It is not our purpose to attempt a comprehensive review of the general literature on proton-conductive polymer electrolyte membranes, for which the reader is directed to the excellent and exhaustive 2017 review article by Kusoglu and Weber [1] as well as prior works correlating structural and chemical properties to performance characteristics [2–4]. Neither do we attempt to consider theories around the morphology and role of ionomer material in the context of the composite structure of catalyst layers, which remains an important open topic of interest, and is discussed elsewhere [5–8]. We will also avoid discussion of atomic-scale theories of the physicochemical structure of materials and instead focus attention on macroscopically observable transport behaviour of the membrane. We will avoid considerations specific to cold start (freezing) conditions, and also exclude the general field of high-temperature PEM devices—our range of consideration here spans conventional low-temperature operating conditions (broadly, 60 °C $\leq T \leq$ 90 °C).

In speaking of "practical simulation methods", we focus our interest upon models predicting the performance and overall electrochemical behaviour of a PEMFC, as well as the fundamental theories that most directly inform the continuum description of the membrane in full cell models. Electrochemical PEMFC models extend from empirical, lumped models to 3D models specifically resolving the dimensions of the various PEMFC components: bipolar plate ("land") design, gas channels, gas diffusion layers (GDL), microporous layers (MPL), catalyst layers (CL), and membrane. The role of the membrane model is to correlate quantitatively the observed overpotentials due to membrane losses, and the membrane's role in the cell water balance, to more fundamental transport laws in the membrane, which depend in turn on the environmental conditions. Such spatially resolved electrochemical simulations are valuable tools for PEMFC stack and component designers as they allow rational design of optimised components and configurations. Equally, they lend insight to researchers investigating the local conditions experienced by different materials and components within operating devices. Higher-dimensional models and those offering greater fidelity of description of the fundamental physical behaviour are expected to be more accurately predictive, providing greater descriptive granularity with respect to changes in operating conditions as well as with respect to exact locations within a cell or stack.

With these goals and restrictions in mind, this paper constitutes a focused review with the intention of acting as a useful digest for the present state-of-the-art in membrane modelling, from the perspective of practicing PEMFC simulation scientists. A number of prior reviews have usefully summarised historical progress and trends, and we recommend these as necessary reading for researchers active in PEMFC theory [7,9–13]; here, we emphasise a synoptic discussion and a survey of the most recent developments with specific focus on the membrane as a feature of the fuel cell.

The primary theoretical literature poses challenges due to inconsistency in notation between different authors, and a lack of traceability of parameterisation. We also recognise that validation of a complex model by means of a polarisation curve alone is often inadequate without further corroborating diagnostics. The detailed discussion of model validation for PEMFCs, as an exercise in general, again exceeds the scope of this review; by presenting theoretical descriptions together in a single source, however, we aim to set the stage for facilitated inter-comparison of models and accelerated implementation of new models for comparison with experimental data. In the Perspective section below, we present a sampled review of recent PEMFC simulations, in which it is demonstrated that even contemporary theoretical works depend heavily on theories and parameterisation established in the early 1990s on then-current membrane materials. For this reason also, we have taken this opportunity to present a collective review of both legacy and recent theoretical developments, as opposed to a selective review considering only the most recent work.

An account of simulation methods requires a summary of the essential transport phenomena to be described by the membrane model (Section 2). We then introduce the most basic level of practical description of the membrane, in the form of lumped and charge transport-only models (Section 3), before proceeding to consider the thermodynamics of membrane hydration through water sorption (Section 4) and the corresponding formulation of models combining charge (proton) transport with water transport (Section 5), optionally including phenomena specific to the interface between the membrane and its environment (Section 6). Thereafter, we describe specialised extensions to the core membrane models: hydraulic transport and membrane mechanics (Section 7); transient phenomena (Section 8); non-isothermal phenomena (Section 9); gas crossover and transport of contaminants (Section 10); and membrane degradation (Section 11). We then provide an outlook on continuing needs for theoretical work (Section 12).

Text abbreviations and symbolic notation used in equations throughout are summarised in the Tables A1 and A2 in Appendix A.

2. Proton-Exchange Membrane: Role and Essential Transport Phenomena

2.1. Role of the Membrane

The purpose of the membrane in a PEMFC is to act as a barrier to gas transport, thereby preventing direct mixing of H_2 and O_2, and preventing electron conduction between the anode and cathode electrodes while acting as an ionic conductor via mobility of protons (H^+). The main active component of PEMFC membranes is the proton-conductive polymer (ionomer) phase. Most commonly, the proton-conductive phase in PEMFCs is a PFSA. These polymers have a perfluorinated hydrophobic backbone connected to hydrophilic sulfonic acid groups that act as strong acids with very labile dissociation of protons. In the presence of water, mobile proton-carrying species (such as hydronium, H_3O^+) form, and the ionic conductivity of the membrane is increased significantly. Membrane hydration is essential for a practically useful proton conductivity to be obtained, so it is common for models to account for the variability of membrane properties with water content, and to describe water transport concurrently with proton transport.

The most well-known example of a PFSA is Nafion, originally developed in the 1960s by DuPont, and now a brand owned by Chemours (Wilmington, DE, USA) [14,15]. Other PFSA materials have been widely used in recent years, with the primary differences being the length and chemistry of the chain linking the backbone and acid groups [16]; these include Aquivion [17,18] (originally developed by Dow as Hyflon and now Solvay) and 3M materials [19,20]. Historically, membranes were made solely of thick extruded sheets of pure proton-conductive polymer; more recently, however, composite membranes with features such as non-conducting polymer reinforcements (e.g., GORE-SELECT materials [21,22]) and radical scavengers [23] (e.g., Nafion XL [24]) have become more common. Figure 1 summarises some of the key developments in membrane technology. While contemporary state-of-the-art commercial membranes bear little resemblance to their 1960s progenitors, either in form or performance, the underlying chemistry and physics remain largely the same.

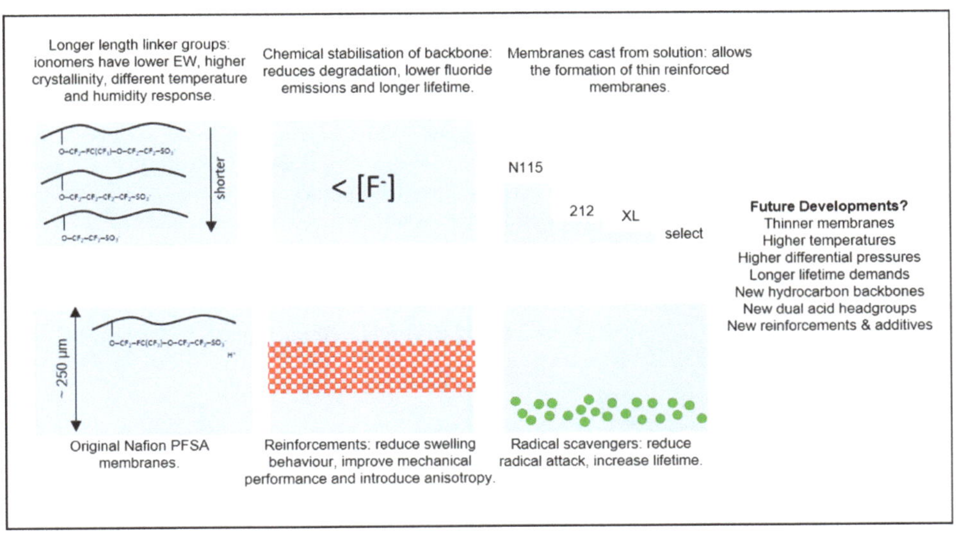

Figure 1. Schematic highlighting the key innovations in polymer electrolyte membrane fuel cell (PEMFC) membranes.

In the PEMFC modelling literature, the significant majority of works address membrane materials in the Nafion family, but other PFSA-based materials, including reinforced membranes, can be treated through similar theoretical approaches, provided experimental data are available. We highlight that it is likely to be insufficient to, for instance, use historical data from experiments on Nafion 115 membranes in models of very thin, reinforced membranes using a different PFSA. Besides one exception in the recent Chinese-language literature [25], we encountered in the literature no instances of models explicitly accounting for the altered properties of thin composite materials used in state-of-the-art devices.

Fundamental research has also considered alternative membrane chemistries: for example, those made with non-fluorinated hydrocarbons [15], with multiple acidic head groups [26] or by the incorporation of new monomers into the backbone [27]. Again, it is likely that such materials can be treated through similar approaches to those developed for Nafion and discussed in this text, provided sufficient experimental data are available. The hydroxide-conducting membranes used in anion exchange membrane fuel cells feature significantly different transport mechanisms [28] and so any adaptation of the models described here to these materials must be made with great caution.

Similar PFSA-based materials to those used in PEMFCs are also used for the membrane in polymer electrolyte membrane water electrolyser (PEMWE) applications, and in alcohol-fuelled proton-exchange fuel cells (direct methanol and direct ethanol fuel cells, DMFC/DEFC). In these devices, one or both faces of the PEM is in contact with liquid water, altering the environmental equilibration of the membrane material compared to the PEMFC case, where both faces of the membrane meet a gas phase (notwithstanding condensation of liquid water in the CLs), and either face of the membrane may be partially humidified depending on operating conditions. We shall draw attention to the applications of the theories reported herein to PEMWE and DMFC/DEFC simulation, selectively and as appropriate.

2.2. Membrane Types and Fundamental Material Properties

Essential properties of the dry PEM material are its density ρ_{dry} and equivalent weight M_{EW}, where the equivalent weight is the mass of polymer per 1 mol of sulfonic acid groups. It is common to report the available ion-exchange capacity (IEC) measured by titration (often as milliequivalents/g of the acidic functional group), which in an ideal condition is the inverse of the equivalent weight (IEC = $1/M_{\mathrm{EW}}$) [29]. IEC can also be measured under specified hydration conditions, in which case it will generally differ from the maximum available IEC.

Nafion 1100 is a standard material with $M_{\mathrm{EW}} \approx 1.100$ kg mol^{-1} and $\rho_{\mathrm{dry}} \approx 2050$ kg m^{-3} [1,30]. Both the Nafion 11x and Nafion 21x series have M_{EW} close to this value, with x denoting the thickness of the manufactured membrane in thousands of an inch (10^{-3} in, "mil"). We note one recent model [31] giving the dry density as 1970 kg m^{-3}, which is equivalent to the basis weight at 5% water content (50% relative humidity, T = 23 °C) given on the Nafion 115 data sheet [32]. In this context it is important to recognise that the basis weight and dry density are not equivalent concepts—there will exist a discrepancy depending on the degree of swelling with water uptake (see also Section 7.2, "Membrane Expansion and Mechanical Constraint", below). Of course, the ideally dry condition is not encountered in the PEMFC context, and so densities of the hydrated membrane have more practical relevance. The role of water sorption on density is discussed further below (Section 4, "Sorption of Water").

The concentration of sulfonic acid groups c_{f} in the dry membrane (or, as an inverse, the molar volume of the dry membrane $\overline{V}_{\mathrm{p}}$) is defined as:

$$c_{\mathrm{f}} = \overline{V}_{\mathrm{p}}^{-1} \equiv \frac{\rho_{\mathrm{dry}}}{M_{\mathrm{EW}}} \tag{1}$$

From the above standard values, $c_{\mathrm{f}} \approx 1850$ mol m^{-3}, $\overline{V}_{\mathrm{p}} \approx 5.35 \times 10^{-4}$ m^{3} mol^{-1}. As a caution, the significant paper on water management by Berg et al. [33] gives c_{f} = 1200 mol m^{-3} which seems to be erroneous for Nafion 1100 if measured against the dry density. Even at high water content, which will lower overall density according to (10) below, this value seems too low.

2.3. Essential Transport Phenomena in the Membrane

Membrane transport phenomena must be described in a PEMFC model to account for the balance across the membrane of the observable quantities of interest (shown schematically in Figure 2) [13]. Spatial variation in these quantities may be of interest: both through the plane of the membrane from anode to cathode, and also in the plane of the membrane, in the case of 2D/3D models that capture spatial variations in the electrode plane due to flow channel design.

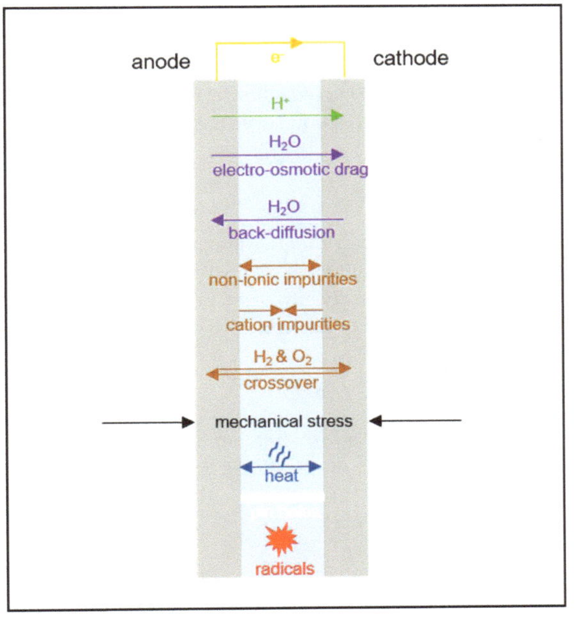

Figure 2. Schematic of different transport phenomena that can be considered in membrane models.

For an electrochemical device, the most common lumped quantity of interest is the electrochemical voltage, with the loss of cell voltage due to the membrane corresponding primarily to resistance to the transport of charge. Since charge is transported in the membrane in the form of protons (or, proton-carrying species) it is accompanied by electroosmotic flow of water; thus it is normally necessary for transport of the following conserved properties to be considered:

- charge;
- proton mass;
- water mass.

In the PEMFC context, it is, therefore, normally considered *essential* to define transport relations for:

- proton flux (current density);
- water flux.

Section 3 below will consider simpler models where the water transport is considered ideal, so that the membrane is uniformly hydrated and a charge transport-only model can be used. Sections 4 and 5 will then consider the quantitative theory of water uptake and describe various models coupling current density to water flux.

If required, the model may be extended by the consideration of other transport phenomena:

- momentum (flow/mechanical stress, discussed in Section 7);
- heat (discussed in Section 9);
- dilute dissolved gas mass, to account for gas crossover (discussed in Section 10.1);
- dissolved ion mass (other than protons, discussed in Section 10.2).

3. Charge Transport-Only Membrane Models

3.1. Zero-Dimensional (0D, Lumped) Resistance Models

The simplest level of description of the electrochemical performance of a PEMFC is an empirical fit to the electrochemical performance as evidenced by a measured polarisation curve, without any physical resolution of the underlying phenomena. Ignoring transport phenomena, a simple fit resolving the kinetic and ohmic regions of the polarisation curve (cell voltage E_{cell} as a function of cell current density i_{cell}) is [34]:

$$E_{\text{cell}} = E_{\text{OCV}} - A_{\text{cat}} \log_{10}\left(\frac{i_{\text{cell}}}{i_{\text{ref}}}\right) - R_{\Omega} i_{\text{cell}} \tag{2}$$

The parameters in this fit are the open circuit voltage E_{OCV}, Tafel slope A_{cat}, reference current density i_{ref}, and ohmic series resistance R_{Ω} ($\Omega \cdot \text{m}^2$). These parameters are determined empirically from the polarisation curve data. R_{Ω} is traditionally attributed primarily to the finite proton conductivity of the membrane (κ). Thus, for a membrane of thickness L_{mem}:

$$R_{\Omega} \approx \frac{L_{\text{mem}}}{\kappa} \tag{3}$$

For the thinnest membranes this approximation is less reliable, since the contributions from proton transport in the CL and from electrical contact resistances in the cell may become proportionally significant.

3.2. Constant Hydration Models

In uncontaminated operating conditions for a PEMFC, there are no dissolved ions in the membrane other than protons: hence, only protons contribute to *mobile* charge in the membrane, with the counter-ions present as the static sulfonate groups. Under these conditions, charge transport and proton mass transport are equivalent phenomena—neither one may take place without the other, and so the parameterisation of the transport of the two properties is inextricable. The current density $\mathbf{i}$ and molar flux of protons $\mathbf{N}_+$ differ only by means of a scaling by the Faraday constant F:

$$\frac{\mathbf{i}}{F} = \mathbf{N}_+ \tag{4}$$

Although water balance is most often included, if a constant hydration condition is assumed then a charge transport-only transport theory results [35,36]. The simplest conductivity model is an Ohm's law model of the current density (5) relating current density to proton conductivity (κ) and membrane-phase electrolyte potential (φ). This can then be combined with a statement of conservation of current in the bulk membrane, (6).

$$\mathbf{i} = -\kappa \nabla \phi \tag{5}$$

$$\nabla \cdot \mathbf{i} = 0 \tag{6}$$

Within a volume-averaged continuum model [37] of the CL, an effective conductivity may be used according to the volume fraction and connectivity of the ionomer within the CL composite [8,38]; also, the electrolyte current balance in the volume-averaged CL will have a source term corresponding to the faradaic current density (and, in principle, capacitive current density) and the corresponding source/sink of protons to the ionomer.

4. Sorption of Water

In a humid or wet environment, the membrane material takes up water through sorption. The water content λ of the polymer is defined as the ratio of moles of sorbed H_2O (n_{H_2O}) to moles of sulfonic acid groups (n_{SO_3}), within a defined reference volume of membrane (V) [1]:

$$\lambda = \frac{n_{H_2O}}{n_{SO_3}} \tag{7}$$

The water content may equivalently be written in terms of a total mass of water (m_w) taken up in the same reference volume:

$$\lambda = \frac{m_w}{M_w c_f V} \tag{8}$$

where the molar mass of water $M_w = 0.018$ kg mol^{-1}.

By rearranging (7) and (8), the volume fraction of water in the hydrated PEM (ϕ_w) is given:

$$\phi_w = \frac{\lambda}{\lambda + \frac{\overline{V}_p}{\overline{V}_w}} \tag{9}$$

where $\overline{V}_w$ is the molar volume of sorbed water $\approx 1.8 \times 10^{-5}$ m^3 mol^{-1} at 25 °C (corresponding to a density of sorbed water $\rho_w \approx 1000$ kg m^{-3}). The total density of hydrated polymer (ρ) approximately obeys a linear relation [1]:

$$\rho = \rho_{dry}(1 - \phi_w) + \rho_w \phi_w \tag{10}$$

From (9),

$$\rho = \frac{M_{EW} + M_w \lambda}{c_f^{-1} + \lambda \overline{V}_w} \tag{11}$$

We define a hygroscopic swelling coefficient β_w to account for membrane volume change under water uptake:

$$\beta_w \equiv \frac{V}{V_{dry}} \tag{12}$$

$$c_w = \frac{c_f}{\beta_w}\lambda \tag{13}$$

Springer et al. proposed the following linear correlation for β_w [39]:

$$\beta_w \approx 1 + 0.0126\lambda \tag{14}$$

4.1. Sorption Isotherms

A sorption isotherm relates the equilibrium water content λ_{eq} to the activity of water a_w in the membrane phase:

$$\lambda_{eq} = \lambda_{eq}(a_w) \tag{15}$$

To reach equilibrium, water may be sorbed from or desorbed to an adjacent phase, which might be either a gas phase containing water vapour at a certain activity, or a liquid phase (aqueous phase). These two cases are referred to as vapour-equilibrated (VE) and liquid-equilibrated (LE) conditions, respectively. It has been widely observed experimentally that the sorbed water uptake to PFSA ionomers is different between VE exposure to saturated water vapour and LE exposure to pure liquid water. Since both saturated water vapour and pure liquid water both have an activity $a_w = 1$, this result is thermodynamically unexpected and is often called "Schröder's paradox" [40]: it is generally explained according to the presence of liquid water promoting an otherwise restricted phase transition of the ionomer that either eliminates the vapour–liquid interface near the membrane surface, or alters the energetics of the ionomer matrix–liquid contact [41–44].

Some experimental studies (notably Jeck et al. [45]) report an absence of Schröder's paradox, but also suggest VE water contents that are higher than presented in other studies and closer to a typical LE value, possibly implicating the presence of a thin water film. The capability of thin water films to maintain bulk LE conditions in membranes with one VE face has also been suggested, based on X-ray tomography evidence [46].

The presence of VE or LE conditions also influences the interfacial resistance to the recovery of the sorption equilibrium, which will be discussed further below (Section 6) in the context of interfacial phenomena. It is relevant to note that whereas a PEMFC may be operated under VE conditions at both electrodes, PEMWE operation is likely to be LE at both electrodes, except possibly at high current density where gas production rate may reduce the extent of wetting. Likewise, liquid-fed DMFCs and similar devices would typically be liquid-equilibrated at the anode face of the membrane in contact with aqueous solution; in this case, however, the activity of water in the liquid phase is not necessarily equal to unity, due to the presence of the concentrated alcohol component.

Under VE conditions, the activity of water in the membrane (a_{w}) can be specified as equal to the activity of water vapour in the equilibrating gas phase ($a_{\mathrm{w,vap}}$):

$$a_{\mathrm{w}} = a_{\mathrm{w,vap}} \text{ (equilibrium)} \tag{16}$$

The activity of water in the vapour phase can in turn be approximated as a function of the partial pressure of water ($p_{\mathrm{w,vap}}$) and the saturated partial pressure of water (vapour pressure, p_{sat}) as a function of temperature (T):

$$a_{\mathrm{w,vap}} \approx \frac{p_{\mathrm{w,vap}}}{p_{\mathrm{sat}}(T)} \tag{17}$$

Equation (17) neglects fugacity corrections, which is normally suitable for PEMFC operating conditions at absolute pressures of a few bar.

The vapour pressure of water used in (17) is conventionally expressed as an empirical function of temperature [47]. Springer et al. used curve fitting on tabulated values for the vapour pressure to give the following standard expression, used also in recent PEMFC models (coefficient data tabulated in Table 1) [39,48]:

$$\log_{10}\left(\frac{p_{\mathrm{sat}}}{p_0}\right) = a_0 + a_1(T - T_0) + a_2(T - T_0)^2 + a_3(T - T_0)^3 \tag{18}$$

Table 1. Parameterisation of the Springer et al. water vapour pressure fit, with p_0 = 1 atm and $T_0 = 0\ ^\circ\mathrm{C}$ [39].

Coefficient	Value
a_0	−2.1794
a_1	+0.02953 K^{-1}
a_2	$-9.1837 \times 10^{-5}\ \mathrm{K}^{-2}$
a_3	$+1.4454 \times 10^{-7}\ \mathrm{K}^{-3}$

Gurau et al. reported an alternative fit given by the American Society of Heating, Refrigerating and Air-Conditioning Engineers (ASHRAE) (coefficient data tabulated in Table 2) [49]:

$$\ln\left(\frac{p_{\mathrm{sat}}}{1\mathrm{Pa}}\right) = \frac{b_{-1}}{(T/1\mathrm{K})} + b_0 + b_1(T/1\mathrm{K}) + b_2(T/1\mathrm{K})^2 + b_3(T/1\mathrm{K})^3 + b_e \ln(T/1\mathrm{K}) \tag{19}$$

Table 2. Parameterisation of the American Society of Heating, Refrigerating and Air-Conditioning Engineers (ASHRAE) water vapour pressure fit reported by Gurau et al. [49].

Coefficient	Value
b_{-1}	-5.8002206×10^3
b_0	1.3914993
b_1	−0.048640239
b_2	4.1764768×10^{-5}
b_3	1.4452093×10^{-8}
b_e	6.5459673

The specification of the sorbed water activity under LE conditions typically depends on extending the range of values a_w to $a_w > 1$, according to a saturation-dependent definition of activity in the presence of liquid water. Simultaneously, the sorption isotherm (15) is extended to the corresponding values of a_w. For instance, Springer et al. defined for the purposes of the sorption equilibrium that [39]:

$$a_w \equiv \frac{x_w p}{p_{sat}} \tag{20}$$

where the mole fraction of water x_w includes both liquid- and gas-phase water. It should be noted that this definition is dependent on the use of a pseudo-two phase description of material transport in the GDL and CL, whereby liquid water is treated a gas-phase species with no independent momentum conservation. More detailed two-phase models, in which liquid water saturation in the porous diffusion media is described through an additional variable with a corresponding transport equation [10,50], may require an alternative specification of the LE sorption condition.

4.2. Empirical Sorption Models for Nafion 1100

A number of empirical sorption isotherms have been established experimentally based on measurements on Nafion 1100 [39,51–54]. All sorption isotherms reported in this subsection were parameterised for this material and, therefore, applications to other related materials—even in the Nafion family—should be undertaken with caution. A selection of the isotherms reported in this subsection are summarised in Figure 3.

The most commonly used sorption isotherm is an empirical polynomial relation due to Springer et al. [39]:

$$\begin{aligned} \lambda_{eq} &= 0.043 + 17.81 a_w - 39.85 a_w^2 + 36.0 a_w^3, \quad 0 \le a_w \le 1 \\ \lambda_{eq} &= 14 + 1.4(a_w - 1), \quad 1 \le a_w \le 3 \end{aligned} \tag{21}$$

The definition for supersaturated conditions ($a_w > 1$) aims to account for Schröder's paradox through the definition (20) given above. The polynomial relation is specific to measurements at 30 °C, although the definition for $a_w > 1$ derives from liquid-equilibration data at 80 °C.

Kusoglu and Weber offered an alternative polynomial fit for VE conditions at 30 °C by averaging a wide range of experimental data (from different authors) [1]:

$$\lambda_{eq} = 0.05 + 20.45 a_w - 42.8 a_w^2 + 36.0 a_w^3, \quad 0 \le a_w \le 1 \tag{22}$$

In spite of a variety of experimental investigations, there exist no definitive data for the temperature dependence of the sorption equilibrium at the higher temperatures more typical of PEMFC operation. Most data, however, suggest a relatively weak temperature dependence up to T = 90 °C: for this reason, it is common in the literature to encounter (21) being used at a typical PEMFC operating temperature also.

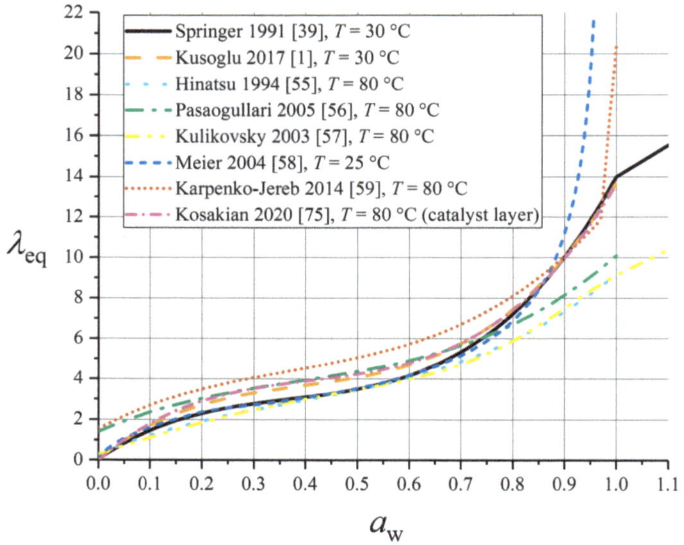

Figure 3. Summary of measured sorption isotherms for Nafion 1100 in the vapour-equilibrated range of a_w.

Hinatsu et al. measured the following sorption isotherm at 80 °C [55]:

$$\lambda_{eq} = 0.3 + 10.8a_w - 16.0a_w^2 + 14.1a_w^3, \quad 0 \le a_w \le 1 \tag{23}$$

while Pasaogullari et al. [56] presented a fit to data from Zawodzinski et al. [53] under VE conditions at 80 °C as:

$$\lambda_{eq} = 1.4089 + 11.263a_w - 18.768a_w^2 + 16.209a_w^3, \quad 0 \le a_w \le 1 \tag{24}$$

Kulikovsky extended the above expression to a super-saturated or LE condition [57]:

$$\lambda_{eq} = 0.3 + 6a_w(1 - \tanh(a_w - 0.5)) + 3.9\sqrt{a_w}\left(1 + \tanh\left(\frac{a_w - 0.89}{0.23}\right)\right) \tag{25}$$

As shown in Figure 3, the Pasaogullari and Kulikovsky isotherms at higher temperature predict overall lower water contents than the Springer isotherm at ambient temperature, especially close to $a_w = 1$.

Meier and Eigenberger proposed a 25 °C isotherm (expressing activity as a function of water content) which is compatible with LE conditions but without exceeding $a_w = 1$ [58]:

$$\begin{aligned} a_w &= 0.01355\lambda_{eq} + 0.03\lambda_{eq}^2, \quad \lambda \le 2.5 \\ a_w &= 1.435 + 0.0022\lambda_{eq} - \tfrac{2.75}{\lambda_{eq}} - 0.13\ln\lambda_{eq}, \quad \lambda > 2.5 \end{aligned} \tag{26}$$

Similarly, Karpenko-Jereb et al. originated an alternative empirical sorption isotherm with a smoothed jump at $0.97 \le a_w \le 1$ to represent Schröder's paradox, wherein the VE sorption isotherm shows no temperature dependence, but a temperature dependence is incorporated in the LE part [59]:

$$\begin{aligned} \lambda_{eq,V} &= 1.55 + 13.71a_w - 24.37a_w^2 + 21.87a_w^3 \\ \lambda_{eq,L} &= 41.83\tfrac{T}{T_0} - 28.31 \\ \lambda_{eq} &= \lambda_{eq,V}, \quad a_w < 0.97 \\ \lambda_{eq} &= \lambda_{eq,V} + \left(\lambda_{eq,L} - \lambda_{eq,V}\right)\left(\tfrac{a_w - 0.97}{0.03}\right), \quad 0.97 \le a_w \le 1 \end{aligned} \tag{27}$$

with $T_0 = 303.15$ K.

Futerko and Hsing gave the temperature dependence of the LE water content with $T_0 = 273$ K as [60]:

$$\lambda_{\text{eq,L}} = 10 + 0.0184(T - T_0) - 9.9 \times 10^{-4}(T - T_0)^2 \tag{28}$$

4.3. Detailed Sorption Models

A number of authors have attempted to construct sorption models based on more fundamental properties of the PEM, rather than from a purely empirical basis. Again, all parameterisation in this subsection is reported for Nafion 1100.

Futerko and Hsing applied a Flory–Huggins model to express the activity of the membrane under VE conditions as [60]:

$$\begin{aligned} a_{\text{w}} &= (1 - \phi_{\text{m}}) \exp\left(\left(1 - \frac{\overline{V}_{\text{w}}}{\overline{V}_{\text{p}}}\right)\phi_{\text{m}} + \chi\phi_{\text{m}}^2\right) \\ \phi_{\text{m}} &= \frac{\overline{V}_{\text{p}} + \overline{V}_{\text{w}}}{\overline{V}_{\text{p}} + \lambda\overline{V}_{\text{w}}} \end{aligned} \tag{29}$$

where the Flory parameter $\chi = 1.936 - (2.18 \text{ kJ mol}^{-1})/RT$ and ϕ_{m} is the effective membrane volume fraction.

Thampan et al. suggested following the Brunauer–Emmett–Teller (BET) adsorption isotherm, expressed as follows [30]:

$$\lambda_{\text{eq}} = \lambda_{\text{eq,Thampan}} = \lambda_{\text{mono}} \frac{\left(K_1 \frac{a_{\text{w}}}{1 - a_{\text{w}}}\right)\left(1 - (n_{\text{w,sat}} + 1)a_{\text{w}}^{n_{\text{w,sat}}} + n_{\text{w,sat}} a_{\text{w}}^{1 + n_{\text{w,sat}}}\right)}{1 + (K_1 - 1)a_{\text{w}} - K_1 a_{\text{w}}^{1 + n_{\text{w,sat}}}} \tag{30}$$

Here, λ_{mono} is the water content corresponding to an effective 'monolayer coverage' within the polymer scaffold of the membrane, which was assumed = 1.8. The other parameters are fitting coefficients to experimental data [52,61] with values given as $K_1 = 150$ and $n_{\text{w,sat}} = 13.5$. The BET isotherm predicts a saturated VE water content in terms of its parameters as [62]:

$$\lambda_{\text{sat}} = \lim_{a_{\text{w}} \to 1} \lambda_{\text{eq,Thampan}} = \lambda_{\text{mono}} \frac{K_1 n_{\text{w,sat}}^2}{1 + K_1 n_{\text{w,sat}}} \tag{31}$$

Klika et al. advocated the Guggenheim–Anderson–de Boer (GAB) isotherm, which extends the BET isotherm with an energy difference between the bulk and multilayer sorbed states of the water represented by the quantity k_{G}, replacing $n_{\text{w,sat}}$ in (30). Their equation is [63]:

$$\lambda_{\text{eq}} = \lambda_{\text{mono}} \frac{K_1 k_{\text{G}} a_{\text{w}}}{(1 - k_{\text{G}} a_{\text{w}})(1 + (K_1 - 1)k_{\text{G}} a_{\text{w}})} \tag{32}$$

with empirical parameters fit to experimental data [39,45] as $\lambda_{\text{mono}} = 1.93$, $K_1 = 44.3$ and $k_{\text{G}} = 0.9$.

Choi and Datta argued that Schröder's paradox can be explained by the restricted morphology of the vapour–liquid interface within a pore, which is resolved by liquid equilibration [41]. Their work derived a rather complicated expression for the sorption isotherm that extends the Thampan isotherm (Equation (30)) for both vapour- and liquid-equilibrated modes, considering bound water as a separate Langmuirian contribution to the isotherm, with a BET model accounting for additional bound water uptake beyond a monolayer, and a Flory–Huggins isotherm for free water. The resulting, rather unwieldy implicit equations relate to (30) as:

$$\begin{aligned} &\left(1 - \lambda_{\text{eq,Thampan}}\right)\left(a_{\text{w}} \exp\left(\frac{\overline{V}_{\text{w}}}{RT}\left(\frac{\kappa_{\text{sorp}}\lambda_{\text{eq,L}}}{\lambda_{\text{eq,L}} + \frac{\overline{V}_{\text{p}}}{\overline{V}_{\text{w}}}}\right)\right) - 1\right)\lambda_{\text{eq,L}} = 1 \\ &\left(1 - \lambda_{\text{eq,Thampan}}\right)\left(a_{\text{w}} \exp\left(\frac{\overline{V}_{\text{w}}}{RT}\left(\frac{\kappa_{\text{sorp}}\lambda_{\text{eq,V}}}{\lambda_{\text{eq,V}} + \frac{\overline{V}_{\text{p}}}{\overline{V}_{\text{w}}}} - a_{\text{pore}}\gamma_{\text{w}} \cos\theta_{\text{c}}\left(1 + \frac{\overline{V}_{\text{p}}}{\overline{V}_{\text{w}}\lambda_{\text{eq,V}}}\right)\right)\right) - 1\right)\lambda_{\text{eq,V}} = 1 \end{aligned} \tag{33}$$

The fitted parameters were as above, except setting $K_1 = 100$ and $n_{\mathrm{w,sat}} = 5$. Additional parameters are defined in Table A2 in Appendix A, and given as $\kappa_{\mathrm{sorp}} = 183$ atm, $a_{\mathrm{pore}} = 2.1 \times 10^8\ \mathrm{m}^{-1}$, $\gamma_{\mathrm{w}} = 0.0721\ \mathrm{N\ m}^{-1}$, $\theta_{\mathrm{c}} = 98°$.

In the context of a DMFC, Meyers and Newman developed an isotherm based on fundamental energetics of the PEM [64]. Like the model of Thampan et al. [30], this model combines an acid-base equilibrium for dissociation of sulfonic acid groups in the membrane with a requirement of electroneutrality. The resulting simultaneous equations that define the isotherm are (coefficient data tabulated in Table 3) [64]:

$$\begin{aligned} a_{\mathrm{w}} &= K_2\left(\lambda_{\mathrm{eq}} - \lambda_+\right)\exp(\phi_2\lambda_+)\exp\left(\phi_3\lambda_{\mathrm{eq}}\right) \\ \lambda_+\exp(\phi_1\lambda_+)\exp\left(\phi_2\lambda_{\mathrm{eq}}\right) &= K_1(1-\lambda_+)\left(\lambda_{\mathrm{eq}} - \lambda_+\right) \end{aligned} \tag{34}$$

where λ_+ is the ratio of moles of hydronium ions to moles of sulfonic acid sites and must be solved for self-consistently. The φ_n coefficients are defined as:

$$\begin{aligned} \phi_1 &= \tfrac{2}{M_{\mathrm{EW}}}(E_{00} - 2E_{++} - 2E_{+0}) \\ \phi_2 &= \tfrac{2}{M_{\mathrm{EW}}}(E_{+0} - 2E_{00}) \\ \phi_3 &= \tfrac{2}{M_{\mathrm{EW}}}E_{00} \end{aligned} \tag{35}$$

Table 3. Parameterisation of the Meyers–Newman sorption model for Nafion 1100 at $T = 30$ °C [52,61].

Coefficient	Value
K_1	100
K_2	0.217
E_{00}	−0.0417 kg mol^{-1}
E_{+0}	−0.052 kg mol^{-1}
E_{++}	−3.7216 kg mol^{-1}

The Meyers–Newman isotherm was applied by Weber and Newman using the above parameters, subject to a further empirical modification to account for inaccuracies of the model at low water uptake [65]:

$$\lambda = \beta\lambda_{\mathrm{eq}}\left(1 + \exp\left(\lambda_{\mathrm{eq,crit}} - \lambda_{\mathrm{eq}}\right)\right) \tag{36}$$

with the scaling coefficient $\beta = 1$ (included for generality, see Section 4.4, "Sorption within the Catalyst Layer" below) and $\lambda_{\mathrm{eq,crit}} = 0.3$. Subsequently, the influence of temperature on this sorption isotherm was accounted for by treating all inputs as temperature-independent except K_2, which varies as [66]:

$$K_2 = 0.217\ \exp\left(\frac{\Delta H_{\mathrm{sorp}}}{R}\left(\frac{1}{T_0} - \frac{1}{T}\right)\right) \tag{37}$$

with $T_0 = 303.15$ K and the enthalpy change of sorption $\Delta H_{\mathrm{sorp}} = +1\ \mathrm{kJ\ mol}^{-1}$.

Murahashi et al. [67] and Eikerling and Berg [44] and have both argued for the influence of size distribution of pores upon the sorption isotherm. Smaller pores may retain water due to the pressure drop of the liquid–vapour interface, even when larger pores become dehydrated [67]. Eikerling and Berg argued that pores with a range of surface charge densities can wet progressively due to more highly charged pores taking up water more slowly, but attaining larger limiting wetted radii [68]. Since the Eikerling–Berg description makes a number of highly specific assumptions about the geometry and governing phenomena of the matrix-liquid-vapour system, it could be viewed as didactic rather than being directly usable for quantitative modelling of membrane sorption in the PEMFC context.

4.4. Sorption within the Catalyst Layer

It has been established experimentally that membrane material within the composite structure of the catalyst layer takes up proportionally less water by sorption than in the bulk membrane [69,70]. This has been attributed to the altered internal morphology of Nafion ionomer when present as a <60 nm film. Recent analysis has indicated that a lamellar structure forms for particularly thin films, and that total water uptake varies non-monotonically with film thickness [71]. This work has also emphasised that for thin films, homogenisation of material properties is typically unreasonable, and interfacial phenomena may dominate.

It has been proposed that the CL water uptake could be described empirically by setting $K_2 = 0.231$, $\beta = 0.342$ (all other parameters identical) in the Meyers–Weber–Newman isotherm (36) [70]. This work also assumed a constant apparent proton conductivity $\kappa_{\text{eff}} = 10^{-4}$ S m^{-1} in the CL. Incorporation into a full PEMFC model was reported to give improved accuracy in the prediction of performance loss due to anode dehydration, under low relative humidity operating conditions. From a thermodynamic point-of-view, it is necessary to recognise that altering the isotherm in the CL specifically implies a phase transition for water between the CL ionomer and the bulk membrane, at constant activity. If no resistance is incorporated for this transfer, reduced water uptake in the CL remains compatible with normal water content and water transfer fluxes in the bulk membrane.

Additionally, studies have suggested that sorbed water uptake of the CL depends upon Pt loading [69], choice of carbon support [38,72], and the solvent used in CL preparation [73]. Using the Pt/C-phase effective electronic conductivity as a probe, Morris et al. showed that sorption/desorption in the CL appeared to be free from hysteresis [74].

Mashio et al. provided a comprehensive model for CL water uptake in which sorption to the membrane was incorporated by means of a Langmuir isotherm [72]:

$$\lambda = \lambda_{\text{sat}} \frac{K_{\text{mem}} a_{\text{w}} p_{\text{sat}}}{1 + K_{\text{mem}} a_{\text{w}} p_{\text{sat}}} \tag{38}$$

with equilibrium constant $K_{\text{mem}} = 3.3 \times 10^{-4}$ Pa^{-1}; this work stresses the role of water adsorption on a variety of materials within the CL, in terms of overall water uptake of this region.

Since the CL has a high volumetric surface area of contact between vapour- or liquid-phase water and the membrane material, the local sorption properties in this region may significantly influence the overall water balance of the membrane. Kosakian et al. have recently presented a dedicated CL isotherm as follows [75]:

$$\begin{aligned} \lambda_{\text{eq}} &= \left(6.932 a_{\text{w}} - 14.53 a_{\text{w}}^2 + 11.82 a_{\text{w}}^3\right) \exp\left(\theta_{\text{sorp}}\left(\tfrac{1}{T_0} - \tfrac{1}{T}\right)\right), \quad 0 \le a_{\text{w}} \le 1 \\ \lambda_{\text{eq}} &= 22, \quad a_{\text{w}} > 1 \end{aligned} \tag{39}$$

with $\theta_{\text{sorp}} = 2509$ K, $T_0 = 303.15$ K. It should be noted that this specification gives a large discontinuity at the transition to liquid equilibration at $a_{\text{w}} = 1$; this is not depicted in Figure 3 which focuses on VE conditions.

5. Coupled Proton-Water Transport

As introduced through the discussion above in Section 2.3 ("Essential Transport Phenomena in the Membrane"), current flow by proton flux through the membrane is always accompanied by water transport due to electroosmotic drag. Therefore, the majority of practical membrane models used in PEMFC simulation are coupled models incorporating both proton and water transport. In this section, we will introduce some essential considerations surrounding the coupling of the two transport processes, and then consider three principal approaches to this coupling and their parameterisation.

First, we will discuss the membrane model originated in the seminal early PEMFC simulation work by Springer, Zawodzinski and Gottesfeld [39] (hereafter "Springer model" for brevity). This widely used model accounts for electroosmotic drag and water diffusion in an essentially empirical manner. Second, the Weber–Newman model [65] will be considered; this model is rooted in a more formal derivation from non-equilibrium thermodynamics of the membrane phase, but remains empirically parameterised. Lastly, the binary friction model developed by Djilali and co-workers will be discussed [76–78].

Weber and Newman have advocated the interpretation of the hydrated membrane as a system with three chemical components: water, protons, and fixed membrane structure [65]. Therefore, there exist no more than three independent transport properties associated with the binary interactions of the three components. If bulk momentum transfer through the membrane is considered, mechanical resistance (friction) may account for a fourth transport property. The three most experimentally accessible choices for the definition of the three independent transport properties, following the Weber–Newman scheme, are [65]:

- proton conductivity κ—the ratio of current density to electrolyte potential gradient for uniform water content;
- electroosmotic drag coefficient ξ—the ratio of water flux to current density for uniform water content;
- water diffusivity D_{w}—the ratio of water flux to water concentration gradient for zero current density.

Parameterisation for each of these intrinsic properties (proton conductivity, water diffusivity, electroosmotic drag coefficient) will be summarised in Section 5.4, Section 5.5, Section 5.6. The rigorous measurement of these properties for Nafion 1100 was initiated in the early 1990s in a series of works by Zawodzinski et al. [52,53,79,80] and Fuller and Newman [81].

Since the fixed membrane structure is considered rigid (there is no mechanical translation of the membrane), the Weber–Newman scheme considers only water and proton fluxes, and each of these fluxes has a conjugate thermodynamic variable whose gradient indicates the direction and magnitude of the flux. Formal treatments define [65,82]:

- membrane-phase electrolyte potential φ as the thermodynamic variable conjugate to the driving force for current flow, under uniform hydration;
- chemical potential of water μ_{w} (expressed as required in terms of the local water content λ) as the thermodynamic variable conjugate to the driving force for water flux, at zero proton current.

From a thermodynamic standpoint, the inclusion of pressure as a local variable, in addition to electrolyte potential and water content, is almost certainly an overdetermination of the system except for the liquid-equilibrated case; it has not found general support [65,78]. Exceptions in the liquid-equilibrated case, where free water channels may be present, will be discussed further in Section 5.7 below.

5.1. Springer Membrane Model

Springer et al. defined the flux of water through the membrane ($\mathbf{N}_{\mathrm{w}}$) phenomenologically as the sum of a Fickian diffusion term in water concentration c_{w}, and an electroosmotic drag term [9,39]:

$$\mathbf{N}_{\mathrm{w}} = -D_{\mathrm{w}} \nabla c_{\mathrm{w}} + \xi \frac{\mathbf{i}}{F} \quad (40)$$

This is compatible with the transport property definitions given above; non-linear transport behaviour is implied if the coefficients D_{w} and ξ are themselves functions of water content. Using (14):

$$\mathbf{N}_{\mathrm{w}} = -D_{\mathrm{w}} c_{\mathrm{f}} \nabla \left(\frac{\lambda}{\beta_{\mathrm{w}}} \right) + \xi \frac{\mathbf{i}}{F} \quad (41)$$

On the basis that β_w is a function of water content only, the Springer water flux formula can be further abbreviated. Springer et al. expressed the water gradient ∇_{dry} with respect to a fixed, dry-membrane coordinate which is undeformed by swelling under water uptake [39]:

$$\mathbf{N}_w = -D_\lambda c_f \nabla_{dry}\lambda + \xi\frac{\mathbf{i}}{F} \quad (42)$$

where the apparent diffusion coefficient with respect to water content (D_λ) is given:

$$D_\lambda = \frac{D_w}{\beta_w^2}\left(1 - \frac{\lambda}{\beta_w}\frac{\partial\beta_w}{\partial\lambda}\right) \quad (43)$$

While the original Springer et al. work considered membrane expansion under swelling, subsequent works have assumed that the compressed membrane has no thickness variation [83]; the role of compression is discussed further below in Section 7.2. Thus one can write simply:

$$\mathbf{N}_w = -D_\lambda c_f \nabla\lambda + \xi\frac{\mathbf{i}}{F} \quad (44)$$

Typically, this diffusivity D_λ is parameterised directly from experimental data, as discussed further in Section 5.5 below. Within the membrane, conservation of water mass requires that, under steady-state conditions:

$$\nabla \cdot \mathbf{N}_w = 0 \quad (45)$$

The widely used Springer model describes coupled proton-water transport in PEMFC membranes by combining Equations (44) and (45) with the Ohm's law expressions (5) and (6). The interaction of the two transport processes is expressed by the electroosmotic drag term in (44) and the water content-dependence of the proton conductivity in (5).

5.2. Weber–Newman Membrane Model

The Weber–Newman model is an alternative to the Springer model that seeks to consider the non-equilibrium thermodynamics of the coupled proton and water transport processes in a more formal and consistent manner [65]. Moreover, the empirically determined transport properties are assumed to have different definitions in the VE and LE regimes.

The substantive difference between the transport equations used for the Weber–Newman model and the Springer model is the presence of a cross-term contribution to the current density expression due to water streaming current, which is non-zero wherever the water chemical potential is non-uniform through the membrane. This phenomenon accounts for the symmetry of the binary proton–water interaction: just as electroosmotic drag describes the motion of water molecules due to proton current, so the streaming current describes the motion of protons due to water diffusion. Specifically:

$$\mathbf{i} = -\kappa\nabla\phi - \frac{\kappa\xi}{F}\nabla\mu_w \quad (46)$$

The water flux is then given as, alongside (46) [65]:

$$\mathbf{N}_{w,m} = -\alpha_{w,m}\nabla\mu_{w,m} + \frac{\xi_m}{F}\mathbf{i} \quad (47)$$

where the subscript m = V or L and indicates VE or LE conditions.

The relation between the intrinsic transport coefficient α_w and the apparent Fick's law diffusion coefficient D_λ has been expressed differently by various authors [65,83]. A simple interpretation for the VE case has been given by Setzler and Fuller, on the basis that the water content is the only parameterising variable for the local state of the hydrated membrane:

$$\begin{aligned} -\alpha_{\mathrm{w,V}}\nabla\mu_{\mathrm{w}} &= -D_\lambda c_{\mathrm{f}}\nabla\lambda \\ &= -D_\lambda c_{\mathrm{f}}\frac{\partial\lambda}{\partial a_{\mathrm{w}}}\frac{\partial a_{\mathrm{w}}}{\partial\mu_{\mathrm{w}}}\nabla\mu_{\mathrm{w}} \\ &= -D_\lambda c_{\mathrm{f}}\frac{\partial\lambda}{\partial a_{\mathrm{w}}}\frac{a_{\mathrm{w}}}{RT}\nabla\mu_{\mathrm{w}} \\ &= -D_\lambda\frac{c_{\mathrm{f}}\lambda}{RT}\frac{\partial\ln\lambda}{\partial\ln a_{\mathrm{w}}}\nabla\mu_{\mathrm{w}} \\ &= -D_\lambda\frac{c_{\mathrm{w}}}{RT}\frac{\partial\ln\lambda}{\partial\ln a_{\mathrm{w}}}\nabla\mu_{\mathrm{w}} \end{aligned} \tag{48}$$

If the self-diffusion coefficient D_μ is defined as given by Springer et al. [39]:

$$\alpha_{\mathrm{w,V}} \equiv D_\mu\frac{c_{\mathrm{w}}}{RT} \tag{49}$$

then the two diffusion coefficients are simply related by the Darken factor [83]:

$$D_\lambda = \frac{\partial\ln a_{\mathrm{w}}}{\partial\ln\lambda}D_\mu \tag{50}$$

However, it should be noted that in their original work, Weber and Newman used instead of the definition (49) the following definition based on the Maxwell–Stefan diffusion coefficient for the membrane-water system as a binary system [65]:

$$\alpha_{\mathrm{w,V}} \equiv D_{\mu,\mathrm{WN}}\frac{c_{\mathrm{w}}}{RT}(1+\lambda) \tag{51}$$

In interpreting the diffusivity data given below in Section 5.5, the inequivalence of (49) and (51) must be borne in mind.

5.3. Binary Friction Model (BFM)

The binary friction model (BFM) is derived beginning from a generalised diffusion equation presented succinctly in the following form [82]:

$$\overline{V}_{\mathrm{w}}\begin{pmatrix}\mathbf{N}_+ \\ \mathbf{N}_{\mathrm{w}}\end{pmatrix} = -\begin{pmatrix}D_{11} & D_{12} \\ D_{21} & D_{22}\end{pmatrix}\begin{pmatrix}\frac{F}{RT}\nabla\phi \\ \nabla\lambda\end{pmatrix} \tag{52}$$

where the generalised diffusion coefficients D_{mn} are functions of λ, T, thereby accounting for non-linearity.

In the BFM developed by Fimrite, Carnes, Struchtrup and Djilali [76,77], the mole fraction of proton carriers is derived using a method originated by Thampan et al. [30] This mole fraction is then applied as a dependent variable in a concentrated solution theory; Shah et al. have also applied the Thampan method previously to dilute solution theory [84]. The Thampan approach assumes that there exists an acid-base equilibrium quantifying the degree of dissociation of the sulfonic acid groups in the fixed membrane structure [30]:

$$\mathrm{SO_3H(fixed)} + \mathrm{H_2O} \overset{K_{\mathrm{a,mem}}}{\rightleftarrows} \mathrm{SO_3^-(fixed)} + \mathrm{H_3O^+} \tag{53}$$

In this case, the total water content is divided between neutral water and charged hydronium species. Electroneutrality requires that the local concentrations of H_3O^+ and $SO_3{}^-$ are equal. Later iterations of the BFM employed the simplifying assumption that sulfonic acid dissociation is complete (strong acid behaviour) [78,82].

In the original BFM, there were five fitted parameters, but no further empirical relations to water content [77]. Fimrite et al. thereafter used the term "BFM2" for a specialised binary friction model specific to the PEMFC device context, in the limit of complete sulfonic acid group dissociation [78,82]. This model has six parameters, which were fitted to a conductivity vs. water content curve derived from alternating current (AC) impedance measurements. (As an aside, the original derivation of the BFM2 model is complicated by two irregular negative multiples: one in the definition of the potential driving force ([82], eqn 3) and then one in the charge of the hydronium ion (set = −1 in [82]). These negatives cancel in the derived equations.) The resulting Weber–Newman transport coefficients are then expressed in terms of a set of binary diffusivities as [82]:

$$\kappa = f_{\text{eff}} c_{\text{f}} \frac{F^2}{RT} \frac{D_{+\text{m}}}{\lambda} \frac{D_{0\text{m}} + D_{+0}\lambda}{D_{+0}\lambda + D_{+\text{m}} + D_{0\text{m}}(\lambda - 1)} \tag{54}$$

$$\xi = \frac{D_{0\text{m}}(\lambda - 1)}{D_{0m} + D_{+0}\lambda} \tag{55}$$

$$D_{\text{w}} = f_{\text{eff}} \frac{D_{0\text{m}}}{\lambda} \frac{D_{+0}}{D_{+0}\lambda + D_{+\text{m}} + D_{0\text{m}}(\lambda - 1)} \tag{56}$$

The three binary diffusivities are expanded further as functions of temperature and/or water content, as:

$$D_{k\text{m}} = D_{+0} A_k \lambda^s \tag{57}$$

$$D_{+0} = D_{+0,\text{ref}} \exp\left(\theta_{\text{diff}}\left(\frac{1}{T_{\text{ref}}} - \frac{1}{T}\right)\right) \tag{58}$$

and the coefficient f_{eff} accounts for the influence of an effective 'membrane porosity' by the relation:

$$f_{\text{eff}} = (\phi_{\text{w}}(\lambda) - \phi_{\text{w}}(\lambda_{\text{min}}))^q \tag{59}$$

The more explicit definition of the functional forms of each of the three orthogonally measurable transport coefficients in terms of the various water-content-independent parameters is suggested as a means to reduce the degree of empiricism in the model setup (coefficients are tabulated in Table 4).

Table 4. Parameterisation of the binary friction model for Nafion 1100 [78,82].

Coefficient	Value
λ_{min}	1.65
D_{+0}	6.5×10^{-9} m^2 s^{-1}
s	0.83
q	1.5
A_+	0.084
A_0	0.5
θ_{diff}	1800 K

Djilali and Sui [85] have argued that the above theory offers improved sensitivity to the variation in membrane behaviour under conditions of low anode humidification, where the empirical correlations from Springer et al. [39] may be less reliable. Conversely, it is unclear whether the BFM will retain applicability in the limit of high water content or liquid-equilibration, where hydraulic transport becomes significant.

5.4. Proton Conductivity as a Function of Water Content

The proton conductivity κ appears in both the Springer model and the Weber–Newman model as an empirical function of water content. Within the Weber–Newman model, proton conductivity is additionally considered to have a different value under LE conditions. At the atomic scale, proton transport in a PEM is understood to occur due to two possible mechanisms: the vehicle mechanism, in which protons are carried in the form of hydronium ions, H_3O^+; and the hopping (Grotthuss) mechanism, in which protons are transported by a hydrogen bond-mediated long-range rearrangement of the bonding network between water molecules and hydronium ions, as in bulk liquid water [42]. Since membranes at lower hydration do not contain connected liquid regions with extensive hydrogen bonding networks, the Grotthuss mechanism is suppressed and the vehicle mechanism is believed to dominate [1,2]. Activation of the Grotthuss mechanism for higher water content, especially when liquid-equilibrated, increases the conductivity as water content rises.

Various empirical relations reported in this subsection for proton conductivity of Nafion 1100 materials are plotted in Figure 4 (T = 30 °C) and Figure 5 (T = 80 °C).

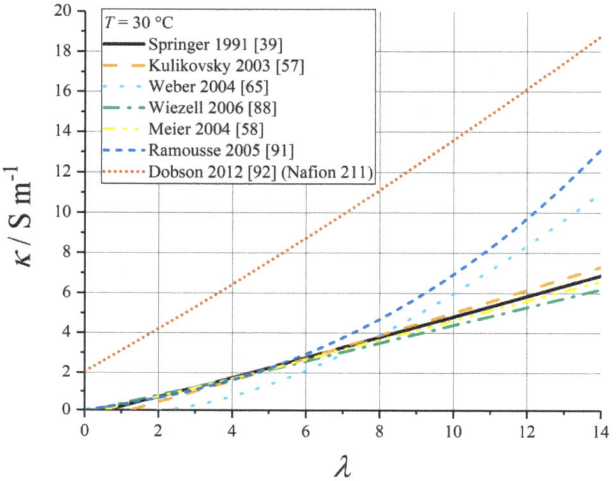

Figure 4. Summary of measured proton conductivity for Nafion 1100 in the vapour-equilibrated range of λ at room temperature (T = 30 °C).

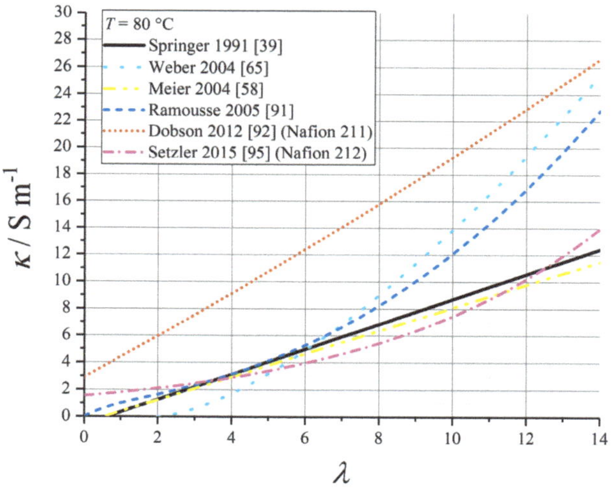

Figure 5. Summary of measured proton conductivity for Nafion 1100 in the vapour-equilibrated range of λ at typical PEMFC operating temperature (T = 80 °C).

The functional dependence of conductivity with respect to water content has been expressed in terms of a polynomial relationship, where there exists some minimum water content λ_0 required for proton conductivity, based on a percolation theory in which there is a need to form connected hydrated channels through which protons can migrate [42,86]. Additionally, an activation energy is included for non-isothermal processes.

$$\begin{aligned} \kappa &= 0, \quad \lambda < \lambda_0 \\ &= \kappa_0(\lambda - \lambda_0)^{n_{\text{cond}}} \exp\left(\theta_{\text{cond}}\left(\tfrac{1}{T_0} - \tfrac{1}{T}\right)\right), \quad \lambda \geq \lambda_0 \end{aligned} \tag{60}$$

For Nafion 1100, the average value for the exponent n_{cond} through a range of studies [1,59] is $\approx$ 1.5, but some empirical models have applied other data values, as summarised in Table 5. In particular, early studies on proton conductivity as a function of water content supported an approximately linear relation [39,52]. Relations for which θ_{cond} is undefined apply only at $T = T_0$.

Table 5. Summary of polynomial proton conductivity relations, with $T_0 = 303.15$ K.

Data Source	κ_0/S m^{-1}	λ_0	θ_{cond}/K	n_{cond}
Springer [39]	0.5139	0.6344	1268	1
van Bussel–Kulikovsky [57,87]	0.5736	1.253	undefined	1
Weber [65]	0.2646	2	1800	1.5
Wiezell [88], fit to Zawodzinski [52]	0.45	0.222	undefined	1
Meier [58], fit to Zawodzinski [79]	0.491 *	0.543	1190	1

* Original value is 0.46 at $T_0 = 298.15$ K.

The lower activation energy ($\theta_{\text{cond}} \approx 1300$ K) was supported by the experimental measurements by Karpenko-Jereb et al. [59]. Moreover, this work sets the dependence strictly in terms of volume fraction rather than water content:

$$\begin{aligned} \kappa &= 0, \quad \phi_{\text{w}} < \phi_{\text{w},0} \\ &= \kappa_{0,\phi}\left(\phi_{\text{w}} - \phi_{\text{w},0}\right)^{n_{\text{cond}}} \exp\left(\theta_{\text{cond}}\left(\tfrac{1}{T_0} - \tfrac{1}{T}\right)\right), \quad \phi_{\text{w}} \geq \phi_{\text{w},0} \end{aligned} \tag{61}$$

This gives approximate equivalence to the tabulated values above when the denominator in (13) is assumed to be constant.

Ju et al. applied the Springer data to a GORE-SELECT membrane with a scaling factor 0.5 to account for tortuosity due to the reinforcement [89].

Ramousse et al. repeated a higher-order polynomial fit due to Neubrand [90,91]:

$$\kappa/\text{Sm}^{-1} = \left(0.2658\lambda + 0.0298\lambda^2 + 0.0013\lambda^3\right)\exp\left(\theta_{\text{cond}}\left(\frac{1}{T_0} - \frac{1}{T}\right)\right) \tag{62}$$

with a water content-dependent activation energy:

$$\theta_{\text{cond}}/\text{K} = 2640\exp(-0.6\lambda) + 1183 \tag{63}$$

Dobson et al. quoted a polynomial expression for Nafion 211 [92]:

$$\kappa/\text{Sm}^{-1} = \left(2.0634 + 1.052\lambda + 0.010125\lambda^2\right)\exp\left(\theta_{\text{cond}}\left(\frac{1}{T_0} - \frac{1}{T}\right)\right) \tag{64}$$

with $\theta_{\text{cond}} = 751.4$ K.

Using the wide data set for proton conductivity compiled by Sone et al. [93], Baschuk and Li tabulated third-order polynomial fit data for the water-content dependence of conductivity, at a range of temperatures from $T = 20$ °C to $T = 70$ °C [94].

Setzler and Fuller used impedance measurements under varying relative humidity conditions to produce the following empirical relation for Nafion 212 at 80 °C [95]:

$$\kappa = \kappa_0 \exp\left(\alpha_\lambda \frac{\lambda}{\lambda_{\text{crit}}}\right) \tag{65}$$

with $\kappa_0 = 1.55$ S m^{-1} and $\alpha_\lambda = 2.2$.

Weber and Newman defined the conductivity under LE and VE conditions (indexed as m = L or m = V below) as depending upon the local volume fraction of water present under the given equilibration condition, without any specification of how the empirical expressions were derived [65]:

$$\kappa_m/\text{Sm}^{-1} = \begin{cases} 10^{-9} & \phi_{\text{w},m} < \phi_{\text{w,crit}} \\ \kappa_0\left(\phi_{\text{w},m} - \phi_{\text{w,crit}}\right)^{\frac{3}{2}} \exp\left(\frac{E_{\text{A,cond}}}{R}\left(\frac{1}{T_0} - \frac{1}{T}\right)\right) & \phi_{\text{w,crit}} \leq \phi_{\text{w},m} \leq \phi_{\text{w,max}} \\ \kappa_0\left(\phi_{\text{w,max}} - \phi_{\text{w,crit}}\right)^{\frac{3}{2}} \exp\left(\frac{E_{\text{A,cond}}}{R}\left(\frac{1}{T_0} - \frac{1}{T}\right)\right) & \phi_{\text{w},m} > \phi_{\text{w,max}} \end{cases} \tag{66}$$

with $\kappa_0 = 50$ S m^{-1} at $T_0 = 298.15$ K, $E_{\text{A,cond}} = +15$ kJ mol^{-1}, and $\varphi_{\text{w,crit}} = 0.06$, $\varphi_{\text{w,max}} = 0.45$.

Kosakian et al. have presented a specific formulation for proton conductivity in the CL ionomer (below) [75]. It should be noted that this expression has no stated correlation to porosity or tortuosity properties of the ionomer in the CL composite, so the absolute magnitude of the conductivity it reports cannot be applied generally to any CL; however, the functional form of the water content-dependence might be considered more widely applicable.

$$\kappa = \left(\sum_{i=0}^{3} a_i \omega^i\right) \exp\left(\frac{E_{\text{A,cond}}}{R}\left(\frac{1}{T_0} - \frac{1}{T}\right)\right) \tag{67}$$

$$\omega = \begin{cases} 100\left(\left(\sum\limits_{i=0}^{3} b_i \lambda^i\right)\right) & 0 < \lambda < 13 \\ 100 & \lambda \geq 13 \end{cases} \tag{68}$$

with $E_{\text{A,cond}} = +15$ kJ mol^{-1} (polynomial data tabulated in Table 6).

Table 6. Parameterisation of the catalyst layer proton conductivity model given by Kosakian et al., $T = 80$ °C [75].

i	a_i	b_i
0	−0.8	−0.1254
1	0.075	0.1832
2	-6.375×10^{-4}	-8.65×10^{-3}
3	1.93×10^{-5}	9.4×10^{-5}

5.5. Water Diffusivity as a Function of Water Content

Implementations of the Springer model typically reference the water diffusivity measured by Zawodzinski et al. using nuclear magnetic resonance (NMR) methods with respect to chemical potential gradients [79], and then solve the sorption isotherm implicitly to convert to a diffusion coefficient with respect to water content. This Fickian diffusion approach has been criticised by Janssen in super-saturated (LE) conditions, because it depends on an extrapolation of the derivative of the sorption isotherm that is defined inexactly in the limit of super-saturation [96]. Eikerling et al.

reported poor correlation of the Springer diffusion data with experimental measurements [97]; however, this appears to have been a comparison made in conjunction with proton conductivity data not matching those used in the Springer work.

The corresponding diffusion coefficient shows a characteristic peak in the range λ = 3 to 4. The original fit used by Springer et al. was reported incompletely [39] and is now deprecated, since it was later refined by Motupally et al. [83]:

$$D_\lambda/\mathrm{m^2s^{-1}} = 3.1\times10^{-7}(\exp(0.28\lambda)-1)\exp\left(-\frac{\theta_{\mathrm{diff}}}{T}\right),\quad \lambda<3 \tag{69}$$

$$D_\lambda/\mathrm{m^2s^{-1}} = 4.17\times10^{-8}(1+161\exp(-\lambda))\exp\left(-\frac{\theta_{\mathrm{diff}}}{T}\right),\quad \lambda\geq3 \tag{70}$$

where θ_{diff} = 2436 K. The Motupally diffusivity model expressed by (69) and (70) was supported by water flux measurements by St-Pierre [98], and has been used in a number of subsequent modelling works.

The data recorded by Okada et al. suggested a constant value of $D_\mathrm{w} = 5\times10^{-10}\ \mathrm{m^2\ s^{-1}}$ [99] which has been used subsequently as $D_\lambda = 3\times10^{-10}\ \mathrm{m^2\ s^{-1}}$ [91]. In a recent work, Kosakian et al. argued from empirical evidence that their own data could be fit accurately by multiplying the Motupally terms by a multiple of 3.2 [75].

The following linear model was measured by Fuller and Newman [100]:

$$D_\lambda = D_{0,\lambda}\lambda\exp\left(-\frac{\theta_{\mathrm{diff}}}{T}\right) \tag{71}$$

with $D_{0,\lambda} = 2.1\times10^{-7}\ \mathrm{m^2\ s^{-1}}$ (as converted by Motupally et al. [83]) and θ_{diff} = 2436 K. Karpenko-Jereb et al. reported a corresponding value $D_{0,\lambda} = 7.84\times10^{-8}\ \mathrm{m^2\ s^{-1}}$ with θ_{diff} = 2383 K [59].

Alternative fits to the data of Springer et al. and Zawodzinski et al. [39,79] have been reported by subsequent authors. For example, Nguyen and White were guided by experimental observations in a PEMFC configuration to scale water diffusivity by the electroosmotic drag coefficient ξ according to the relation [101]:

$$D_\mathrm{w} = D_{0,\mathrm{w}}\xi\exp\left(-\frac{\theta_{\mathrm{diff}}}{T}\right) \tag{72}$$

with $D_{0,\mathrm{w}} = 1.6\times10^{-7}\ \mathrm{m^2\ s^{-1}}$ and θ_{diff} = 2416 K.

Mazumder reported an alternative fit as follows [102]:

$$D_\lambda/\mathrm{m^2s^{-1}} = 2.9\times10^{-7}f(\lambda)\exp\left(-\frac{\theta_{\mathrm{diff}}}{T}\right) \tag{73}$$

$$\begin{aligned} f(\lambda) &= 1,\quad \lambda\leq2\\ f(\lambda) &= 1+2(\lambda-2),\quad 2<\lambda\leq3\\ f(\lambda) &= 3-1.38(\lambda-3),\quad 3<\lambda\leq4\\ f(\lambda) &= 2.563-0.33\lambda+0.0264\lambda^2-0.000671\lambda^3,\quad \lambda\geq4 \end{aligned} \tag{74}$$

Gurau et al. applied the Mazumder polynomial fit (for $\lambda \geq 4$) across the full water content range [49]. This water content dependence has also been applied to models of GORE-SELECT membranes, but with an altered pre-factor to give approximately half the diffusivity of pure Nafion 1100, due to tortuosity from the reinforcement [103].

Data recorded by van Bussel et al. [87] have been used by models developed by Kulikovsky [57] as well as Wu et al. [104]. These data lack any variation of membrane thickness and so may be perturbed by interfacial phenomena: they also lack the characteristic peak at low λ given by (69) and (70). The following fit is given by Kulikovsky, at T = 80 °C [57]:

$$D_\mathrm{w}/\mathrm{m^2s^{-1}} = 4.1\times10^{-10}\left(\frac{\lambda}{25}\right)^{0.15}\left(1+\tanh\left(\frac{\lambda-2.5}{1.4}\right)\right) \tag{75}$$

Figure 6 plots various correlations for diffusion coefficient at operating temperature, assuming in the case of (75) that $D_w = D_\lambda$ (that is, ignoring swelling corrections according to (43)). It should be noted that the various reported equations do not give close agreement, and differ by over an order of magnitude in the limit of high water content. One possible reason for this could be the unreliable extrapolation of data measured close to room temperature to much higher temperatures, but the extent of inconsistency merits further controlled measurements on contemporary state-of-the-art materials.

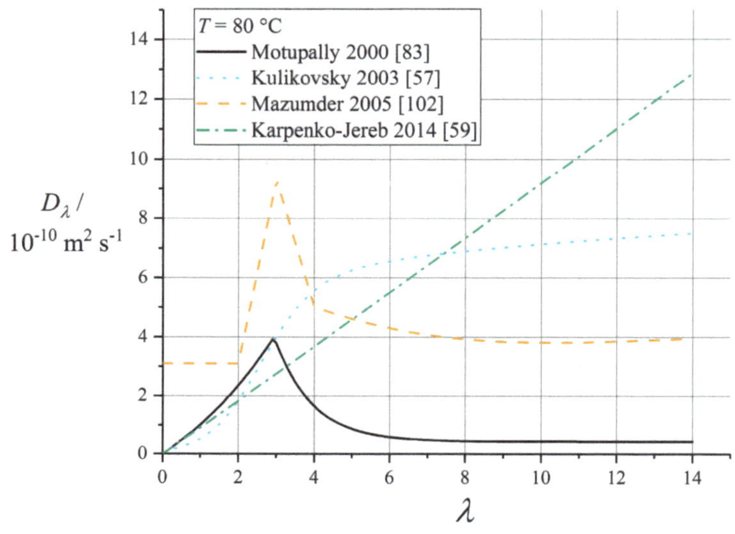

Figure 6. Summary of Fick's law diffusivities of water in Nafion 1100, in the vapour-equilibrated range of λ at T = 80 °C.

Weber and Newman parameterised the self-diffusion coefficient for Nafion 1100 as [65]:

$$D_{\mu,\mathrm{WN}} = D_{\mu,0}\phi_\mathrm{w}\exp\left(-\theta_\mathrm{diff}\left(\frac{1}{T_0}-\frac{1}{T}\right)\right) \tag{76}$$

with $D_{\mu,0}$ = 1.8 × 10^{-9} m^2 s^{-1}, T_0 = 303.15 K and θ_diff = 2400 K. For use in the CL specifically, Kosakian et al. gave correspondingly [75]:

$$D_\mu = D_{\mu,0}\phi_\mathrm{w}\exp\left(-\theta_\mathrm{diff}\left(\frac{1}{T_0}-\frac{1}{T}\right)\right) \tag{77}$$

with $D_{\mu,0}$ = 5.44 × 10^{-9} m^2 s^{-1} and other parameters the same. Here, the substantial difference in magnitude of the pre-factor presumably arises from the different definitions of self-diffusion coefficient in each case (see above, Section 5.2).

5.6. Electroosmotic Drag Coefficient

Electroosmotic drag is known to increase with higher water content (liquid-equilibrated membrane), and with temperature under LE conditions; measurements under different conditions or with different experimental methods have yielded significant disparity [105]. The need for careful control of hydration conditions has been emphasised [106]; however, due to interfacial water transport resistances (see Section 6.2 below), electroosmotic drag itself may induce a water content gradient, even in configurations where the humidity on each face of the membrane is rigorously controlled.

Therefore, care is always needed in interpreting experimental data. Also, the literature is often unclear due to the occasional use of "electroosmotic drag coefficient" to indicate the phenomenological property of net number of water molecules transferred per proton transferred, K_{drag}:

$$K_{drag} \equiv F\frac{|\mathbf{N}_w|}{|\mathbf{i}|} \tag{78}$$

This quantity should not be confused with the intrinsic transport property ξ appearing in the water transport Equations (42) and (47).

Some of the parameterisations reported in this subsection for the electroosmotic drag coefficient in Nafion 1100 are presented in Figure 7.

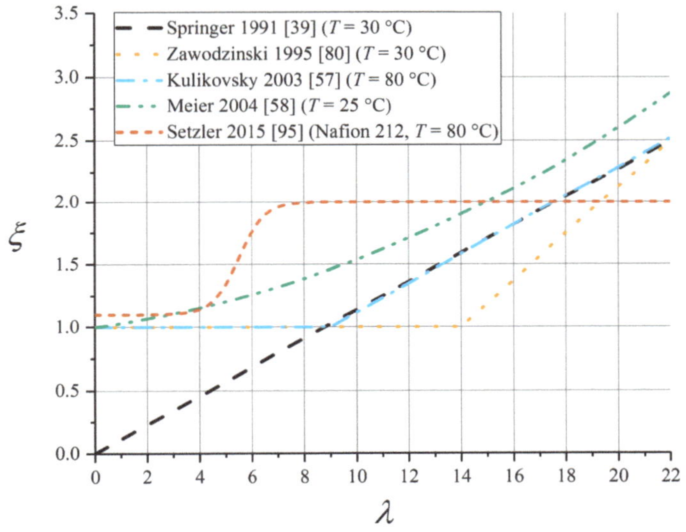

Figure 7. Summary of common parameterisations of electroosmotic drag coefficient for Nafion 1100.

The original parameterisation of the Springer model used a linear relation in λ, tending to zero for zero water content [39]:

$$\xi = a_{\xi,\lambda}\frac{\lambda}{\lambda_{sat}} \tag{79}$$

with $a_{\xi,\lambda} = 2.5$, $\lambda_{sat} = 22$ for Nafion 1100. This was supported by subsequent data from Okada et al. [99]. However, several later studies have presented evidence that the electroosmotic drag coefficient is constant and very close to unity up to a critical water content associated with the liquid-equilibration phase transition, whereafter it rises linearly (data tabulated in Table 7) [80,87,103,107]:

$$\begin{aligned} &\xi = 1, \quad \lambda < \lambda_{crit} \\ &\xi = 1 + \alpha_{\xi,\lambda}(\lambda - \lambda_{crit}), \quad \lambda \geq \lambda_{crit} \end{aligned} \tag{80}$$

Table 7. Parameterisation of linear expressions for the electroosmotic drag coefficient in the liquid-equilibrated regime.

Data Source	λ_{crit}	$\alpha_{\xi,\lambda}$
Zawodzinski et al. [80] (room temperature)	14	0.1875
van Bussel–Kulikovsky (T = 80 °C) [57,87]	9	0.117

Nonetheless, many modelling studies have continued to use the original linear relation given by Equation (79).

In spite of evidence from the aforementioned studies that a liquid-equilibrated state elevates the electroosmotic drag coefficient, some simulations assumed $\xi = 1$ uniformly [57,108]. Quoted values of ξ_L for the liquid-equilibrated membrane vary across a wide range from 2 to 5 [105]. Zawodzinski et al. measured $\xi_V = 1$ and $\xi_L = 2.5$ at $T = 30$ °C [80], which has been subsequently applied in the Weber–Newman model implementation [65]. A polynomial fit spanning both VE and LE conditions was used by Meier and Eigenberger [58]:

$$\xi = 1 + 0.028\lambda + 0.0026\lambda^2 \tag{81}$$

A non-linear correlation has recently been presented for Nafion 212 (a related material in the Nafion 1100 family) [95]:

$$\xi = 1.1 + \frac{0.9}{1 + \exp(-2(\lambda - 5.5))} \tag{82}$$

The experimental study by Ge et al. gave a detailed polynomial fit for electroosmotic drag coefficient as a function of water content [109]. The data were derived from experiment using a model assuming a given explicit diffusivity (measured separately [110]) that lacked a detailed water content dependence, together with an interfacial resistance. Since any imprecision in the chosen diffusivity expression would carry forwards into the water content dependence of the electroosmotic drag coefficient, these data seem uncertain.

Data from both Ge et al. [109] (water flux measurements) and Ise et al. [111] (electrophoretic NMR) suggested a linear temperature dependence of electroosmotic drag coefficient in the fully liquid-equilibrated limit, and negligible temperature dependence otherwise (coefficient data tabulated in Table 8):

$$\xi_{\lambda \to \lambda_{max}} = \xi_0 + \alpha_{\xi,T}(T - T_0) \tag{83}$$

Table 8. Parameterisation of temperature dependence of the electroosmotic drag coefficient for Nafion 1100, with $T_0 = 303.15$ K.

Data Source	ξ_0	$\alpha_{\xi,T}$/K^{-1}
Ge et al. [109]	1.984	0.0126
Ise et al. [111]	1.812	0.014

These temperature-dependent expressions for electroosmotic drag coefficient are plotted in Figure 8.

Benziger et al. presented data from a hydrogen–hydrogen cell showing that even when interfacial or bulk transport limits the current density under high overpotential (e.g., due to limited hydrogen availability for reaction), water flow can continue to grow with greater applied voltage [112]. The implications of these results are not yet clear, especially as might apply to a PEMFC context rather than a hydrogen–hydrogen cell. The original work does not justify them in detail theoretically, but these observations do merit further investigation by comparison to non-equilibrium thermodynamic theories such as the Weber–Newman model.

We note in this context that the interpretation of water transport presented by Berning et al. [113] is incorrect: the authors mistakenly argue that there is no contribution from electroosmotic drag to water transport in the bulk of the membrane, on the basis that a constant electroosmotic drag coefficient allows an equation for water transport to be written that contains only the diffusional term. Even in this case, electroosmotic drag will still contribute to net water flux, and hence to the rate of sorption/desorption in the catalyst layers; thus electroosmotic drag retains a role in determining the overall water content profile and it is not correct to argue that diffusion is the only active water transport process in the bulk membrane.

Recently, Berg and Stornes [114] combined a random pore network model with the detailed swelling model of Eikerling and Berg [44] to predict a variety of apparent experimental water flux to proton flux ratios, and suggested that a 'consistent' thermodynamic model following Dreyer et al. [115] predicts that this experimental ratio of water flux to proton flux should tend to zero in the limit of a membrane consisting of sub-nanopores of negligible size. As far as we are aware, no continuum model has been developed to test the origin of this prediction.

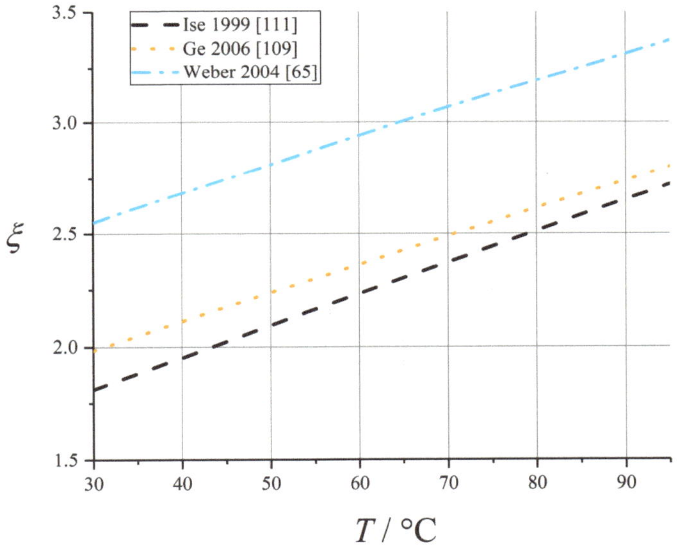

Figure 8. Temperature-dependent parameterisation of the liquid-equilibrated electroosmotic drag coefficient for Nafion 1100.

5.7. Water Transport: Liquid-Equilibrated Conditions

Given the multiphase nature of the membrane, there has been historical disagreement over whether the definition of water content from (7) is relevant or sufficient for describing the process of diffusion down a chemical potential gradient for dissolved water. Some models account also for hydraulic flow of liquid water under a pressure gradient, in addition to diffusion down the water concentration gradient; these are discussed further below under Section 7.1, "Hydraulic Transport of Water (Flow)". One early example is the Bernardi–Verbrugge model in which Schlögl's equation is used to express the water flux as the sum of an electroosmotic term and a hydraulic term, with no diffusional contribution [116]; a later assessment of this work recognises that the water transport model "formulated to simulate the cathode and its gas diffusion layer" was extended to the membrane model in identical form simply "for the sake of integrity" [11], in spite of the different governing water transport phenomena in the different PEMFC regions. Weber and Newman [65] as well as Wu and

Berg [104] argue that models implicating a pressure gradient become relevant only in the LE mode where free liquid water is present in the membrane.

Weber and Newman assumed that the LE conductivity becomes constant at its maximum VE value; that is:

$$\kappa_{\mathrm{L}} = \kappa_{\mathrm{V},a_{\mathrm{w}}=1} \tag{84}$$

Instead of (83), Weber and Newman suggested an Arrhenius relation for the electroosmotic drag coefficient of the liquid-equilibrated membrane [65]:

$$\xi_{\mathrm{L}} = 2.55 \exp\left(\theta_{\mathrm{EOD}}\left(\frac{1}{T_0} - \frac{1}{T}\right)\right) \tag{85}$$

with $T_0 = 303.15$ K and $\theta_{\mathrm{EOD}} = 481.1$ K. This relation is plotted in Figure 8.

6. Interface-Specific Phenomena

6.1. Interfacial Proton Transport Resistance

Interfacial proton transport resistance between the CL and membrane is normally considered to be negligible. Pivovar and Kim reported an experimental measurement of the ionic resistance associated with the CL–membrane contact, for a directly painted Nafion-based membrane-electrode assembly (MEA), as being at least 8 mΩ cm^2 [117]; however, results from commercially manufactured MEAs were not reported and it has subsequently been suggested by the same authors that commercial MEAs are likely free from such resistances, which were attributed to poorly correlated swelling magnitudes under water uptake between the CL and bulk membrane [118]. Interfacial proton transport and electrical contact resistances may also increase at low water content due to a reduced membrane thickness causing mechanical decohesion of the membrane from the electrode.

The presence of an interfacial proton transport resistance even in the presence of liquid equilibration has been identified by Tsampas et al., who raise the possibility that the measured ohmic resistance of the membrane is, in fact, dominated by an interfacial membrane resistance for proton transfer across the liquid water-membrane boundary [119]. It is not clear that this study has been considered in any subsequent continuum modelling work, or whether its conclusions still apply to a membrane-impregnated CL, where other resistances may be appreciable. Evidence for significant interfacial contributions to membrane conductivity measurements has also been presented by Rangel-Cárdenas and Koper [120].

6.2. Interfacial Water Transport Resistance: Vapour-Equilibrated Conditions

In principle, there may be a kinetic barrier to equilibration of the membrane with adjacent phases (vapour or liquid) which would manifest itself as an interfacial resistance to water transport at the surfaces of the membrane. Interfacial transport resistances are typically measured by performing water transport measurements at a range of membrane thicknesses and then extrapolating the measured resistance to zero thickness: the intercept gives the interfacial resistance while the gradient gives intrinsic properties [121,122].

Many studies of PEMFCs have ignored interfacial transport resistance and instead assumed that the membrane surfaces equilibrate instantaneously according to the sorption isotherm [39,57,82,94,96,108,123–125]. This equilibrium assumption enables a numerically advantageous mathematical transformation that simplifies the overall water conservation equation by using an effective (fictitious) equivalent gas-phase concentration (as predicted from the sorption isotherm) to represent water content throughout the membrane. Within an equilibrium model, Janssen proposed that under super-saturated conditions in the membrane, both a vapour phase and a liquid phase must also be present for bulk water [96].

Also, in its original formulation, the Weber–Newman model defined the equilibrium with the vapour environment (with water activity $a_{\mathrm{w,vap}}$) as [65]:

$$\mu_{\mathrm{w}} - \mu_{0,\mathrm{w}} = RT \ln a_{\mathrm{w}} = RT \ln a_{\mathrm{w,vap}} \tag{86}$$

where the second equality implies that interfacial resistance is ignored. For the case of membranes where one face is vapour-equilibrated but the other is liquid-equilibrated, interfacial effects under a humidity gradient have been argued to dominate pressure gradient effects in other works [122,126,127]. This case implicitly includes alcohol-fuelled devices, also.

Some experimental studies have suggested that relaxation of a membrane to a true equilibrium state when in equilibrium with vapour may take days to weeks, which would not be compatible with experimental timescales: the apparent equilibria maintained in PEMFC experiments might be more properly understood as quasi-equilibria, whose properties may nonetheless be determined [1]. Magnetic resonance imaging (MRI) measurements by Teranishi et al. showed a cell water content response towards equilibrium occurring on the order of 100 s after cell startup, which while much shorter than the above is still a significant duration compared to typical timescales of interest for transient PEMFC phenomena [69,128]. This is comparable to membrane and CL water uptake timescales indicated by electrochemical measurements [52,69,129] as well as to membrane sorption equilibration times [130]. Overall membrane resistance and current density were shown to respond to a potential step on the order of 1 h by Cheah et al. [121]; it is worth noting, however, that this work described an MEA based on Nafion 115, whose dimensions and total water content (by mass per unit electrode area) greatly exceed those in more modern, thinner membranes, for which equilibration following galvanostatic steps or rapid changes in inlet humidity is empirically expected to be significantly more rapid ($\leq$1 min, see e.g., [131] with some variation reported between wetting and drying).

A general approach used to account for interfacial resistance to water transport is the inclusion of a linear mass transfer coefficient relation for the water flux at the vapour-membrane boundary:

$$\mathbf{N}_{\mathrm{w}} \cdot \mathbf{n} = k_{\mathrm{int}}\left(a_{\mathrm{w}} - a_{\mathrm{w,vap}}\right) \tag{87}$$

where $\mathbf{n}$ is the outward normal unit vector from the membrane towards the adjacent vapour phase, and k_{int} is an interfacial mass transfer coefficient (mol m^{-2} s^{-1}). This expression was used for dehydration of the super-saturated membrane by Futerko et al., while assuming that condensation under the vapour or liquid phase maintained equilibrium [60]. In an experimental measurement at zero current, Monroe et al. measured $k_{\mathrm{v}} = k_{\mathrm{int}}(RT/p_{\mathrm{sat}}) \approx 6.3 \times 10^{-3}$ m s^{-1} at T = 50 °C for the VE membrane [132]. Klika has suggested $k_{\mathrm{int}} = 4.4 \times 10^{-3}$ mol m^{-2} s^{-1}, which seems compatible in order of magnitude.

Alternatively, some treatments have required that the water uptake derives directly from faradaic proton current density $\mathbf{i}_{\mathrm{far}}$ across the boundary, without any chemical equilibration [101]:

$$\mathbf{N}_{\mathrm{w}} \cdot \mathbf{n} = \frac{\alpha}{F}\mathbf{i}_{\mathrm{far}} \cdot \mathbf{n} \tag{88}$$

where α is a number of water molecules transferred per proton generated through the reaction, which is not necessarily equal to ξ.

Okada et al. considered the direct sum of an active uptake (electroosmotic) and a passive uptake (independent of current density), such that [99]:

$$\mathbf{N}_{\mathrm{w}} \cdot \mathbf{n} = \frac{\xi}{F}\mathbf{i}_{\mathrm{far}} \cdot \mathbf{n} + k_{\mathrm{int}}\left(a_{\mathrm{w}} - a_{\mathrm{w,vap}}\right) \tag{89}$$

The presence or absence of the active uptake term, given directly from the electroosmotic drag coefficient, has not generally been agreed upon in the literature, and merits further study. One prior study has suggested loosely but without clear justification that the electroosmotic drag term should be absent at the anode but present at the cathode [133]; for a case in which the electroosmotic drag coefficient is *excluded* from the interfacial sorption rate as specified in (89) in both CLs, cathode drying has been predicted under conditions of a low volumetric interfacial area in the CL, leading to a low effective value of k_{int} [134].

A similar model due to Berg et al. defined an uptake model for the VE case as linear in the difference of water contents, rather than activities [33,135]:

$$\mathbf{N}_{\text{w}} \cdot \mathbf{n} = \frac{\xi}{F}\mathbf{i}_{\text{far}} \cdot \mathbf{n} + k_{\text{int},\lambda} c_{\text{f}} \big(\lambda - \lambda_{\text{eq}}\big) \tag{90}$$

The best experimental fit by Berg et al. gave $k_{\text{int},\lambda} \approx 5 \times 10^{-6}$ m s^{-1} [33]. Ge et al. offered measurements as a function of water content and discovered an asymmetry between absorption and desorption rates, while suggesting that $k_{\text{int},\lambda} \approx 10^{-5}$ m s^{-1} to the closest order of magnitude and that both absorption and desorption rate can be treated as linearly dependent upon water volume fraction (determined from (9)) [110]. The temperature-dependence of $k_{\text{int},\lambda}$ can be expressed by an Arrhenius relation with activation energy in the range 25–31 kJ mol^{-1} [136].

Alternatively, a volumetric rate constant can be expressed as:

$$k_{\text{vol},\lambda} = k_{\text{int},\lambda} a_{\text{vol}} \tag{91}$$

where a_{vol} (m^{-1}) is the specific surface area of the membrane-vapour contact in the CL. Reported values include: $k_{\text{vol},\lambda} = 1.3$ s^{-1} [137]; $k_{\text{vol},\lambda} = 5.7$ s^{-1} [135]; $k_{\text{vol},\lambda} = 1$ s^{-1} [36]; 50 s^{-1} $\leq k_{\text{vol},\lambda} \leq$ 80 s^{-1} [35]. Kosakian et al. presented fitted data at T = 30 °C with $k_{\text{vol},\lambda} \approx 0.1$ s^{-1} for sorption and ≈ 1 s^{-1} for desorption; this work also took the rate as proportional to the volume fraction of the vapour phase, and assumed an activated process with $E_{\text{A,sorp}}$ = +20 kJ mol^{-1} [75]. Vorobev et al. performed a phenomenological study of the role of this rate constant within a CL model [138].

Klika et al. have clearly demonstrated [63] that absorption-desorption asymmetries [110,136] in fits to (90) can be explained by understanding that the true thermodynamic driving force is the activity difference as expressed in (89); the nonlinearity of the sorption isotherm then explains the resulting asymmetry when expressed in terms of water content or concentration difference. This criticism, as supported by experimental evidence, suggests that (89) should be preferred [110,132].

Kienitz et al. (originally published under the lead author name "Kientiz [*sic*] et al.") explored the idea that interfacial resistance is itself a humidity-dependent quantity, with an investigation on Nafion 21x membranes, and attributed the liquid–vapour interface at a hydrophobic surface to interfacial resistances [122]. To account for this, they proposed the following expression:

$$k_{\text{int}} / \text{ mol m}^{-2} \text{ s}^{-1} = 1.04 \times 10^{-3} \exp^{4.48\times10^{-4} a_{\text{w}}} \tag{92}$$

6.3. Interfacial Water Transport Resistance: Liquid-Equilibrated Conditions

Experiments suggest that LE conditions cause negligible water transport resistance at the liquid contact to the membrane [1,122,132]. The Weber–Newman model argues that there exist two membrane morphologies depending on whether or not liquid water is present [42]. In the absence of liquid water, linked clusters are connected by 'collapsed channels', accounting for electroosmotic drag at $\xi = 1$ due to the transport of H_3O^+ as the active proton carrier ion. In the presence of liquid water, channels open so that water can flow under a pressure gradient. For the intermediate regime, Weber and Newman defined S_{L} as a fraction of expanded channels and then used this quantity to take a linear average between the predictions of the VE and LE transport models, where the VE and LE cases have different transport properties: the consequent doubling of the number of required inputs has been identified as a limitation to the use of this model by the review of Jiao and Li [10]. The corresponding transport equations are (subscripts V and L denoting VE and LE membrane proportion, respectively):

$$\begin{aligned} \mathbf{i} &= (1 - S_L)\mathbf{i}_V + S_L\mathbf{i}_L \\ \mathbf{N}_w &= (1 - S_L)\mathbf{N}_{w,V} + S_L\mathbf{N}_{w,L} \end{aligned} \tag{93}$$

where the fluxes under each case are given by (46) and (47), with transport coefficients taking different definitions for VE or LE membrane conditions.

For the liquid-equilibrated membrane, the water content is defined directly as:

$$\lambda = \lambda_{max,V} + S_L(\lambda_{max,L} - \lambda_{max,V}) \tag{94}$$

The fraction of expanded channels is evaluated by assuming a specific pore-size distribution of hydrophobic channels in the membrane, such that the proportion of expanded channels can be expressed from the hydraulic pressure. An integration is performed across a log-normal pore-size distribution with the lower bound given by a pressure-dependent critical radius which, as developed in a later refinement of the model, has hydrophobic properties parameterised by a general energetic parameter $\Gamma = 4 \times 10^{-5}$ N m^{-1} [65]:

$$S_L = \int_{r_{crit}}^{\infty} V(r)dr = \frac{1}{2}\left(1 - \mathrm{erf}\left(\frac{\ln(r_{crit}/\mathrm{nm}) - \ln 1.25}{0.3\sqrt{2}}\right)\right) \tag{95}$$

$$r_{crit} = \frac{\Gamma}{p_L} \tag{96}$$

In the critical radius Equation (96), the liquid water pressure at the membrane-vapour interface (p_L) is then determined by mass conservation: the specific form will depend also upon the GDL model, which was developed by later authors after some simplistic assumptions in early works [139]. The work by Weber and Newman is somewhat vague about how the degree of saturation is defined when a membrane is liquid-equilibrated only in one spatial region: any implementation of the above approach in such a case seems to require additional assumptions [36,140].

Meng et al. implemented the liquid saturation by means of an adapted Springer model [141]:

$$\lambda_{eq} = 14 + 2.8S_L, \quad S_L > 0 \tag{97}$$

noting that the saturation extent here references an equilibrium water content rather than an actual water content. This work argues that once liquid water channels are opened in the membrane, they can accommodate additional free liquid water which is not bonded, while the membrane accommodates a given quantity of water per (97)—puzzlingly, this approach seems to count the additional water present due to liquid-equilibration twice over, because of the presence of the saturation in (97). In turn the liquid saturation evolves according to a capillary diffusion equation:

$$\nabla \cdot \left(D_{cap}\nabla S_L\right) - \nabla \cdot \left(\frac{S_L^3 \kappa_p}{v_w}\nabla p\right) = R_w \tag{98}$$

where D_{cap} is capillary diffusivity (set as a constant = 2×10^{-5} kg m^{-1} s^{-1}) and κ_p is a permeability for the membrane (set = 1.8×10^{-18} m^2 following Bernardi and Verbrugge [142]). The water source R_w is expressed in terms of the balance of condensation of vapour to liquid water and bonding of water to the membrane according to the sorption equilibrium (97):

$$R_w = k_{vap}M_w\left(p_{vap} - p_{sat}\right) - k_{bond}M_w c_f\left(\lambda_{eq} - \lambda\right) \tag{99}$$

with $k_{bond} = 1\ s^{-1}$ [36]. The value for k_{vap} is not clearly specified in the original work but for a typical specific surface area of vapour–liquid contact, the work of Wu et al. implies a value of the order of $7000\ s\ m^{-2}\ kg^{-1}$ [36].

Hwang et al. employed the van Genuchten model for the degree of saturation in partially saturated porous media to describe the membrane as governed by capillary hydrodynamics (detailed formulas are given in the referenced work) [143]. This assumed a van Genuchten parameter $n = 3.56$ for Nafion, without any real justification. At the boundary between porous media properties, this work asserted continuity of liquid and vapour pressure, and hence discontinuity of saturation. Due to the lack of morphological similarity between Nafion and the porous rocks on which the van Genuchten model is based, this approach seems doubtful, and it does not appear to have been followed subsequently.

The works of Wu et al. and Falcão et al. distinguished between the anode as being exclusively vapour-equilibrated, and the cathode side, where the water produced through the cathode reaction (oxygen reduction) is initially membrane-dissolved, as liquid-equilibrated [36,144]. The frequent observation of net cathode-to-anode water transport in the presence of a water-saturated anode fuel has been described as a manifestation of Schröder's paradox due to the liquid equilibration of the cathode [127]. Adachi et al. demonstrated experimentally that allowing liquid equilibration of one face of the membrane, while maintaining vapour equilibration on the other, greatly increased the permeation rate compared to a vapour–vapour membrane, due to the elimination of one interfacial resistance [126].

7. Mechanical Phenomena

7.1. Hydraulic Transport of Water (Flow)

Hydraulic transport of water is dependent upon a continuous liquid water phase within the membrane, which according to the physical model of Weber and Newman only arises under LE conditions [42]. Models that focus on conditions of vapour equilibration, especially those based on the Springer model and its transient extension, may ignore pressure gradients altogether [39,124,145]. The Weber–Newman theory argues with support from Janssen that liquid water content, as distinct from water content dissolved in the membrane, must be modelled explicitly to allow a pressure gradient to drive liquid water flow alongside the chemical potential gradient [9,96]; this point is analogous to the analysis by Kreuer et al. which showed that the liquid water chemical potential gradient must vanish in the presence of connected channels containing bulk liquid [2]. A significant increase in water transport rate under LE conditions has been demonstrated experimentally [130].

The quasi-empirical Schlögl equation used for water transport in the works of Bernardi and Verbrugge assumes the presence of liquid water filling the membrane pores, and sets [116,142]:

$$\frac{\mu M_w}{\rho_w}\mathbf{N}_w = -\kappa_p \nabla p - \kappa_\phi c_f F \nabla \phi \tag{100}$$

where μ is viscosity, κ_p and κ_φ are the hydraulic and electroosmotic permeabilities respectively, and p is pressure. This equation is unsuitable in the VE regime where the assumption of the presence of liquid water is not valid [65,96]. The original data (at T = 80 °C) are: $\mu = 3.56 \times 10^{-4}\ kg\ m^{-1}\ s^{-1}$; $\kappa_p = 1.8 \times 10^{-18}\ m^2$; $(\kappa_\varphi\ c_f) = 8.616 \times 10^{-17}\ mol\ m^{-1}$. Hydraulic permeability was given as $5 \times 10^{-19}\ m^2$ by Nam et al. [146]. Meier and Eigenberger used experimental data to measure the following hydraulic permeability at 25 °C:

$$\kappa_p/m^2 = 10^{-20}\left(0.38 + 0.04\lambda + 0.014\lambda^2\right) \tag{101}$$

The works of the Djilali group have criticised features of the "dusty fluid" model combining hydraulic and diffusive transport, as developed by Thampan et al. [30], as being physically unreasonable for the PEMFC context [77,82]. In particular, this model is criticised for double-accounting flux contributions by imposing Schlögl's equation for viscous phenomena convective velocity on top of a Maxwell–Stefan equation, which already considers all contributions to the velocity of each component species [78]. The similar non-equilibrium thermodynamic model presented by Rama et al. gave independent diffusion and hydraulic pressure gradient terms in the water flux expression [147]; this would contradict the same principles of fundamental multi-component transport theory, and this model has not been taken up by other researchers. Eikerling et al. [97] argued for the predominance of convective transport according to (100), but emphasised that sorption equilibrium meant that the pressure could be expressed directly in terms of water content, and hence the pressure gradient term is not distinguishable from an effective diffusivity expressed in terms of water content, as in the Springer model.

In the Weber–Newman model, the liquid-equilibrated gradient of water chemical potential in (47) is expressed as [65]:

$$\nabla \mu_{\mathrm{w,L}} = \overline{V}_{\mathrm{w}} \nabla p \tag{102}$$

Hence, the LE water flux becomes a hydraulic flow [65]:

$$\mathbf{N}_{\mathrm{w,L}} = -\alpha_{\mathrm{w,L}} \nabla p + \frac{\xi_{\mathrm{L}}}{F} \mathbf{i} \tag{103}$$

The mass transfer coefficient is assumed to take the value:

$$\alpha_{\mathrm{w,L}} = \frac{\kappa_{\mathrm{p}}}{\mu \overline{V}_{\mathrm{w}}^{2}} \left(\frac{\phi_{\mathrm{w}}}{\phi_{\mathrm{w,max}}} \right)^2 \tag{104}$$

In the above, the viscosity μ should be understood to be its bulk value for liquid water, which is a general function of temperature as reported in standard engineering data sources.

7.2. Membrane Expansion and Mechanical Constraint

Membrane materials are known to undergo swelling as their water content increases. In the unconstrained material, the inelastic volumetric expansion can be expressed using (12). In a PEMFC, however, the membrane is constrained through compression, and so is not free to expand; consequently, membrane strain in a PEMFC depends on the overall mechanical properties of the device, and the extent of hydration. It has also been reported that membrane expansion is likely to compress the GDL of the operating PEMFC due to the higher stiffness of the membrane compared to the GDL [13,148]. Investigation of the complex structural interactions between laminated MEA components has indicated that stress–strain measurements on free membranes are unlikely to be representative of in situ behaviour [149].

Experimental measurements of the inelastic expansion strain due to membrane swelling suggest a magnitude of the order of 0.01 per unit λ in unreinforced membranes [1,39,150]; corresponding measurements on reinforced expanded polytetrafluoroethylene (ePTFE) GORE-SELECT membranes gave inelastic expansion strains about 5 times lower [151]. Both experimental and theoretical studies have implicated the load due to membrane swelling in a lowering of the effective water content of the membrane [128,148,152], with an experimental cap on water uptake of $\lambda_{\mathrm{max}} = 6.5$ reported in one case [128].

The formulation of mechanical constitutive relations (e.g., hyperelastic and/or viscoelastic-plastic constitutive models) for hydrated Nafion materials has been a subject of several works but, since such models are seldom if ever combined with practical electrochemical device models, a detailed discussion falls outside the scope of this review; the interested reader is directed to some key publications for further information [150,153–155]. It has also been established that the CL and interfacial properties may have a significant role in overall structural behaviour of the membrane [156].

To assess the influence of mechanical stress on water transport in an operating PEMFC, Weber and Newman proposed that a membrane within a PEMFC MEA can be described by a degree of constraint

χ_c, such that the membrane volume change under hydration is zero with $\chi_c = 1$ and equals its free, unconstrained value with $\chi_c = 0$ [148]:

$$\overline{V}_{\mathrm{mem}} = \overline{V}_{\mathrm{p}} + \lambda \overline{V}_{\mathrm{w}}(1 - \chi_{\mathrm{c}}) \tag{105}$$

Considering that the constraint introduces an associated stress that will contribute to the chemical potential of the water, this work assumed a balance between the chemical potential for water inside and outside the membrane in terms of membrane bulk modulus, and yielded a comparison to the unconstrained case as [148]:

$$\frac{\lambda_{\mathrm{constrained}}}{\lambda_{\mathrm{free}}} = \left(\frac{\overline{V}_{\mathrm{mem}}}{\overline{V}_{\mathrm{p}} + \lambda \overline{V}_{\mathrm{w}}} \right)^{\frac{Y_{\mathrm{mem}} \overline{V}_{\mathrm{w}}}{3RT}} \tag{106}$$

where Y_{mem} is the Young's modulus of Nafion 1100, expressed as:

$$Y_{\mathrm{mem}} = Y_{0,\mathrm{mem}} \frac{T_0}{T} \exp\left(-0.1655\left(12 - 10\left(M_{\mathrm{EW}} / \mathrm{kgmol}^{-1}\right) + \frac{\lambda M_{\mathrm{w}}}{M_{\mathrm{EW}}}\right)\right) \tag{107}$$

where $Y_{0,\mathrm{mem}}$ = 275 MPa and T_0 = 303.15 K. This work then used an iterative simulation procedure, beginning from the unconstrained case, in order to predict the self-consistent value of the constrained water content from (106).

Kusoglu et al. used a Flory–Huggins model for the sorption thermodynamics together with the Mori-Tanaka model for bulk modulus of a two-phase material, in order to provide an overall sorption isotherm under pressure, with experimental corroboration [152]. Both Kusoglu et al. and Klika et al. have argued that the influence of pressure on contact resistances will have a greater practical impact than variation in uptake or transport due to swelling phenomena [63,152]. Hasan et al. incorporated in isotropic swelling model combined with a viscoelastic-plastic mechanical model into an electrochemical analysis: this allowed strain to be predicted as a function of membrane hydration, but no feedback from the mechanical response was considered to the water transport and uptake [31].

8. Transient Response of the Membrane

As noted in Section 6.2 above, sorption and water transport phenomena may not equilibrate rapidly compared to relevant experimental timescales. Additionally, intrinsically dynamic experimental methods such as electrochemical impedance spectroscopy (EIS) are of interest for PEMFC characterisation, and require corresponding simulation development for interpretation of their results [75,88,95,157]. For this reason, some authors have explored transient extensions to the quasistatic membrane models discussed thus far.

In general, interfacial resistances and dynamics of sorption may dominate the membrane response to a perturbation in its surroundings [1]. Therefore, diffusion coefficients reported from mass-uptake experimental methods must be treated with great care [130]. The consistent incorporation of interfacial phenomena and liquid equilibration effects described in Section 6.2 is the best strategy to allow for a correct transient prediction.

The first significant development of a transient continuum membrane model was the extension of the model established by Um et al. [123] to a transient model [124]. This defined a transient water content balance equation for the membrane phase as:

$$c_{\mathrm{f}} \frac{\partial \lambda}{\partial t} + \nabla \cdot \mathbf{N}_{\mathrm{w}} = 0 \tag{108}$$

This equation applies for the pure membrane region where no electrochemical reaction takes place; in the CL, a source term could apply due to the rate of sorption/desorption. Transient volume changes associated with swelling are not considered self-consistently in the form (108).

In general it is common to consider that the relaxation timescales for proton conductivity are significantly more rapid than other transport processes, and so a quasistatic proton conduction model can be used in conjunction with transient transport models for water content and for species transport in other regions of the PEMFC [158]. Thus, the stationary proton conduction Equation (6) is applied alongside (108) in the typical transient formulation of the Springer model [124,159]. Ziegler et al. extended the Weber–Newman model to consider transient systems [160], using the explicit Thampan isotherm (Equation (30)).

9. Non-Isothermal Phenomena

Because of the thin spatial dimension of the membrane, it is standard to assume that heat transfer is dominated by thermal conduction, and maintains a quasi-steady state. The following equations then apply to describe the heat flux $\mathbf{q}$:

$$\nabla \cdot \mathbf{q} = Q \tag{109}$$

$$\mathbf{q} = -k\nabla T \tag{110}$$

Thermal conductivities (k) for Nafion 1100 have been measured variously in the range 0.1–1 W m^{-1} K^{-1}; data have been reported as a function of temperature and water content [12,161]. The net heat source Q arises within the membrane due to resistive heating [84]; for an Ohm's law treatment (Equation (5), as used in the Springer model), the corresponding Joule heat source is:

$$Q = \frac{\mathbf{i}^2}{\kappa} \tag{111}$$

Temperature dependence of the transport coefficients was discussed above in Section 5. The proton conductivity and water diffusivity are generally agreed to obey Arrhenius equations in the range 50 °C < T < 80 °C [1,59,65].

For a more general non-isothermal case, it is necessary to consider thermoosmosis—that is, the transport of water under a temperature gradient. Dai et al. (writing in 2009) suggested that thermoosmotic transport of water in the membrane was not well understood [162], but could contribute appreciably due to internal temperature gradients, even in the presence of good thermal balance between bipolar plates.

A standard equation expresses this additional contribution in terms of a thermoosmotic diffusion coefficient $D_{\mathrm{w},T}$:

$$\mathbf{N}_{\mathrm{w}} = \mathbf{N}_{\mathrm{w,const}T} - D_{\mathrm{w},T}\nabla T \tag{112}$$

Even the direction of this effect is uncertain: the sign of $D_{\mathrm{w},T}$ should depend on the relative hydrophilicity or hydrophobicity of the membrane. Some experimental studies have suggested $D_{\mathrm{w},T} < 0$ [163] while others have suggested $D_{\mathrm{w},T} > 0$ [164]. The former measurements also gave an Arrhenius behaviour for $D_{\mathrm{w},T}$ with an activation energy comparable to that for mass diffusion of water in Nafion 1100 (as Equation (72)); this was incorporated into a recent full cell study as [75]:

$$D_{\mathrm{w},T} = D_{\mathrm{w},T,0} \exp\left(-\frac{\theta_{\mathrm{diff},T}}{T}\right) \tag{113}$$

with $D_{\mathrm{w},T,0} = -1.04 \times 10^{-5}$ kg m^{-1} s^{-1} K^{-1} and $\theta_{\mathrm{diff},T} = 2362$ K.

Within the same experimental studies, it has been argued that thermoosmosis is sufficiently negligible to make it reasonable to approximate that $D_{\mathrm{w},T} = 0$ [163,164]. In this theory, the perceived contribution to mass flux due to temperature gradients in fact arises due to other mechanisms: (a) 'heat piping' due to the difference in water saturation pressure on the two faces of the membrane driving condensation and evaporation at cold and hot faces, respectively; (b) temperature dependence of the sorption isotherm, introducing a diffusive driving force at constant water content. To account for the latter, Fu et al. introduced the following modification to (97) [164]:

$$\lambda_{eq} = 9.2 + (0.18 + 0.138(T - T_0))S_L, \quad 0 < S_L \leq 1 \tag{114}$$

10. Transport of Other Chemical Species

10.1. Dilute Gas Transport

In device models, it is most common to treat the membrane as strictly gas-impermeable. As membranes have become thinner, however, gas crossover has become an increasingly important phenomenon to understand quantitatively. For PEMWEs, gas crossover effectively limits the extent to which the membrane can be thinned; for PEMFCs, it is important when considering the gas composition in the anode recirculation loop, and for investigating radical formation.

Wherever an explicit description of gas crossover is required, the flux of dilute dissolved gas ($\mathbf{N}_{gas}$) across the membrane can be expressed empirically using permeation coefficients (ψ_{gas}) and the difference in partial pressure on the two faces of the membrane, as related to the membrane thickness:

$$\mathbf{N}_{gas} = -\psi_{gas} \nabla p_{gas} \tag{115}$$

In such gas crossover models, it is common to assume an infinitely rapid reaction of H_2 at the cathode and O_2 at the anode, such that the concentrations of the dissolved gases go to zero at the respective boundaries [146].

Values of ψ_{gas} are typically of the order 10^{-15} to 10^{-14} mol m^{-1} s^{-1} Pa^{-1}, with values in pure water closer to 10^{-13} mol m^{-1} s^{-1} Pa^{-1} [1]. For the common PEMFC gases, $\psi_{H2} > \psi_{O2} > \psi_{N2}$. Weber suggested the following approximate relative relations [165]:

$$\begin{aligned} \psi_{O_2} = \psi_{N_2} &= \tfrac{2}{3}\psi_{H_2} \\ \psi_{H_2O} &= \tfrac{8}{9}\psi_{H_2} \end{aligned} \tag{116}$$

This work argued that the threshold for significant performance impact arises at 10^{-13} mol m^{-1} s^{-1} Pa^{-1}, which is appreciably higher than the measured permeation coefficients in wet Nafion [165]. Kundu et al. measured $\psi_{H2} \approx 7.4 \times 10^{-14}$ mol m^{-1} s^{-1} Pa^{-1} in a GORE PRIMEA series 5510 MEA [166].

Zhang et al. measured $\psi_{O2} \approx 1.6 \times 10^{-14}$ mol m^{-1} s^{-1} Pa^{-1} through Nafion 117 at T = 80 °C and gave an activation energy for permeation of 23 kJ mol^{-1} in the range 40 °C < T < 100 °C [167]. Earlier measurements had suggested a comparable activation energy for O_2 permeation of the order 30 kJ mol^{-1} [168]. Various studies have reported Nafion 1100 permeation data additionally for CO_2 and N_2, including the functional relationship with relative humidity and temperature [169,170]. Studies have indicated particularly low permeability for N_2 (<10^{-15} mol m^{-1} s^{-1} Pa^{-1}) both through Nafion 1100 and GORE PRIMEA catalyst-coated membranes [170,171].

Weber and Newman compiled then-available experimental data to give the following expressions for H_2 and O_2 permeability under VE and LE conditions, with water content-dependence in the former case (coefficient data are tabulated in Table 9, and permeation coefficients are plotted against temperature in Figure 9) [65]:

$$\psi_{i,V} = (\psi_{w,i}\phi_w + \psi_{V0,i}) \exp\left(\frac{E_{A,DV,i}}{R}\left(\frac{1}{T_0} - \frac{1}{T}\right)\right) \tag{117}$$

$$\psi_{i,L} = \psi_{L0,i} \exp\left(\frac{E_{A,DL,i}}{R}\left(\frac{1}{T_0} - \frac{1}{T}\right)\right) \tag{118}$$

Table 9. Gas permeability data in Nafion 1100 as summarised by Weber and Newman, T_0 = 303.15 K [65].

Coefficient	H_2	O_2
ψ_{V0}/mol m^{-1} s^{-1} Pa^{-1}	2.9×10^{-15}	1.1×10^{-15}
ψ_{w}/mol m^{-1} s^{-1} Pa^{-1}	2.2×10^{-14}	1.9×10^{-14}
ψ_{L0}/mol m^{-1} s^{-1} Pa^{-1}	1.8×10^{-14}	1.2×10^{-14}
$E_{A,DV}$/kJ mol^{-1}	21	22
$E_{A,DL}$/kJ mol^{-1}	18	20

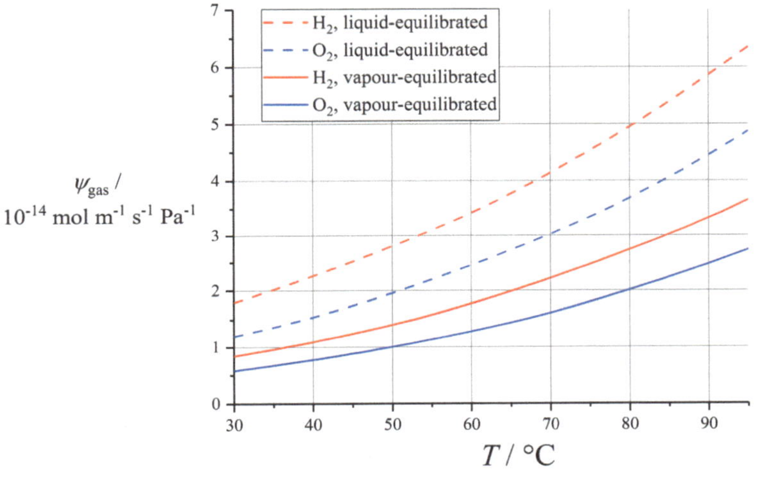

Figure 9. Permeation coefficients for H_2 and O_2 as a function of temperature, using data summarised by Weber and Newman [65].

The permeation coefficient is a convenient measure due to its combination of the solubility of the gas and the diffusivity of dissolved gas in the membrane into a single empirical quantity [9]. Since these effects often have opposing temperature dependences, with solubility falling with temperature while diffusivity rises, permeability is only weakly dependent on temperature, albeit still positively increasing.

Where a diffusion model is required for the dissolved gas, it is common to assume that the dissolved gas concentration (c_{gas}) is low enough that Fick's law can be applied [9]:

$$\mathbf{N}_{gas} = -D_{gas}\nabla c_{gas} \tag{119}$$

Here the diffusion coefficient D_{gas} relates to the permeation coefficient as:

$$D_{gas} = \frac{\psi_{gas}}{K_{H,gas}} \tag{120}$$

where $K_{H,gas}$ is the Henry's law coefficient such that, at equilibrium:

$$c_{soln} = \frac{p_{gas}}{K_{H,gas}} \tag{121}$$

Wong and Kjeang compiled solubilities from a variety of prior sources as follows (coefficient data are tabulated in Table 10 and Henry's law coefficients are plotted against temperature in Figure 10) [137]:

$$K_{\mathrm{H,gas}} = K_{\mathrm{H,gas,0}} \exp\left(-\frac{\theta_{\mathrm{soln,gas}}}{T}\right) \tag{122}$$

Table 10. Henry's law data for various gases in Nafion 1100 as summarised by Wong and Kjeang [137].

	$K_{\mathrm{H,gas,0}}$/Pa m^3 mol^{-1}	$\theta_{\mathrm{soln,gas}}$/K
H_2	2.584×10^3	−170
O_2	1.348×10^5	666
HF	4.149×10^8	7400
H_2O_2	6.83×10^7	7379

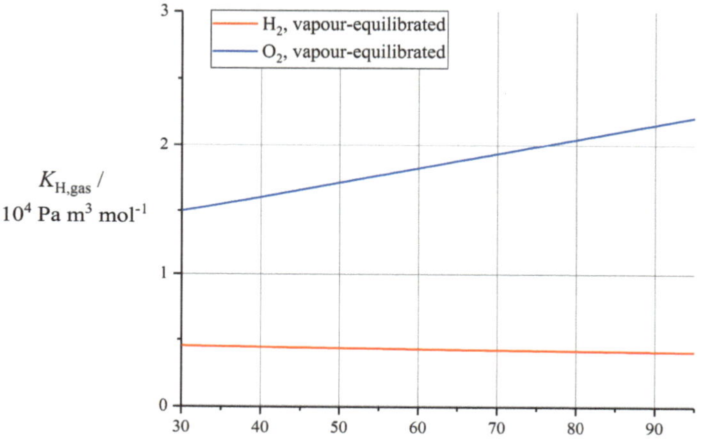

Figure 10. Henry's law coefficients for H_2 and O_2 as a function of temperature, using data summarised by Wong and Kjeang [137].

Bernardi and Verbrugge expressed the gas diffusion coefficients directly, in the following temperature-dependent form (coefficient data are tabulated in Table 11) [142]:

$$D_{\mathrm{gas}} = D_{\mathrm{gas,0}} \exp\left(\frac{-\theta_{\mathrm{diff,gas}}}{T}\right) \tag{123}$$

Table 11. Gas diffusion coefficients in Nafion 1100 as reported by Bernardi and Verbrugge [142].

Gas	$D_{\mathrm{gas,0}}$/m^2 s^{-1}	$\theta_{\mathrm{diff,gas}}$/K
H_2	4.1×10^{-7}	2602
O_2	3.1×10^{-7}	2736

Within a degradation model, Wong et al. gave $D_{\mathrm{H2O2}} = D_{\mathrm{HF}} = 1.5 \times 10^{-10}$ m^2 s^{-1} (at an unspecified temperature) [137]. It has been suggested that N_2 can be treated with identical solubility and diffusion coefficient properties as O_2 [172]. In a study of carbon corrosion, Hu et al. used $D_{\mathrm{O2}} = 10^{-9}$ m^2 s^{-1}, which is much higher than values proposed elsewhere [173].

Rangel-Cárdenas and Koper included H_2 permeation of the membrane in a non-equilibrium thermodynamic model, but without any further application of the equations derived thereby [120].

10.2. Transport of Other Ions

The presence of mobile ions other than protons in the membrane complicates the description of transport phenomena discussed above. Contaminating ions may arise from membrane manufacture, degradation of PEMFC components during operation or reactant impurities. Since if multiple ionic species are present in the membrane, proton transport and current density are no longer equivalent, relation (4) does not hold; furthermore, the addition of other ionic components will mean that binary quantities such as the conductivity and diffusivity values reported above are no longer valid [174].

The simplest approach is to assume that contaminating ionic species are present in very dilute concentration, such that the Nernst–Planck equations can be used for their transport while considering the current density and apparent membrane potential to be dominated by proton transport, so that (5) or (46) still holds. Then, the flux of a dilute ionic species i with charge number z_i is:

$$\mathbf{N}_i = -D_i \nabla c_i - \frac{z_i F}{RT} D_i c_i \nabla \phi \tag{124}$$

Weber and Delacourt extended a concentrated electrolyte solution theory to consider the presence of a single contaminating cation [175]. Burlatsky et al. described Pt^{2+} transport with $D_{Pt} \approx 10^{-10}$ m^2 s^{-1} at unspecified temperature [176]. Fe^{2+} and Fe^{3+} ions as contaminants have been described with diffusivities $D_{Fe2+} \approx 4 \times 10^{-10}$ m^2 s^{-1} and $D_{Fe3+} \approx 4 \times 10^{-11}$ m^2 s^{-1} at $T = 95$ °C [177,178]. The presence of metallic cations in the membrane can accelerate membrane degradation [16] both through the formation of radical species and through the precipitation of solid bands of platinum inside the membrane; chemical degradation models are discussed further below (Section 11.2), but more detailed discussion of the specific contaminated membrane case in the work of Burlatsky et al. [176] exceeds the scope of this review.

11. Membrane Degradation

For nearly all applications, the durability of PEMFCs is critical. For instance the US Department of Energy has fuel cell targets in the automotive sector requiring no more than a 10% loss in rated power over operating times of 8000 h for passenger cars and >25,000 h for heavy-duty vehicles [179]. Performance may be especially impacted by membrane degradation, particularly through reduced proton conductivity and through enhanced hydrogen crossover due to loss of membrane thickness or formation of pinholes. These degradation pathways may also cause other components to degrade more rapidly.

Due to the central role of the membrane in cell degradation, predictive modelling can support industrial development by facilitating prediction of the rate of degradation, and/or correlating metrics of degradation to cell performance in a more fundamental, physical manner. The relevant mechanisms of degradation are complex and varied; a detailed discussion is beyond the scope of this review. The most important phenomena can broadly be classed as either chemical aging of the membrane by radicals, or mechanical aging by repeated dimensional change of the membrane [16].

The simplest approach to incorporating general membrane degradation in a practical device model is to use a purely empirical specification of degradation rate, without correlation to other physical features of the model. Without correlation to a specifically mechanical or chemical origin of degradation, Karpenko-Jereb et al. followed this approach by specifying a constant proportional decrease in proton conductivity (and, concurrently, c_f) at 5.64×10^{-4} h^{-1}, and a constant proportional decrease in membrane thickness at 3.71×10^{-4} h^{-1} [172]. Once the membrane thickness reaches a critical thickness (defined as 10% of its initial thickness by the authors), pinholes are simulated by increasing gas crossover rates by an arbitrary but empirically motivated multiple of 100.

11.1. Mechanical Degradation Models

Mechanical degradation of the membrane may consist of pinhole and microcrack formation, membrane creep and even delamination of the CL from the membrane surface [180]. In general, models to date focus either on empirical description of the impact of degradation, or prediction of degradation rate; there are no coupled models for ongoing mechanical degradation prediction alongside performance prediction.

Weber described a membrane pinhole with radius r_{hole} according to a localised volume fraction $\varepsilon_{\text{hole}}$ within a degraded area of the membrane A_{deg} [165]:

$$\varepsilon_{\text{hole}} = \frac{\pi r_{\text{hole}}^2}{A_{\text{deg}}} \tag{125}$$

Within this region, the membrane effective properties are scaled by (1-$\varepsilon_{\text{hole}}$) while the Maxwell–Stefan diffusion equations from the adjacent GDLs are extended through the membrane with porosity $\varepsilon_{\text{hole}}$ and unit tortuosity (assuming a straight, cylindrical pinhole). Based on this theory, the study explored the impacts of pinholes with different sizes and frequencies along a channel length, revealing especially significant performance degradation in the case of single pinholes occupying a volume fraction $\varepsilon_{\text{hole}} > 0.002$. In a similar approach, membrane-electrode delamination over a prescribed area has also been described empirically, simply by applying an infinite contact resistance at the membrane-CL interface [181].

Burlatsky and co-workers developed a cyclic stress model in order to predict damage accrual due to fatigue from repeated hydration and dehydration of the membrane [182,183]. However, they did not provide any model of performance deterioration or the impact upon chemical properties of this degradation route. In general, membrane fatigue models have been used for durability analysis [184] but have not yet been applied to a direct prediction of performance deterioration. As mechanical stress on the membrane is closely coupled to dimensional changes, membranes with reinforcements are expected to show very different mechanical aging behaviour, and care should be taken when parameterising models using historical data.

11.2. Chemical Degradation Models

The dominant chemical mechanism considered in chemical degradation models of Nafion membranes is via the generation of H_2O_2 through the crossover reactions of dissolved H_2 at the cathode and/or dissolved O_2 at the anode. In the presence of Fe^{2+}, which is generally present as a dispersed membrane contaminant, the Fenton reaction generates $^{\bullet}OH$ and $^{\bullet}OOH$ radicals, which lead to membrane degradation by side-chain and main-chain scission of the perfluorosulfonate membrane material. The F^- byproduct of the degradation reaction is a common experimental tracer of degradation rate and as such can be applied to model validation [166].

Some semi-empirical models have assumed that the rate of $^{\bullet}OH$ radical release is in direct proportion to the crossover flux of the contributing species (H_2 or O_2), and membrane degradation rate is in turn proportional to $^{\bullet}OH$ radical generation rate. Hence, Kundu et al. gave the variation in membrane thickness L_{mem} as [166]:

$$\frac{dL_{\text{mem}}}{dt} = -k_{\text{deg}} \psi_{H_2} p_{H_2,\text{ano}} \tag{126}$$

with $k_{\text{deg}} = 1.8 \times 10^{-8}$ m^2 mol^{-1} for a GORE PRIMEA series 5510 MEA.

Chandesris et al. gave a similar semi-empirical expression, including potential and temperature dependence of k_{deg} and assuming thickness-dependent degradation (assuming constant reaction rate with crossover flux) [185]:

$$\frac{L_{mem}}{L_{mem,0}} \frac{dL_{mem}}{dt} = -k_{deg}(U_{ano}, T)\psi_{O_2} p_{O_2,cat} \tag{127}$$

Pinhole formation is predicted within this model as thinner regions of the membrane degrade more rapidly, amplifying any initially present non-uniformities.

These models were extended by Shah et al. to incorporate the full chemical mechanism for the Fenton reactions and consequent 'unzipping' of the membrane structure [186]. In this model, independent concentrations are assigned to several constituent repeated moieties within the polymer structure (carboxylic acid, weak polymer end groups, side chains, CF_2); as these concentrations evolve under the degradation model, they allow a prediction of spatial and temporal evolution of degradation sites. This model did not consider the consequences of degradation on relevant performance properties of the membrane, however. Subsequently, Futter et al. incorporated the role of Fe^{2+} and Fe^{3+} ion transport explicitly [178]. Burlatsky and co-workers have also considered the role of Pt reprecipitation on membrane degradation, since reprecipitated Pt from Pt^{2+} transport can agglomerate within the membrane and act as a catalytic site for peroxide-driven chemical degradation; these membrane-scale Pt reprecipitation models have not yet been combined with a full electrochemical model for the PEMFC [176,187]. We recommend that these papers be consulted directly for the detailed specification and parameterisation of the reaction mechanism.

More recent works have correlated the extent of chemical degradation predicted from a mechanistically detailed Fenton reaction model to increased gas crossover and performance deterioration. The principal approach is the definition of an 'effective porosity' for the membrane which varies with the concentrations of the membrane chemical moieties [188,189]. By considering c_f to be a variable depending on the sulfonate and total polymer concentrations, Wong and Kjeang defined an altered $\overline{V}_p$ for the degraded membrane, and used this to update the membrane porosity in the BFM description (using (9) and (59)) [137]. This model also altered membrane thickness directly in proportion to total mass loss, and has recently been incorporated into a 3D full cell performance model [190]. Similarly, Quiroga et al. gave a specific expression for degraded membrane molar volume according to the side-chain concentration [191]:

$$\overline{V}_p = \frac{1}{c_{side-chain}(1 + \beta_w \lambda)^3} \tag{128}$$

The membrane conductivity was in turn correlated to $\overline{V}_p$ according to an effective medium model; the latter was then based on a coarse-grained molecular dynamics database.

In recent years there have been advances in membranes that reduce their susceptibility to chemical degradation. These are not always explicitly indicated in theoretical descriptions of membranes and, therefore, great care should be used when parameterising degradation models using literature data from older materials. Post-fluorination of PFSA end groups and the introduction of radical scavengers into the membrane have been particularly effective, as has the use of mechanical reinforcement to reduce chemical-mechanical aging mechanisms [16].

12. Perspective

The ability to simulate fuel cells has been key to their recent advancement. We expect device simulation to play an increasingly important role in the rational design of new fuel cell technologies and in understanding the behaviour of these complex devices. As PEMFC technology matures, ever greater optimisation will be required to achieve meaningful performance improvements, and higher fidelity models will, therefore, be required to continue to design better fuel cells. Modelling will also play a key role in extending the operating life of PEMFCs. With widespread fuel cell deployment needed urgently to contribute to climate change targets, it is impossible to experimentally test new prototype PEMFCs for all operating conditions or end-use applications when the required lifetimes are in excess of 30,000 h (>3 years). Simulation will, therefore, be key in providing designers and users with confidence that new fuel cell designs will have the high longevity needed for use in heavy-duty applications such as trucks, trains, ships and aeroplanes. It will also be crucial in the development of the diagnostic and prognostic measurements that will be needed for the operation of PEMFCs and for the application of new measurement-integrated digitisation technologies such as 'digital twins'.

On consulting the literature, it is clear that even models developed in the last 5 years depend extensively on heritage parameterisation from the 1990s, in spite of significant developments in industrially relevant PEMFC membrane materials since this time. While many fundamental modelling concepts continue to apply to contemporary PFSA-based membranes, parameterisation based on industrially outdated materials like Nafion 117 is increasingly irrelevant. As an instructive example, we investigated, randomly and without any *a priori* pre-selection, 8 papers published since 2019 that contain 2D or 3D PEMFC device simulations, all from different research groups [75,178,192–197]. Of these, 6 use the Springer model and 2 use the Weber–Newman model. In works using the Springer model, parameterisation of the membrane properties is exclusively drawn from four experimental works [39,52,53,87], all dating to 1998 and earlier, and all measured on Nafion 117. Data correction in CLs was limited to scaling according to an effective tortuosity. In works using the Weber–Newman model, a recent measurement was used for the proton conductivity in one case, but otherwise the data (where reported) drew from the same 1990s experimental measurements, with some rescaling for fitting purposes in one case.

On this basis, we note a general absence of models and/or data in the recent literature that explicitly address the following features of state-of-the-art PEMFC membranes:

- The composite nature of reinforced membranes, including its effect on conductivity, diffusivity, water uptake, and mechanical coupling to transport phenomena.
- The introduction of radical scavengers or other membrane 'additives'.
- Parameterisation at relevant operating conditions (>60 °C), as opposed to at ambient conditions.
- The impact of variations in different membrane chemistries/side-chain length.
- The use of very thin membranes (<50 μm).
- In-plane inhomogeneities in electrode profiles.

We highlight these industry-relevant, contemporary membrane features as priorities for further theoretical study.

13. Conclusions

This review has summarised 30 years of development in the macroscopic theory of transport phenomena of polymer electrolyte membranes, as applied to practical models of polymer electrolyte membrane fuel cells. A spatial model of proton current from Ohm's law is achievable under the assumption of uniform membrane hydration. However, the water content dependence of conductivity, as well as the intrinsic role of membrane water transport phenomena (electroosmotic drag and water diffusion) in cell water balance, encourage the vast majority of spatially-resolved PEMFC simulations to incorporate a coupled proton-water transport model in the membrane. The semi-empirical Springer and Weber–Newman models have been the most popular of these, and allow relatively straightforward extension to transient and/or non-isothermal conditions, as well as to account for interfacial resistance to water uptake or loss in the CLs. During the last decade in particular, an increasing number of predictive models have also been developed to consider the commercially relevant concerns of gas crossover, impurity transport and membrane degradation.

Accurate physical models of the charge, mass and heat transport phenomena of the membrane are essential to high-fidelity prediction of PEMFC performance and localised behaviour. In collating models with their parameterisation data in this review, and by appraising collectively the heritage of work back to the early 1990s rather than focusing only on the last few years of progress, we aim to instigate greater efforts to compare existing models to new models. In particular, such comparisons may aid in demarcating the range of conditions in which particular models are most suitable, and the tasks to which they can be applied.

The theoretical description of PFSA membranes is challenging in terms of its physics, but also represents a moving target for PEMFC modelling and its underlying experimental parameterisation. Membrane chemistries and structures will continue to advance and, considering the increasing lifetime and performance demands for PEMFCs, more accurate modelling of 'edge-case' transport and degradation mechanisms, such as poisoning by ions leached from bipolar plates, is likely to become ever more important. This review, therefore, highlights the need for higher-fidelity models and the high-quality fundamental experimental data on state-of-the-art materials needed to parameterise them.

Our outlook in the Perspective section underlined a present disconnection between the experimental state-of-the art and the parameterisation most often employed in practical PEMFC simulation. Thus, the task of developing PEMFC membrane models is far from complete. We strongly encourage the community to continue to develop useful, practical models of these vitally important and fascinating materials.

Author Contributions: Conceptualisation, E.J.F.D.; investigation, E.J.F.D. and G.S.; writing—original draft preparation, E.J.F.D. and G.S.; writing—review and editing, E.J.F.D. and G.S. All authors have read and agreed to the published version of the manuscript.

Funding: This work was funded by the National Measurement System of the UK Department of Business, Energy and Industrial Strategy (BEIS).

Acknowledgments: Luís Castanheira (formerly NPL, now Symbio, Grenoble), Andy Wain (NPL) and Gareth Hinds (NPL) provided helpful comments on the text.

Conflicts of Interest: The authors declare no conflict of interest.

Appendix A

Tables of Abbreviations and Symbols.

Table A1. Table of abbreviations.

Abbreviation	Definition
BET	Brunauer–Emmett–Teller
BFM	binary friction model
CL	catalyst layer
DEFC	direct ethanol fuel cell
DMFC	direct methanol fuel cell
EIS	electrochemical impedance spectroscopy
ePTFE	expanded polytetrafluoroethylene
GAB	Guggenheim–Anderson–de Boer
GDL	gas diffusion layer
IEC	ion-exchange capacity
LE	liquid-equilibrated
MEA	membrane-electrode assembly
MPL	microporous layer
MRI	magnetic resonance imaging
NMR	nuclear magnetic resonance
PEM	polymer electrolyte membrane
PEMFC	polymer electrolyte membrane fuel cell
PEMWE	polymer electrolyte membrane water electrolyser
PFSA	perfluorosulfonic acid
VE	vapour-equilibrated

Table A2. Table of symbols.

Symbol	Unit	Definition
A_k	1	Empirical coefficient, binary friction model
A_{cat}	V	Tafel slope (cathode)
A_{deg}	m^2	Degraded area
a_k	various	Polynomial coefficients, various models
a_{pore}	m^{-1}	Effective pore surface area per unit volume, Choi–Datta sorption isotherm
a_{vol}	m^{-1}	Volumetric surface area, membrane-vapour contact in CL
$a_{w,vap}$	1	Activity, water vapour in gas phase
a_w	1	Activity, sorbed water
b_k	various	Coefficients, various models
c_f	mol m^{-3}	Sulfonic acid group concentration
$c_{side\text{-}chain}$	mol m^{-3}	Side chain concentration, Quiroga degradation model
c_w	mol m^{-3}	Concentration, sorbed water
D_i	m^2 s^{-1}	Fick's law diffusion coefficient, dilute species *i*
D_{cap}	kg m^{-1} s^{-1}	Capillary diffusion coefficient
D_w	m^2 s^{-1}	Fick's law diffusion coefficient, sorbed water
$D_{w,T}$	kg m^{-1} s^{-1} K^{-1}	Thermal diffusion coefficient
D_λ	m^2 s^{-1}	Effective diffusion coefficient with respect to water content gradient, sorbed water
D_μ	m^2 s^{-1}	Self-diffusion coefficient, sorbed water
$D_{\mu,WN}$	m^2 s^{-1}	Self-diffusion coefficient, sorbed water (binary Weber-Newman definition)
$E_{A,m}$	J mol^{-1}	Activation energy, process *m*
E_{cell}	V	Cell voltage
E_{OCV}	V	Open-circuit cell voltage
E_{mn}	kg mol^{-1}	Empirical coefficients, Meyers sorption isotherm
F	C mol^{-1}	Faraday constant
f	1	Mazumder diffusion function
f_{eff}	1	Membrane porosity, binary friction model
ΔH_m	J mol^{-1}	Enthalpy change, process *m*
$\mathbf{i}$	A m^{-2}	Current density
i_{far}	A m^{-2}	Equivalent current density of proton flux from faradaic reaction
i_{cell}	A m^{-2}	Cell current density
i_{ref}	A m^{-2}	Reference current density
$K_{a,mem}$	mol m^{-3}	Acidity constant, sulfonic acid groups
K_{drag}	1	Phenomenological ratio of water flux to proton flux
$K_{H,i}$	Pa m^3 mol^{-1}	Henry's law coefficient, species *i*
K_{mem}	Pa^{-1}	Langmuir equilibrium constant, Mashio sorption isotherm
K_1	1	Empirical coefficient, Thampan BET isotherm / Meyers sorption isotherm
K_2	1	Empirical coefficient, Meyers isotherm
k	W m^{-1} K^{-1}	Thermal conductivity
k_{bond}	s^{-1}	Water bonding rate constant, Meng water uptake model
k_{deg}	various	Degradation rate constant
k_G	1	Empirical coefficient, Klika GAB isotherm
k_{int}	m s^{-1}	Interfacial water transport coefficient
k_{vol}	s^{-1}	Interfacial water transport coefficient (volumetric)
k_{vap}	s m^{-2} kg^{-1}	Sorption rate constant, Meng model
L_{mem}	m	Membrane thickness

Table A2. *Cont.*

Symbol	Unit	Definition
M_{EW}	kg mol^{-1}	Equivalent weight (polymer mass per 1 mol sulfonic acid groups)
M_{w}	kg mol^{-1}	Molar mass, water
m_{w}	kg	Mass of water in reference volume V
$\mathbf{N}_{\mathrm{w}}$	mol m^{-2} s^{-1}	Molar flux, sorbed water
$\mathbf{N}_{+}$	mol m^{-2} s^{-1}	Molar flux, protons
n_i	mol	Content of species in reference volume V
n_{cond}	1	Scaling exponent with water content, proton conductivity
$n_{\mathrm{w,sat}}$	1	Empirical coefficient, Thampan BET isotherm
p	Pa	Pressure
$p_{i,k}$	Pa	Partial pressure, species i in phase k
p_0	Pa	Reference pressure
p_{L}	Pa	Absolute pressure in liquid phase
p_{sat}	Pa	Saturation vapour pressure (water)
$p_{\mathrm{w,vap}}$	Pa	Partial pressure, water vapour in gas phase
Q	W m^{-3}	Volumetric heat source
$\mathbf{q}$	W m^{-2}	Heat flux
q	1	Empirical coefficient, binary friction model
R_{Ω}	Ω	Ohmic cell series resistance
R	J K^{-1} mol^{-1}	Gas constant
R_{w}	kg m^{-3} s^{-1}	Liquid water source
r_{crit}	m	Critical maximum/minimum radius of liquid saturation
r_{hole}	m	Pinhole radius
S_{L}	1	Fraction of available membrane channels expanded by liquid water
s	1	Empirical coefficient, binary friction model
T	K	Temperature
T_0	K	Reference temperature
t	s	Time
U_{ano}	V	Anode operating potential
V	m^3	Reference volume
$V(r)$	m^{-1}	Relative volume probability distribution function
V_{dry}	m^3	Reference volume, dry conditions
$\overline{V}_{\mathrm{mem}}$	m^3 mol^{-1}	Molar volume, hydrated membrane
$\overline{V}_{\mathrm{p}}$	m^3 mol^{-1}	Molar volume, sulfonic acid groups
$\overline{V}_{\mathrm{w}}$	m^3 mol^{-1}	Molar volume, sorbed water
Y_{mem}	Pa	Young's modulus, membrane
z_i	1	Charge number, species i
α	1	Net number of water molecules transferred per proton
α_{λ}	1	Empirical coefficient, Setzler–Fuller conductivity model
α_{w}	mol^2 m^{-1} s^{-1} J^{-1}	Diffusion coefficient, Weber–Newman model
β	1	Scaling coefficient, Meyers isotherm
β_{w}	1	Swelling coefficient, water sorption
γ_{w}	N m^{-1}	Surface tension, water-gas interface
ε_{ij}	1	Strain tensor
$\varepsilon_{\mathrm{deg}}$	1	Void proportion of degraded membrane
θ_{c}	1	Contact angle of water-gas interface in membrane

Table A2. *Cont.*

Symbol	Unit	Definition
θ_m	K	Characteristic temperature, process m
κ	S m^{-1}	Proton conductivity
κ_p	m^2	Permeability
κ_sorp	Pa	Pore elasticity, Choi–Datta sorption isotherm
κ_k	S m^{-1}	Proton conductivity, equilibration condition k
κ_ϕ	m^2	Electroosmotic permeability, Schlögl equation
κ_0	S m^{-1}	Proton conductivity, reference conditions
λ	1	Water content
λ_crit	1	Critical water content, electroosmotic drag models
λ_eq	1	Water content, at equilibrium with a defined water activity
$\lambda_\mathrm{eq,crit}$	1	Critical water content, Meyers isotherm
$\lambda_\mathrm{eq,L}$	1	Water content, liquid-equilibrated conditions
$\lambda_\mathrm{eq,V}$	1	Water content, vapour-equilibrated conditions
$\lambda_{\mathrm{max},k}$	1	Maximum water content, equilibration condition k, Weber–Newman model
λ_min	1	Minimum water content, binary friction model
λ_mono	1	Water content, effective monolayer
λ_sat	1	Saturated water content
λ_+	1	Proton content
λ_0	1	Reference water content
μ	Pa s	Dynamic viscosity, sorbed water
μ_w	J mol^{-1}	Chemical potential, sorbed water
ξ	1	Electroosmotic drag coefficient
ρ	kg m^{-3}	Membrane density
ρ_dry	kg m^{-3}	Membrane density (dry conditions)
ρ_w	kg m^{-3}	Density of sorbed water
ϕ	V	Electrolyte potential, membrane phase
ϕ_m	1	Effective membrane volume fraction, Flory–Huggins model
ϕ_w	1	Water volume fraction, membrane phase
$\phi_\mathrm{w,crit}$	1	Critical volume fraction, Weber–Newman conductivity model
$\phi_\mathrm{w,max}$	1	Conductivity-maximum volume fraction, Weber–Newman conductivity model
ϕ_k	1	Empirical coefficient, Meyers isotherm
χ	1	Flory–Huggins parameter
χ_c	1	Degree of constraint, Weber membrane mechanical model
ψ_i	mol m^{-1} s^{-1} Pa^{-1}	Permeation coefficient, species i
ω	1	Empirical coefficient, Kosakian conductivity model

References

1. Kusoglu, A.; Weber, A.Z. New Insights into Perfluorinated Sulfonic-Acid Ionomers. *Chem. Rev.* **2017**, *117*, 987–1104. [CrossRef]
2. Kreuer, K.D.; Paddison, S.J.; Spohr, E.; Schuster, M. Transport in proton conductors for fuel-cell applications: Simulations, elementary reactions, and phenomenology. *Chem. Rev.* **2004**, *104*, 4637–4678. [CrossRef]
3. Hickner, M.A.; Pivovar, B.S. The chemical and structural nature of proton exchange membrane fuel cell properties. *Fuel Cells* **2005**, *5*, 213–229. [CrossRef]
4. Peighambardoust, S.J.; Rowshanzamir, S.; Amjadi, M. Review of the proton exchange membranes for fuel cell applications. *Int. J. Hydrogen Energy* **2010**, *35*, 9349–9384. [CrossRef]

5. Yuan, X.-Z.; Wang, H. PEM Fuel Cell Fundamentals. In *PEM Fuel Cell Electrocatalysts and Catalyst Layers: Fundamentals and Applications*; Zhang, J., Ed.; Springer: London, UK, 2008; pp. 1–87.
6. Andreaus, B.; Eikerling, M. Catalyst Layer Operation in PEM Fuel Cells: From Structural Pictures to Tractable Models. In *Device and Materials Modeling in PEM Fuel Cells*; Paddison, S.J., Promislow, K.S., Eds.; Springer: New York, NY, USA, 2009; pp. 41–90.
7. Weber, A.Z.; Borup, R.L.; Darling, R.M.; Das, P.K.; Dursch, T.J.; Gu, W.B.; Harvey, D.; Kusoglu, A.; Litster, S.; Mench, M.M.; et al. A Critical Review of Modeling Transport Phenomena in Polymer-Electrolyte Fuel Cells. *J. Electrochem. Soc.* **2014**, *161*, F1254–F1299. [CrossRef]
8. Huang, J.; Li, Z.; Zhang, J.B. Review of characterization and modeling of polymer electrolyte fuel cell catalyst layer: The blessing and curse of ionomer. *Front. Energy* **2017**, *11*, 334–364. [CrossRef]
9. Weber, A.Z.; Newman, J. Modeling transport in polymer-electrolyte fuel cells. *Chem. Rev.* **2004**, *104*, 4679–4726. [CrossRef]
10. Jiao, K.; Li, X.G. Water transport in polymer electrolyte membrane fuel cells. *Prog. Energy Combust. Sci.* **2011**, *37*, 221–291. [CrossRef]
11. Nasef, M.M.; Aly, A.A. Water and charge transport models in proton exchange membranes: An overview. *Desalination* **2012**, *287*, 238–246. [CrossRef]
12. Zamel, N.; Li, X.G. Effective transport properties for polymer electrolyte membrane fuel cells - With a focus on the gas diffusion layer. *Prog. Energy Combust. Sci.* **2013**, *39*, 111–146. [CrossRef]
13. Balasubramanian, S.; Weber, A.Z. Continuum, Macroscopic Modeling of Polymer-Electrolyte Fuel Cells. In *Physical Multiscale Modeling and Numerical Simulation of Electrochemical Devices for Energy Conversion and Storage: From Theory to Engineering to Practice*; Franco, A.A., Doublet, M.L., Bessler, W.G., Eds.; Springer: London, UK, 2016; pp. 91–149.
14. Costamagna, P. Transport phenomena in polymeric membrane fuel cells. *Chem. Eng. Sci.* **2001**, *56*, 323–332. [CrossRef]
15. Smitha, B.; Sridhar, S.; Khan, A.A. Solid polymer electrolyte membranes for fuel cell applications—A review. *J. Membr. Sci.* **2005**, *259*, 10–26. [CrossRef]
16. Zaton, M.; Roziere, J.; Jones, D.J. Current understanding of chemical degradation mechanisms of perfluorosulfonic acid membranes and their mitigation strategies: A review. *Sustain. Energy Fuels* **2017**, *1*, 409–438. [CrossRef]
17. Gebert, M.; Ghielmi, A.; Merlo, L.; Corasaniti, M.; Arcella, V. AQUIVION(TM)–The Short-Side-Chain and Low-EW PFSA for Next-Generation PEFCs Expands Production and Utilization. *ECS Trans.* **2010**, *26*, 279–283. [CrossRef]
18. Solvay. Available online: https://www.solvay.com/en/brands/aquivion-ion-conducting-polymers (accessed on 20 October 2020).
19. Emery, M.; Frey, M.; Guerra, M.; Haugen, G.; Hintzer, K.; Lochhaas, K.H.; Pham, P.; Pierpont, D.; Schaberg, M.; Thaler, A.; et al. The Development of New Membranes for Proton Exchange Membrane Fuel Cells. *ECS Trans.* **2007**, *11*, 3–14. [CrossRef]
20. 3M Company. Available online: https://www.3m.com/3M/en_US/design-and-specialty-materials-us/ionomers/ (accessed on 20 October 2020).
21. Penner, R.M.; Martin, C.R. Ion Transporting Composite Membranes: I. Nafion–Impregnated Gore–Tex. *J. Electrochem. Soc.* **1985**, *132*, 514–515. [CrossRef]
22. Bahar, B.; Hobson, A.R.; Kolde, J.A. Integral Composite Membrane. U.S. Patent US5599614A, 4 February 1997.
23. Endoh, E. Highly Durable MEA for PEMFC Under High Temperature and Low Humidity Conditions. *ECS Trans.* **2006**, *3*, 9–18. [CrossRef]
24. Shi, S.W.; Weber, A.Z.; Kusoglu, A. Structure/property relationship of Nafion XL composite membranes. *J. Membr. Sci.* **2016**, *516*, 123–134. [CrossRef]
25. Wang, Z.J.; Tan, J.T.; Zhan, Z.G.; Pan, M. Water transport law of fuel cell membranes at different current densities. *Chin. Sci. Bull. Chin.* **2019**, *64*, 2254–2261. [CrossRef]
26. Yandrasits, M.A.; Lindell, M.J.; Hamrock, S.J. New directions in perfluoroalkyl sulfonic acid-based proton-exchange membranes. *Curr. Opin. Electrochem.* **2019**, *18*, 90–98. [CrossRef]
27. Rolfi, A.; Oldani, C.; Merlo, L.; Facchi, D.; Ruffo, R. New perfluorinated ionomer with improved oxygen permeability for application in cathode polymeric electrolyte membrane fuel cell. *J. Power Sources* **2018**, *396*, 95–101. [CrossRef]

28. Varcoe, J.R.; Atanassov, P.; Dekel, D.R.; Herring, A.M.; Hickner, M.A.; Kohl, P.A.; Kucernak, A.R.; Mustain, W.E.; Nijmeijer, K.; Scott, K.; et al. Anion-exchange membranes in electrochemical energy systems. *Energy Environ. Sci.* **2014**, *7*, 3135–3191. [CrossRef]
29. The Chemours Company FC, LLC. *Product Bulletin P-11: Nafion(TM) NR211 and NR212 Ion Exchange Materials: Solution Cast Membranes*; The Chemours Company FC, LLC.: Wilmington, DE, USA, 2017.
30. Thampan, T.; Malhotra, S.; Tang, H.; Datta, R. Modeling of conductive transport in proton-exchange membranes for fuel cells. *J. Electrochem. Soc.* **2000**, *147*, 3242–3250. [CrossRef]
31. Hasan, M.; Goshtasbi, A.; Chen, J.X.; Santare, M.H.; Ersal, T. Model-Based Analysis of PFSA Membrane Mechanical Response to Relative Humidity and Load Cycling in PEM Fuel Cells. *J. Electrochem. Soc.* **2018**, *165*, F3359–F3372. [CrossRef]
32. The Chemours Company FC, LLC. *Product Bulletin P-12: Nafion(TM) N115, N117 and N1110 Ion Exchange Materials: Extrusion Cast Membranes*; The Chemours Company FC, LLC.: Wilmington, DE, USA, 2016.
33. Berg, P.; Promislow, K.; St Pierre, J.; Stumper, J.; Wetton, B. Water management in PEM fuel cells. *J. Electrochem. Soc.* **2004**, *151*, A341–A353. [CrossRef]
34. Kim, J.; Lee, S.M.; Srinivasan, S.; Chamberlin, C.E. Modeling of Proton Exchange Membrane Fuel Cell Performance with an Empirical Equation. *J. Electrochem. Soc.* **1995**, *142*, 2670–2674. [CrossRef]
35. Gerteisen, D.; Heilmann, T.; Ziegler, C. Modeling the phenomena of dehydration and flooding of a polymer electrolyte membrane fuel cell. *J. Power Sources* **2009**, *187*, 165–181. [CrossRef]
36. Wu, H.; Li, X.G.; Berg, P. On the modeling of water transport in polymer electrolyte membrane fuel cells. *Electrochim. Acta* **2009**, *54*, 6913–6927. [CrossRef]
37. Newman, J.; Tiedemann, W. Porous-electrode theory with battery applications. *AIChE J.* **1975**, *21*, 25–41. [CrossRef]
38. Iden, H.; Sato, K.; Ohma, A.; Shinohara, K. Relationship among Microstructure, Ionomer Property and Proton Transport in Pseudo Catalyst Layers. *J. Electrochem. Soc.* **2011**, *158*, B987–B994. [CrossRef]
39. Springer, T.E.; Zawodzinski, T.A.; Gottesfeld, S. Polymer Electrolyte Fuel Cell Model. *J. Electrochem. Soc.* **1991**, *138*, 2334–2342. [CrossRef]
40. Von Schroeder, P. Über Erstarrungs- und Quellugserscheinungen von Gelatine. *Z. Phys. Chem.* **1903**, *45*, 75–117. [CrossRef]
41. Choi, P.H.; Datta, R. Sorption in proton-exchange membranes—An explanation of Schroeder's paradox. *J. Electrochem. Soc.* **2003**, *150*, E601–E607. [CrossRef]
42. Weber, A.Z.; Newman, J. Transport in polymer-electrolyte membranes—I. Physical model. *J. Electrochem. Soc.* **2003**, *150*, A1008–A1015. [CrossRef]
43. Freger, V. Hydration of Ionomers and Schroeder's Paradox in Nafion. *J. Phys. Chem. B* **2009**, *113*, 24–36. [CrossRef]
44. Eikerling, M.H.; Berg, P. Poroelectroelastic theory of water sorption and swelling in polymer electrolyte membranes. *Soft Matter* **2011**, *7*, 5976–5990. [CrossRef]
45. Jeck, S.; Scharfer, P.; Kind, M. Absence of Schroeder's paradox: Experimental evidence for water-swollen Nafion (R) membranes. *J. Membr. Sci.* **2011**, *373*, 74–79. [CrossRef]
46. Hwang, G.S.; Parkinson, D.Y.; Kusoglu, A.; MacDowell, A.A.; Weber, A.Z. Understanding Water Uptake and Transport in Nafion Using X-ray Microtomography. *ACS Macro Lett.* **2013**, *2*, 288–291. [CrossRef]
47. Bridgeman, O.C.; Aldrich, E.W. Vapor Pressure Tables for Water. *J. Heat Transf.* **1964**, *86*, 279–286. [CrossRef]
48. Jiao, K.; Li, X.G. Three-dimensional multiphase modeling of cold start processes in polymer electrolyte membrane fuel cells. *Electrochim. Acta* **2009**, *54*, 6876–6891. [CrossRef]
49. Gurau, V.; Zawodzinski, T.A.; Mann, J.A. Two-phase transport in PEM fuel cell cathodes. *J. Fuel Cell Sci. Technol.* **2008**, *5*. [CrossRef]
50. Andersson, M.; Beale, S.B.; Espinoza, M.; Wu, Z.; Lehnertbe, W. A review of cell-scale multiphase flow modeling, including water management, in polymer electrolyte fuel cells. *Appl. Energy* **2016**, *180*, 757–778. [CrossRef]
51. Springer, T.E.; Wilson, M.S.; Gottesfeld, S. Modeling and Experimental Diagnostics in Polymer Electrolyte Fuel Cells. *J. Electrochem. Soc.* **1993**, *140*, 3513–3526. [CrossRef]
52. Zawodzinski, T.A.; Derouin, C.; Radzinski, S.; Sherman, R.J.; Smith, V.T.; Springer, T.E.; Gottesfeld, S. Water Uptake by and Transport Through Nafion® 117 Membranes. *J. Electrochem. Soc.* **1993**, *140*, 1041–1047. [CrossRef]

53. Zawodzinski, T.A.; Springer, T.E.; Uribe, F.; Gottesfeld, S. Characterization of polymer electrolytes for fuel cell applications. *Solid State Ion.* **1993**, *60*, 199–211. [CrossRef]
54. Broka, K.; Ekdunge, P. Modelling the PEM fuel cell cathode. *J. Appl. Electrochem.* **1997**, *27*, 281–289. [CrossRef]
55. Hinatsu, J.T.; Mizuhata, M.; Takenaka, H. Water Uptake of Perfluorosulfonic Acid Membranes from Liquid Water and Water Vapor. *J. Electrochem. Soc.* **1994**, *141*, 1493–1498. [CrossRef]
56. Pasaogullari, U.; Wang, C.Y.; Chen, K.S. Two-phase transport in polymer electrolyte fuel cells with bilayer cathode gas diffusion media. *J. Electrochem. Soc.* **2005**, *152*, A1574–A1582. [CrossRef]
57. Kulikovsky, A.A. Quasi-3D modeling of water transport in polymer electrolyte fuel cells. *J. Electrochem. Soc.* **2003**, *150*, A1432–A1439. [CrossRef]
58. Meier, F.; Eigenberger, G. Transport parameters for the modelling of water transport in ionomer membranes for PEM-fuel cells. *Electrochim. Acta* **2004**, *49*, 1731–1742. [CrossRef]
59. Karpenko-Jereb, L.; Innerwinkler, P.; Kelterer, A.M.; Sternig, C.; Fink, C.; Prenninger, P.; Tatschl, R. A novel membrane transport model for polymer electrolyte fuel cell simulations. *Int. J. Hydrogen Energy* **2014**, *39*, 7077–7088. [CrossRef]
60. Futerko, P.; Hsing, I.M. Two-dimensional finite-element method study of the resistance of membranes in polymer electrolyte fuel cells. *Electrochim. Acta* **2000**, *45*, 1741–1751. [CrossRef]
61. Morris David, R.; Sun, X. Water-sorption and transport properties of Nafion 117 H. *J. Appl. Polym. Sci.* **1993**, *50*, 1445–1452. [CrossRef]
62. Steinkamp, K.; Schumacher, J.O.; Goldsmith, F.; Ohlberger, M.; Ziegler, C. A nonisothermal PEM fuel cell model including two water transport mechanisms in the membrane. *J. Fuel Cell Sci. Technol.* **2008**, *5*. [CrossRef]
63. Klika, V.; Kubant, J.; Pavelka, M.; Benziger, J.B. Non-equilibrium thermodynamic model of water sorption in Nafion membranes. *J. Membr. Sci.* **2017**, *540*, 35–49. [CrossRef]
64. Meyers, J.P.; Newman, J. Simulation of the Direct Methanol Fuel Cell: I. Thermodynamic Framework for a Multicomponent Membrane. *J. Electrochem. Soc.* **2002**, *149*, A710–A717. [CrossRef]
65. Weber, A.Z.; Newman, J. Transport in polymer-electrolyte membranes—II. Mathematical model. *J. Electrochem. Soc.* **2004**, *151*, A311–A325. [CrossRef]
66. Weber, A.Z.; Newman, J. A Combination Model for Macroscopic Transport in Polymer-Electrolyte Membranes. In *Device and Materials Modeling in PEM Fuel Cells*; Paddison, S.J., Promislow, K.S., Eds.; Springer: New York, NY, USA, 2009; pp. 157–198.
67. Murahashi, T.; Naiki, M.; Nishiyama, E. Water transport in the proton exchange-membrane fuel cell: Comparison of model computation and measurements of effective drag. *J. Power Sources* **2006**, *162*, 1130–1136. [CrossRef]
68. Eikerling, M. Water management in cathode catalyst layers of PEM fuel cells—A structure-based model. *J. Electrochem. Soc.* **2006**, *153*, E58–E70. [CrossRef]
69. Kusoglu, A.; Kwong, A.; Clark, K.T.; Gunterman, H.P.; Weber, A.Z. Water Uptake of Fuel-Cell Catalyst Layers. *J. Electrochem. Soc.* **2012**, *159*, F530–F535. [CrossRef]
70. Jung, C.Y.; Yi, S.C. Influence of the water uptake in the catalyst layer for the proton exchange membrane fuel cells. *Electrochem. Commun.* **2013**, *35*, 34–37. [CrossRef]
71. DeCaluwe, S.C.; Baker, A.M.; Bhargava, P.; Fischer, J.E.; Dura, J.A. Structure-property relationships at Nafion thin-film interfaces: Thickness effects on hydration and anisotropic ion transport. *Nano Energy* **2018**, *46*, 91–100. [CrossRef]
72. Mashio, T.; Sato, K.; Ohma, A. Analysis of Water Adsorption and Condensation in Catalyst Layers for Polymer Electrolyte Fuel Cells. *Electrochim. Acta* **2014**, *140*, 238–249. [CrossRef]
73. Kim, T.-H.; Yoo, J.H.; Maiyalagan, T.; Yi, S.-C. Influence of the Nafion agglomerate morphology on the water-uptake behavior and fuel cell performance in the proton exchange membrane fuel cells. *Appl. Surf. Sci.* **2019**, *481*, 777–784. [CrossRef]
74. Morris, D.R.P.; Liu, S.P.; Gonzalez, D.V.; Gostick, J.T. Effect of Water Sorption on the Electronic Conductivity of Porous Polymer Electrolyte Membrane Fuel Cell Catalyst Layers. *ACS Appl. Mater. Interfaces* **2014**, *6*, 18609–18618. [CrossRef]
75. Kosakian, A.; Padilla Urbina, L.; Heaman, A.; Secanell, M. Understanding single-phase water-management signatures in fuel-cell impedance spectra: A numerical study. *Electrochim. Acta* **2020**, *350*. [CrossRef]

76. Fimrite, J.; Struchtrup, H.; Djilali, N. Transport phenomena in polymer electrolyte membranes—I. Modeling framework. *J. Electrochem. Soc.* **2005**, *152*, A1804–A1814. [CrossRef]
77. Fimrite, J.; Carnes, B.; Struchtrup, H.; Djilali, N. Transport phenomena in polymer electrolyte membranes—II. Binary friction membrane model. *J. Electrochem. Soc.* **2005**, *152*, A1815–A1823. [CrossRef]
78. Fimrite, J.; Carnes, B.; Struchtrup, H.; Djilali, N. Coupled Proton and Water Transport in Polymer Electrolyte Membranes. In *Device and Materials Modeling in PEM Fuel Cells*; Paddison, S.J., Promislow, K.S., Eds.; Springer: New York, NY, USA, 2009; pp. 123–155.
79. Zawodzinski, T.A.; Neeman, M.; Sillerud, L.O.; Gottesfeld, S. Determination of water diffusion coefficients in perfluorosulfonate ionomeric membranes. *J. Phys. Chem.* **1991**, *95*, 6040–6044. [CrossRef]
80. Zawodzinski, T.A.; Davey, J.; Valerio, J.; Gottesfeld, S. The water content dependence of electro-osmotic drag in proton-conducting polymer electrolytes. *Electrochim. Acta* **1995**, *40*, 297–302. [CrossRef]
81. Fuller, T.F.; Newman, J. Experimental Determination of the Transport Number of Water in Nafion 117 Membrane. *J. Electrochem. Soc.* **1992**, *139*, 1332–1337. [CrossRef]
82. Carnes, B.; Djilali, N. Analysis of coupled proton and water transport in a PEM fuel cell using the binary friction membrane model. *Electrochim. Acta* **2006**, *52*, 1038–1052. [CrossRef]
83. Motupally, S.; Becker, A.J.; Weidner, J.W. Diffusion of water in Nafion 115 membranes. *J. Electrochem. Soc.* **2000**, *147*, 3171–3177. [CrossRef]
84. Shah, A.A.; Kim, G.S.; Gervais, W.; Young, A.; Promislow, K.; Li, J.; Ye, S. The effects of water and microstructure on the performance of polymer electrolyte fuel cells. *J. Power Sources* **2006**, *160*, 1251–1268. [CrossRef]
85. Djilali, N.; Sui, P.C. Transport phenomena in fuel cells: From microscale to macroscale. *Int. J. Comput. Fluid Dyn.* **2008**, *22*, 115–133. [CrossRef]
86. Li, J.Y.; Nemat-Nasser, S. Micromechanical analysis of ionic clustering in Nafion perfluorinated membrane. *Mech. Mater.* **2000**, *32*, 303–314. [CrossRef]
87. Van Bussel, H.P.L.H.; Koene, F.G.H.; Mallant, R.K.A.M. Dynamic model of solid polymer fuel cell water management. *J. Power Sources* **1998**, *71*, 218–222. [CrossRef]
88. Wiezell, K.; Gode, P.; Lindbergh, G. Steady-state and EIS investigations of hydrogen electrodes and membranes in polymer electrolyte fuel cells I. Modeling. *J. Electrochem. Soc.* **2006**, *153*, A749–A758. [CrossRef]
89. Ju, H.C.; Wang, C.Y.; Cleghorn, S.; Beuscher, U. Nonisothermal modeling of polymer electrolyte fuel cells—I. Experimental validation. *J. Electrochem. Soc.* **2005**, *152*, A1645–A1653. [CrossRef]
90. Neubrand, W. *Modellbildung und Simulation von Elektromembranverfahren*; Logos Verlag Berlin: Berlin, Germany, 1999.
91. Ramousse, J.; Deseure, J.; Lottin, O.; Didierjean, S.; Maillet, D. Modelling of heat, mass and charge transfer in a PEMFC single cell. *J. Power Sources* **2005**, *145*, 416–427. [CrossRef]
92. Dobson, P.; Lei, C.; Navessin, T.; Secanell, M. Characterization of the PEM Fuel Cell Catalyst Layer Microstructure by Nonlinear Least-Squares Parameter Estimation. *J. Electrochem. Soc.* **2012**, *159*, B514–B523. [CrossRef]
93. Sone, Y.; Ekdunge, P.; Simonsson, D. Proton Conductivity of Nafion 117 as Measured by a Four–Electrode AC Impedance Method. *J. Electrochem. Soc.* **1996**, *143*, 1254–1259. [CrossRef]
94. Baschuk, J.J.; Li, X.G. Modeling of ion and water transport in the polymer electrolyte membrane of PEM fuel cells. *Int. J. Hydrogen Energy* **2010**, *35*, 5095–5103. [CrossRef]
95. Setzler, B.P.; Fuller, T.F. A Physics-Based Impedance Model of Proton Exchange Membrane Fuel Cells Exhibiting Low-Frequency Inductive Loops. *J. Electrochem. Soc.* **2015**, *162*, F519–F530. [CrossRef]
96. Janssen, G.J.M. A phenomenological model of water transport in a proton exchange membrane fuel cell. *J. Electrochem. Soc.* **2001**, *148*, A1313–A1323. [CrossRef]
97. Eikerling, M.; Kharkats, Y.I.; Kornyshev, A.A.; Volfkovich, Y.M. Phenomenological Theory of Electro–osmotic Effect and Water Management in Polymer Electrolyte Proton–Conducting Membranes. *J. Electrochem. Soc.* **1998**, *145*, 2684–2699. [CrossRef]
98. St-Pierre, J. Simple mathematical model for water diffusion in Nafion membranes. *J. Electrochem. Soc.* **2007**, *154*, B88–B95. [CrossRef]
99. Okada, T.; Xie, G.; Tanabe, Y. Theory of water management at the anode side of polymer electrolyte fuel cell membranes. *J. Electroanal. Chem.* **1996**, *413*, 49–65. [CrossRef]

100. Fuller, T.F.; Newman, J. Water and Thermal Management in Solid–Polymer–Electrolyte Fuel Cells. *J. Electrochem. Soc.* **1993**, *140*, 1218–1225. [CrossRef]
101. Nguyen, T.V.; White, R.E. A Water and Heat Management Model for Proton–Exchange–Membrane Fuel Cells. *J. Electrochem. Soc.* **1993**, *140*, 2178–2186. [CrossRef]
102. Mazumder, S. A generalized phenomenological model and database for the transport of water and current in polymer electrolyte membranes. *J. Electrochem. Soc.* **2005**, *152*, A1633–A1644. [CrossRef]
103. Ye, X.H.; Wang, C.Y. Measurement of water transport properties through membrane-electrode assemblies. *J. Electrochem. Soc.* **2007**, *154*, B676–B682. [CrossRef]
104. Wu, H.; Berg, P.; Li, X.G. Non-isothermal transient modeling of water transport in PEM fuel cells. *J. Power Sources* **2007**, *165*, 232–243. [CrossRef]
105. Pivovar, B.S. An overview of electro-osmosis in fuel cell polymer electrolytes. *Polymer* **2006**, *47*, 4194–4202. [CrossRef]
106. Peng, Z.; Morin, A.; Huguet, P.; Schott, P.; Pauchet, J. In-Situ Measurement of Electroosmotic Drag Coefficient in Nafion Membrane for the PEMFC. *J. Phys. Chem. B* **2011**, *115*, 12835–12844. [CrossRef] [PubMed]
107. Wang, X.H.; McClure, J.P.; Fedkiw, P.S. Transport properties of proton- and hydroxide-exchange membranes for fuel cells. *Electrochim. Acta* **2012**, *79*, 126–132. [CrossRef]
108. Um, S.; Wang, C.Y. Computational study of water transport in proton exchange membrane fuel cells. *J. Power Sources* **2006**, *156*, 211–223. [CrossRef]
109. Ge, S.H.; Yi, B.L.; Ming, P.W. Experimental determination of electro-osmotic drag coefficient in Nafion membrane for fuel cells. *J. Electrochem. Soc.* **2006**, *153*, A1443–A1450. [CrossRef]
110. Ge, S.H.; Li, X.G.; Yi, B.L.; Hsing, I.M. Absorption, desorption, and transport of water in polymer electrolyte membranes for fuel cells. *J. Electrochem. Soc.* **2005**, *152*, A1149–A1157. [CrossRef]
111. Ise, M.; Kreuer, K.D.; Maier, J. Electroosmotic drag in polymer electrolyte membranes: An electrophoretic NMR study. *Solid State Ion.* **1999**, *125*, 213–223. [CrossRef]
112. Benziger, J.B.; Cheah, M.J.; Klika, V.; Pavelka, M. Interfacial Constraints on Water and Proton Transport Across Nafion Membranes. *J. Polym. Sci. Part. B-Polym. Phys.* **2015**, *53*, 1580–1589. [CrossRef]
113. Berning, T. On water transport in polymer electrolyte membranes during the passage of current. *Int. J. Hydrogen Energy* **2011**, *36*, 9341–9344. [CrossRef]
114. Berg, P.; Stornes, M. Towards a Consistent Interpretation of Electro-osmotic Drag in Polymer Electrolyte Membranes. *Fuel Cells* **2016**, *16*, 715–724. [CrossRef]
115. Dreyer, W.; Guhlke, C.; Müller, R. Overcoming the shortcomings of the Nernst-Planck model. *Phys. Chem. Chem. Phys.* **2013**, *15*, 7075–7086. [CrossRef] [PubMed]
116. Bernardi, D.M.; Verbrugge, M.W. Mathematical model of a gas diffusion electrode bonded to a polymer electrolyte. *AIChE J.* **1991**, *37*, 1151–1163. [CrossRef]
117. Pivovar, B.S.; Kim, Y.S. The membrane-electrode interface in PEFCs. *J. Electrochem. Soc.* **2007**, *154*, B739–B744. [CrossRef]
118. Kim, Y.S.; Pivovar, B.S. The Membrane-Electrode Interface in PEFCs IV. The origin and implications of interfacial resistance. *J. Electrochem. Soc.* **2010**, *157*, B1616–B1623. [CrossRef]
119. Tsampas, M.N.; Pikos, A.; Brosda, S.; Katsaounis, A.; Vayenas, C.G. The effect of membrane thickness on the conductivity of Nafion. *Electrochim. Acta* **2006**, *51*, 2743–2755. [CrossRef]
120. Rangel-Cárdenas, A.L.; Koper, G.J.M. Transport in Proton Exchange Membranes for Fuel Cell Applications—A Systematic Non-Equilibrium Approach. *Materials* **2017**, *10*, 576. [CrossRef]
121. Cheah, M.J.; Kevrekidis, I.G.; Benziger, J. Effect of Interfacial Water Transport Resistance on Coupled Proton and Water Transport Across Nafion. *J. Phys. Chem. B* **2011**, *115*, 10239–10250. [CrossRef] [PubMed]
122. Kientiz, B.; Yamada, H.; Nonoyama, N.; Weber, A.Z. Interfacial Water Transport Effects in Proton-Exchange Membranes. *J. Fuel Cell Sci. Technol.* **2011**, *8*. [CrossRef]
123. Um, S.; Wang, C.Y.; Chen, K.S. Computational fluid dynamics modeling of proton exchange membrane fuel cells. *J. Electrochem. Soc.* **2000**, *147*, 4485–4493. [CrossRef]
124. Wang, Y.; Wang, C.Y. Transient analysis of polymer electrolyte fuel cells. *Electrochim. Acta* **2005**, *50*, 1307–1315. [CrossRef]
125. Meng, H. A three-dimensional PEM fuel cell model with consistent treatment of water transport in MEA. *J. Power Sources* **2006**, *162*, 426–435. [CrossRef]

126. Adachi, M.; Navessin, T.; Xie, Z.; Frisken, B.; Holdcroft, S. Correlation of In Situ and Ex Situ Measurements of Water Permeation Through Nafion NRE211 Proton Exchange Membranes. *J. Electrochem. Soc.* **2009**, *156*, B782–B790. [CrossRef]
127. Yau, T.C.; Cimenti, M.; Bi, X.T.T.; Stumper, J. Water transport and Schroder's Paradox in fuel cell membrane electrode assemblies. *J. Power Sources* **2013**, *224*, 285–289. [CrossRef]
128. Teranishi, K.; Tsushima, S.; Hirai, S. Analysis of water transport in PEFCs by magnetic resonance imaging measurement. *J. Electrochem. Soc.* **2006**, *153*, A664–A668. [CrossRef]
129. Liu, X.L.; Lou, G.F.; Wen, Z. Three-dimensional two-phase flow model of proton exchange membrane fuel cell with parallel gas distributors. *J. Power Sources* **2010**, *195*, 2764–2773. [CrossRef]
130. Majsztrik, P.W.; Satterfield, M.B.; Bocarsly, A.B.; Benziger, J.B. Water sorption, desorption and transport in Nafion membranes. *J. Membr. Sci.* **2007**, *301*, 93–106. [CrossRef]
131. Davey, J.; Mukundan, R.; Spendelow, J.; Hussey, D.S.; Jacobson, D.; Arif, M.; Borup, R.L. Water Dynamics in a PEM Fuel Cell: Effect of Current and Humidity Transients. *ECS Trans.* **2008**, *16*, 329–340. [CrossRef]
132. Monroe, C.W.; Romero, T.; Merida, W.; Eikerling, M. A vaporization-exchange model for water sorption and flux in Nafion. *J. Membr. Sci.* **2008**, *324*, 1–6. [CrossRef]
133. Gurau, V.; Edwards, R.V.; Mann, J.A.; Zawodzinski, T.A. A look at the multiphase mixture model for PEM fuel cell simulations. *Electrochem. Solid State Lett.* **2008**, *11*, B132–B135. [CrossRef]
134. Berning, T.; Odgaard, M.; Kær, S.K. Water balance simulations of a polymer-electrolyte membrane fuel cell using a two-fluid model. *J. Power Sources* **2011**, *196*, 6305–6317. [CrossRef]
135. Berg, P.; Novruzi, A.; Promislow, K. Analysis of a cathode catalyst layer model for a polymer electrolyte fuel cell. *Chem. Eng. Sci.* **2006**, *61*, 4316–4331. [CrossRef]
136. Satterfield, M.B.; Benziger, J.B. Non-fickian water vapor sorption dynamics by nafion membranes. *J. Phys. Chem. B* **2008**, *112*, 3693–3704. [CrossRef] [PubMed]
137. Wong, K.H.; Kjeang, E. Macroscopic In-Situ Modeling of Chemical Membrane. Degradation in Polymer Electrolyte Fuel Cells. *J. Electrochem. Soc.* **2014**, *161*, F823–F832. [CrossRef]
138. Vorobev, A.; Zikanov, O.; Shamim, T. A computational model of a PEM fuel cell with finite vapor absorption rate. *J. Power Sources* **2007**, *166*, 92–103. [CrossRef]
139. Ziegler, C.; Gerteisen, D. Validity of two-phase polymer electrolyte membrane fuel cell models with respect to the gas diffusion layer. *J. Power Sources* **2009**, *188*, 184–191. [CrossRef]
140. Weber, A.Z.; Newman, J. Transport in polymer-electrolyte membranes—III. Model validation in a simple fuel-cell model. *J. Electrochem. Soc.* **2004**, *151*, A326–A339. [CrossRef]
141. Meng, H.; Han, B.; Ruan, B. Numerical modeling of liquid water transport inside and across membrane in PEM fuel cells. *Asia-Pac. J. Chem. Eng.* **2013**, *8*, 104–114. [CrossRef]
142. Bernardi, D.M.; Verbrugge, M.W. A Mathematical Model of the Solid–Polymer–Electrolyte Fuel Cell. *J. Electrochem. Soc.* **1992**, *139*, 2477–2491. [CrossRef]
143. Hwang, G.S.; Kaviany, M.; Nam, J.H.; Kim, M.H.; Son, S.Y. Pore-Water Morphological Transitions in Polymer Electrolyte of a Fuel Cell. *J. Electrochem. Soc.* **2009**, *156*, B1192–B1200. [CrossRef]
144. Falcão, D.S.; Oliveira, V.B.; Rangel, C.M.; Pinho, C.; Pinto, A.M.F.R. Water transport through a PEM fuel cell: A one-dimensional model with heat transfer effects. *Chem. Eng. Sci.* **2009**, *64*, 2216–2225. [CrossRef]
145. Jung, C.Y.; Lee, C.S.; Yi, S.C. Computational analysis of transport phenomena in proton exchange membrane for polymer electrolyte fuel cells. *J. Membr. Sci.* **2008**, *309*, 1–6. [CrossRef]
146. Nam, J.; Chippar, P.; Kim, W.; Ju, H. Numerical analysis of gas crossover effects in polymer electrolyte fuel cells (PEFCs). *Appl. Energy* **2010**, *87*, 3699–3709. [CrossRef]
147. Rama, P.; Chen, R. Polymer Electrolyte Fuel Cell Transport Mechanisms: A Universal Approach to Multilayer Two-Phase Modeling Through the General Transport Equation. *J. Fuel Cell Sci. Technol.* **2010**, *7*. [CrossRef]
148. Weber, A.Z.; Newman, J. A theoretical study of membrane constraint in polymer-electrolyte fuel cells. *AIChE J.* **2004**, *50*, 3215–3226. [CrossRef]
149. Goulet, M.-A.; Arbour, S.; Lauritzen, M.; Kjeang, E. Water sorption and expansion of an ionomer membrane constrained by fuel cell electrodes. *J. Power Sources* **2015**, *274*, 94–100. [CrossRef]
150. Tang, Y.L.; Karlsson, A.M.; Santare, M.H.; Gilbert, M.; Cleghorn, S.; Johnson, W.B. An experimental investigation of humidity and temperature effects on the mechanical properties of perfluorosulfonic acid membrane. *Mater. Sci. Eng. A-Struct. Mater. Prop. Microstruct. Process.* **2006**, *425*, 297–304. [CrossRef]

151. Tang, Y.L.; Kusoglu, A.; Karlsson, A.M.; Santare, M.H.; Cleghorn, S.; Johnson, W.B. Mechanical properties of a reinforced composite polymer electrolyte membrane and its simulated performance in PEM fuel cells. *J. Power Sources* **2008**, *175*, 817–825. [CrossRef]
152. Kusoglu, A.; Kienitz, B.L.; Weber, A.Z. Understanding the Effects of Compression and Constraints on Water Uptake of Fuel-Cell Membranes. *J. Electrochem. Soc.* **2011**, *158*, B1504–B1514. [CrossRef]
153. Tang, Y.L.; Santare, M.H.; Karlsson, A.M.; Cleghorn, S.; Johnson, W.B. Stresses in proton exchange membranes due to hygro-thermal loading. *J. Fuel Cell Sci. Technol.* **2006**, *3*, 119–124. [CrossRef]
154. Kusoglu, A.; Tang, Y.L.; Lugo, M.; Karlsson, A.M.; Santare, M.H.; Cleghorn, S.; Johnson, W.B. Constitutive response and mechanical properties of PFSA membranes in liquid water. *J. Power Sources* **2010**, *195*, 483–492. [CrossRef]
155. Khattra, N.S.; Karlsson, A.M.; Santare, M.H.; Walsh, P.; Busby, F.C. Effect of time-dependent material properties on the mechanical behavior of PFSA membranes subjected to humidity cycling. *J. Power Sources* **2012**, *214*, 365–376. [CrossRef]
156. Khorasany, R.M.H.; Goulet, M.-A.; Sadeghi Alavijeh, A.; Kjeang, E.; Wang, G.G.; Rajapakse, R.K.N.D. On the constitutive relations for catalyst coated membrane applied to in-situ fuel cell modeling. *J. Power Sources* **2014**, *252*, 176–188. [CrossRef]
157. Futter, G.A.; Gazdzicki, P.; Friedrich, A.; Latz, A.; Jahnke, T. Physical modeling of polymer-electrolyte membrane fuel cells: Understanding water management and impedance spectra. *J. Power Sources* **2018**, *391*, 148–161. [CrossRef]
158. Franco, A.A.; Schott, P.; Jallut, C.; Maschke, B. A multi-scale dynamic mechanistic model for the transient analysis of PEFCs. *Fuel Cells* **2007**, *7*, 99–117. [CrossRef]
159. Shah, A.A.; Kim, G.S.; Sui, P.C.; Harvey, D. Transient non-isothermal model of a polymer electrolyte fuel cell. *J. Power Sources* **2007**, *163*, 793–806. [CrossRef]
160. Ziegler, C.; Yu, H.M.; Schumacher, J.O. Two-phase dynamic modeling of PEMFCs and simulation of cyclo-voltammograms. *J. Electrochem. Soc.* **2005**, *152*, A1555–A1567. [CrossRef]
161. Khandelwal, M.; Mench, M.M. Direct measurement of through-plane thermal conductivity and contact resistance in fuel cell materials. *J. Power Sources* **2006**, *161*, 1106–1115. [CrossRef]
162. Dai, W.; Wang, H.J.; Yuan, X.Z.; Martin, J.J.; Yang, D.J.; Qiao, J.L.; Ma, J.X. A review on water balance in the membrane electrode assembly of proton exchange membrane fuel cells. *Int. J. Hydrogen Energy* **2009**, *34*, 9461–9478. [CrossRef]
163. Kim, S.; Mench, M.M. Investigation of temperature-driven water transport in polymer electrolyte fuel cell: Thermo-osmosis in membranes. *J. Membr. Sci.* **2009**, *328*, 113–120. [CrossRef]
164. Fu, R.S.; Preston, J.S.; Pasaogullari, U.; Shiomi, T.; Miyazaki, S.; Tabuchi, Y.; Hussey, D.S.; Jacobson, D.L. Water Transport Across a Polymer Electrolyte Membrane under Thermal Gradients. *J. Electrochem. Soc.* **2011**, *158*, B303–B312. [CrossRef]
165. Weber, A.Z. Gas-crossover and membrane-pinhole effects in polymer-electrolyte fuel cells. *J. Electrochem. Soc.* **2008**, *155*, B521–B531. [CrossRef]
166. Kundu, S.; Fowler, M.W.; Simon, L.C.; Abouatallah, R.; Beydokhti, N. Degradation analysis and modeling of reinforced catalyst coated membranes operated under OCV conditions. *J. Power Sources* **2008**, *183*, 619–628. [CrossRef]
167. Zhang, J.X.; Gasteiger, H.A.; Gu, W.B. Electrochemical Measurement of the Oxygen Permeation Rate through Polymer Electrolyte Membranes. *J. Electrochem. Soc.* **2013**, *160*, F616–F622. [CrossRef]
168. Zhang, L.; Ma, C.S.; Mukerjee, S. Oxygen permeation studies on alternative proton exchange membranes designed for elevated temperature operation. *Electrochim. Acta* **2003**, *48*, 1845–1859. [CrossRef]
169. Ahluwalia, R.K.; Wang, X. Buildup of nitrogen in direct hydrogen polymer-electrolyte fuel cell stacks. *J. Power Sources* **2007**, *171*, 63–71. [CrossRef]
170. Giacinti Baschetti, M.; Minelli, M.; Catalano, J.; Sarti, G.C. Gas permeation in perflurosulfonated membranes: Influence of temperature and relative humidity. *Int. J. Hydrogen Energy* **2013**, *38*, 11973–11982. [CrossRef]
171. Baik, K.D.; Kim, M.S. Characterization of nitrogen gas crossover through the membrane in proton-exchange membrane fuel cells. *Int. J. Hydrogen Energy* **2011**, *36*, 732–739. [CrossRef]
172. Karpenko-Jereb, L.; Sternig, C.; Fink, C.; Tatschl, R. Membrane degradation model for 3D CFD analysis of fuel cell performance as a function of time. *Int. J. Hydrogen Energy* **2016**, *41*, 13644–13656. [CrossRef]

173. Hu, J.W.; Sui, P.C.; Kumar, S.; Djilali, N. Modelling and simulations of carbon corrosion during operation of a Polymer Electrolyte Membrane fuel cell. *Electrochim. Acta* **2009**, *54*, 5583–5592. [CrossRef]
174. Sundaram, N.; Peppas, N.A. Friction coefficient analysis of multicomponent solute transport through polymer membranes. *J. Appl. Polym. Sci.* **1996**, *60*, 95–101. [CrossRef]
175. Weber, A.Z.; Delacourt, C. Mathematical Modelling of Cation Contamination in a Proton-exchange Membrane. *Fuel Cells* **2008**, *8*, 459–465. [CrossRef]
176. Burlatsky, S.F.; Gummalla, M.; Atrazhev, V.V.; Dmitriev, D.V.; Kuzminyh, N.Y.; Erikhman, N.S. The Dynamics of Platinum Precipitation in an Ion Exchange Membrane. *J. Electrochem. Soc.* **2011**, *158*, B322–B330. [CrossRef]
177. Wong, K.H.; Kjeang, E. Mitigation of Chemical Membrane Degradation in Fuel Cells: Understanding the Effect of Cell Voltage and Iron Ion Redox Cycle. *ChemSusChem* **2015**, *8*, 1072–1082. [CrossRef] [PubMed]
178. Futter, G.A.; Latz, A.; Jahnke, T. Physical modeling of chemical membrane degradation in polymer electrolyte membrane fuel cells: Influence of pressure, relative humidity and cell voltage. *J. Power Sources* **2019**, *410*, 78–90. [CrossRef]
179. U.S. Department of Energy. 3.4. Fuel Cell. In *Fuel Cell Technologies Office Multi-Year Research, Development, and Demonstration Plan*; U.S. Department of Energy: Washington, DC, USA, 2017.
180. Qiu, D.K.; Peng, L.F.; Lai, X.M.; Ni, M.; Lehnert, W. Mechanical failure and mitigation strategies for the membrane in a proton exchange membrane fuel cell. *Renew. Sustain. Energy Rev.* **2019**, *113*. [CrossRef]
181. Kim, S.; Khandelwal, M.; Chacko, C.; Mench, M.M. Investigation of the Impact of Interfacial Delamination on Polymer Electrolyte Fuel Cell Performance. *J. Electrochem. Soc.* **2009**, *156*, B99–B108. [CrossRef]
182. Burlatsky, S.F.; Gummalla, M.; O'Neill, J.; Atrazhev, V.V.; Varyukhin, A.N.; Dmitriev, D.V.; Erikhman, N.S. A mathematical model for predicting the life of polymer electrolyte fuel cell membranes subjected to hydration cycling. *J. Power Sources* **2012**, *215*, 135–144. [CrossRef]
183. Atrazhev, V.V.; Astakhova, T.Y.; Dmitriev, D.V.; Erikhman, N.S.; Sultanov, V.I.; Patterson, T.; Burlatsky, S.F. The Model of Stress Distribution in, Polymer Electrolyte Membrane. *J. Electrochem. Soc.* **2013**, *160*, F1129–F1137. [CrossRef]
184. Khorasany, R.M.H.; Singh, Y.; Sadeghi Alavijeh, A.; Rajapakse, R.K.N.D.; Kjeang, E. In-situ simulation of membrane fatigue in polymer electrolyte fuel cells. *Int. J. Hydrogen Energy* **2017**, *42*, 11838–11844. [CrossRef]
185. Chandesris, M.; Vincent, R.; Guetaz, L.; Roch, J.-S.; Thoby, D.; Quinaud, M. Membrane degradation in PEM fuel cells: From experimental results to semi-empirical degradation laws. *Int. J. Hydrogen Energy* **2017**, *42*, 8139–8149. [CrossRef]
186. Shah, A.A.; Ralph, T.R.; Walsh, F.C. Modeling and Simulation of the Degradation of Perfluorinated Ion-Exchange Membranes in PEM Fuel Cells. *J. Electrochem. Soc.* **2009**, *156*, B465–B484. [CrossRef]
187. Gummalla, M.; Atrazhev, V.V.; Condit, D.; Cipollini, N.; Madden, T.; Kuzminyh, N.Y.; Weiss, D.; Burlatsky, S.F. Degradation of Polymer-Electrolyte Membranes in Fuel Cells II. Theoretical model. *J. Electrochem. Soc.* **2010**, *157*, B1542–B1548. [CrossRef]
188. Coulon, R.; Bessler, W.G.; Franco, A.A. Modeling Chemical Degradation of a Polymer Electrolyte Membrane and its Impact on Fuel Cell Performance. *ECS Trans.* **2010**, *25*, 259–273. [CrossRef]
189. Singh, R.; Sui, P.C.; Wong, K.H.; Kjeang, E.; Knights, S.; Djilali, N. Modeling the Effect of Chemical Membrane Degradation on PEMFC Performance. *J. Electrochem. Soc.* **2018**, *165*, F3328–F3336. [CrossRef]
190. Fink, C.; Gößling, S.; Karpenko-Jereb, L.; Urthaler, P. CFD Simulation of an Industrial PEM Fuel Cell with Local Degradation Effects. *Fuel Cells* **2020**, *20*, 431–452. [CrossRef]
191. Quiroga, M.A.; Malek, K.; Franco, A.A. A Multiparadigm Modeling Investigation of Membrane Chemical Degradation in PEM Fuel Cells. *J. Electrochem. Soc.* **2016**, *163*, F59–F70. [CrossRef]
192. Carcadea, E.; Varlam, M.; Marinoiu, A.; Raceanu, M.; Ismail, M.S.; Ingham, D.B. Influence of catalyst structure on PEM fuel cell performance—A numerical investigation. *Int. J. Hydrogen Energy* **2019**, *44*, 12829–12841. [CrossRef]
193. Penga, Z.; Bergbreiter, C.; Barbir, F.; Scholta, J. Numerical and experimental analysis of liquid water distribution in PEM fuel cells. *Energy Convers. Manag.* **2019**, *189*, 167–183. [CrossRef]
194. Rizvandi, O.B.; Yesilyurt, S. A pseudo three-dimensional, two-phase, non-isothermal model of proton exchange membrane fuel cell. *Electrochim. Acta* **2019**, *302*, 180–197. [CrossRef]
195. Wang, Y.L.; Wang, S.X.; Liu, S.C.; Li, H.; Zhu, K. Three-dimensional simulation of a PEM fuel cell with experimentally measured through-plane gas effective diffusivity considering Knudsen diffusion and the liquid water effect in porous electrodes. *Electrochim. Acta* **2019**, *318*, 770–782. [CrossRef]

196. Zhang, G.B.; Xie, X.; Xie, B.A.; Du, Q.; Jiao, K. Large-scale multi-phase simulation of proton exchange membrane fuel cell. *Int. J. Heat Mass Transf.* **2019**, *130*, 555–563. [CrossRef]
197. Zhao, J.; Li, X.G. Oxygen transport in polymer electrolyte membrane fuel cells based on measured electrode pore structure and mass transport properties. *Energy Convers. Manag.* **2019**, *186*, 570–585. [CrossRef]

membranes

MDPI

Review

Insights into Advancements and Electrons Transfer Mechanisms of Electrogens in Benthic Microbial Fuel Cells

Mohammad Faisal Umar [1], Syed Zaghum Abbas [2,*], Mohamad Nasir Mohamad Ibrahim [3], Norli Ismail [1] and Mohd Rafatullah [1,*]

1 Division of Environmental Technology, School of Industrial Technology, Universiti Sains Malaysia, Penang 11800, Malaysia; faisalumar@student.usm.my (M.F.U.); norlii@usm.my (N.I.)
2 Biofuels Institute, School of Environment, Jiangsu University, Zhenjiang 212013, China
3 School of Chemical Sciences, Universiti Sains Malaysia, Penang 11800, Malaysia; mnm@usm.my
* Correspondence: Zabbas@ujs.edu.cn (S.Z.A.); mohd_rafatullah@yahoo.co.in or mrafatullah@usm.my (M.R.); Tel.: +60-4-6532111 (M.R.); Fax: +60-4-656375 (M.R.)

Received: 7 August 2020; Accepted: 19 August 2020; Published: 28 August 2020

check for updates

Abstract: Benthic microbial fuel cells (BMFCs) are a kind of microbial fuel cell (MFC), distinguished by the absence of a membrane. BMFCs are an ecofriendly technology with a prominent role in renewable energy harvesting and the bioremediation of organic pollutants through electrogens. Electrogens act as catalysts to increase the rate of reaction in the anodic chamber, acting in electrons transfer to the cathode. This electron transfer towards the anode can either be direct or indirect using exoelectrogens by oxidizing organic matter. The performance of a BMFC also varies with the types of substrates used, which may be sugar molasses, sucrose, rice paddy, etc. This review presents insights into the use of BMFCs for the bioremediation of pollutants and for renewable energy production via different electron pathways.

Keywords: bioremediation; renewable energy; organic pollutants; electrogens; wastewater

1. Introduction

Different environmental pollutants, such as organic- and inorganic-based contaminants, remain a severe challenge to the sustainability of water resources [1,2]. This poses a serious threat to living organisms, including human beings and marine organisms [3]. Due to the depletion of natural water resources, there is an imbalance in the natural ecosystem, but simultaneously the commutability of renewable pure water resources has been enhanced. There is a plethora of potential sources of pollution in water bodies (e.g., oceans, lakes, rivers and reservoirs) stemming from human activity, and notably the chemical and oil filtration industries. The chemical substances emitted from these industries contain very harmful and potentially carcinogenic inorganic and organic pollutants [4]. These pollutants have a severe impact on living organisms and pose a serious threat to the environment.

Several techniques exist for the treatment of wastewater prior to irrigation, such as lagoon ponds, constructed wetlands, conventional wastewater treatment plants, membrane bioreactors and membrane filtration. Although these techniques have been shown to be effective, disadvantages remain, i.e., they require a large area for operation, along with high economic stability [5]. Recently, a novel approach was introduced for the treatment of wastewater: the microbial fuel cell. Microbial fuel cells (MFCs) are devices which utilize microbial activity to produce electricity from chemical energy stored in an organic substrate. Thus, MFCs are a promising technique for wastewater bioremediation and for generating electricity in an economical way.

Organic pollutant compounds are oxidized by microorganisms and the transfer of electrons to the anode of the MFC via exoelectrogens [6,7]. A new type of MFC, the benthic microbial fuel cell (BMFC), was designed to generate electricity from organic matter present in wastewater. As a result, like with MFCs, chemical energy is converted into electrical energy with exoelectrogens working as a catalyst, i.e., electrons (e^-) and protons (H^+) are released. In this way, a potential difference exists between the anode and cathode. Here, we present information regarding recent developments using exoelectrogens on the anode by direct and indirect processes.

2. Benthic Microbial Fuel Cell (BMFC)

There is a need for sustainable and clean energy sources to meet growing energy demands. In 2014, the global percentage of electricity generated via the consumption of fossil fuels was 66%; however, only 11% of this was utilized together with renewable energy [8,9]. Organic substrates are used as bio sediments, and they protect the microbial ecosystem in various regions and provide a suitable environment for the bioremediation of accumulated pollutants via the electron donor–acceptor mechanism [10]. Currently, physiochemical processes, such as dredging, ozonation and electrochemical degradation, are used for the bioremediation of pollutants. These techniques are effective but require a lot of energy and are costly, limiting their application. Usually, the accumulation of reductive substances and the lack of electron acceptors are the main limitations for the remediation of sediment under anaerobic conditions.

In recent years, microbial fuel cells (MFC) have been considered as an alternative, cheap approach to the bioremediation of toxic organic pollutants via power generation. Recently, BMFCs have attracted the attention of many researchers due to their nonaggressive and easily controllable nature. BMFCs consist of an anode, which is embedded in organic matter, and a cathode, which is placed in the overlying water. The air diffuser provides a constant supply of oxygen which plays a vital role in the transfer of electrons and protons from the anode to cathode via an external circuit, where electrons react with oxygen and produce water [11,12].

Reimers et al. [13] were the first to employ BMFCs; their approach included a platinum mesh for the anode and carbon fiber for the cathode. A unique feature of the BMFC is its membrane-less assembly; this is possible thanks to the boundary organic substrate used as a substrate, which itself acts as a pseudo membrane. Nowadays, many researchers are working on improving ecofriendly systems, including BMFCs [14]. The prototype of a double chamber BMFC is shown in Figure 1.

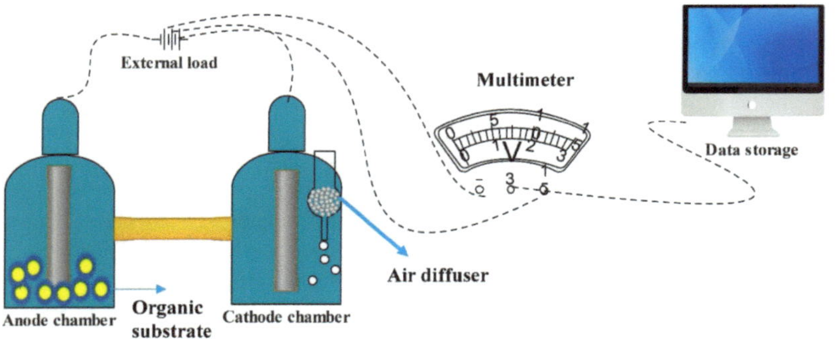

Figure 1. General prototype scheme of a benthic microbial fuel cell.

An air cathode in the overlaying water connected with a benthic-integrating anode is the most common BMFC model. In a saline environment, conductivity is normally high, so the overpotential limits the BMFC performance; this is not the case in freshwater [15]. Under the latter scenario, the efficiency of the anode decreases because of anodic contamination, i.e., the accumulation of waste substrate in the anodic region. BMFCs are usually restricted in terms of the proximity of the electrode by the naturally forming spatial separation of oxic and anoxic zones [16]. The tubular air cathode designs along with the cathodic fabric assembly structure suggest that only low-cost fabric would separate the electrodes. In this configuration, the cathode catalytic layer was exposed to air and would allow a hydrogen oxidation reaction [17]. However, as this setup requires long tubes for air exposure, the BMFC's setup cannot operate in deep-water environment. If the BMFC can adapt the cathode carbon cloth, then embedded cathode in the organic substrate can also be used optionally [2]. In the simple design of the BMFC, though, electrodes can be constructed from both graphite felt or carbon cloth.

3. Degradation of Organic Matter by BMFC

Like bio-electrochemical systems, BMFCs too have been shown to boost the organic compounds biodegradation, i.e., total petroleum hydrocarbons, total organic carbon, ignition loss and polycyclic aromatic hydrocarbons present in the wastewater, as shown in Figure 2. BMFC takes some time for the formation of a biofilm on the anode, which is the main requirement for the removal of the organic contents [18,19]. The anodic biofilm consists of two types of bacteria, the fermentative bacteria and the exoelectrogens. Fermentative bacteria are primarily involved in the complex organic matter hydrolysis and transform the products of hydrolysis into ethanol, H_2, volatile organic acids and CO_2 by acid-forming fermentation [20]. Ethanol, into which lactic acid can easily be converted, is volatile and readily escapes, allowing the reaction to proceed easily. CO_2 is the other product, but is weakly acidic and even more volatile than ethanol. H_2 is a substrate for methanogens and sulfate reducers, which keep the concentration of hydrogen low and favor the production of such an energy-rich compound, but hydrogen gas at a fairly high concentration can nevertheless be formed.

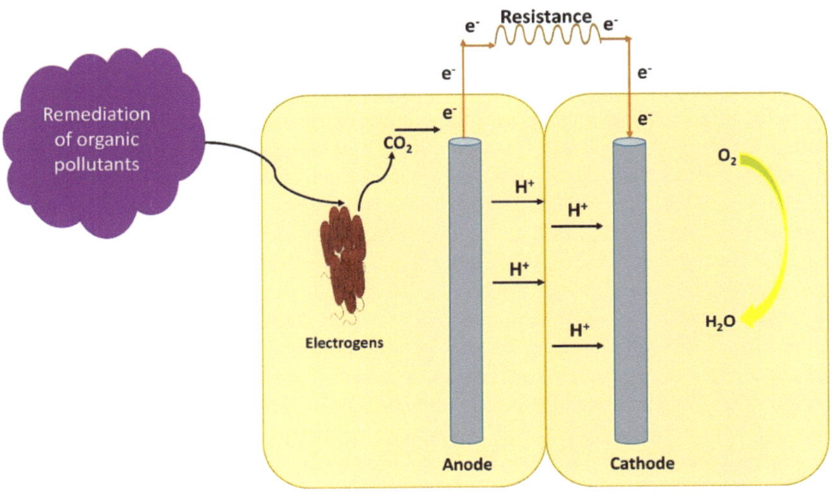

Figure 2. Overview of organic pollutants removal by benthic microbial fuel cell.

The metabolites of fermentative bacteria used by electrogenic bacteria as substrates, which produce electrons, CO_2 and protons by oxidation, are shown in Equation (1). The protons are shifted to the overlying cathodic water and transfer few electrons towards the anode, which can be seen in Equation (2). These electrons are passed to the cathode through an external circuit and a redox reaction occurs that generates protons and dissolves oxygen, as mentioned in Equation (3) [21]. The existence of these electrodes has established a new microbial mechanism for metabolism, and to some degree it alters anodic microbial communities too. Recently, it has been reported that BMFCs alone cannot efficiently remove the organic pollutants. Wu et al. [18] reported that zero-valent iron (ZVI) has a high reducing ability (E0 = −0.44 V) and could react with the oxidizing contaminants. The hydroxyl radical formed through this method is a very durable oxidative degradation of bio-refractory organics, which allows for the common use of ZVI technology in the treatment of dyes, complexing agents, chlorinated organic compounds and preservatives. ZVI can also alter the metabolic pathways and redox capacity, regulate acidification and promote extracellular electron transfer. Estevezcanales et al. [22] cultivated *Geobacter sulfurreducense* with an iron-free substratum and found an abruptly reduced cytochrome *c*, which showed a limited capacity of outer membrane electrons transport. However, using ZVI alone, the desired effect cannot be guaranteed, particularly the final removal of certain refractory contaminants. The combined use of ZVI and BMFC technologies offers an enhanced substitute approach for eliminating organic contaminants.

Anode: (oxidation)

$$a(OP) + bH_2O \rightarrow cCO_2 + ne^- + dH^+ \tag{1}$$

Cathode: (reduction)

$$eO_2 + dH^+ + ne^- \rightarrow bH_2O \tag{2}$$

Overall reaction: (redox reaction)

$$a(OP) + eO_2 \rightarrow cCO_2 + bH_2O \tag{3}$$

a = number of organic pollutants (OP) molecules, b = number of water molecules, c = number of carbon dioxide molecules, d = number of protons, e = number of oxygen molecules and n = number of electrons.

The removal of organic contaminants from BMFC is the foremost priority for organic contents remediation. Many hydrocarbons, such as those consisting of nitro and chlorine aromatic compounds, can be employed as substrates in BMFC. For bioremediation, these compounds need bioreduction [23]. The amalgamation of bioremediation and the electrochemical system forms a synergistic connection among electrodes and bacteria and enables the bioreduction of perchloroethane and polycyclic aromatic hydrocarbons. The in-situ generation of oxygen and hydrogen can be employed for intermediates reduction. The energy efficiency and removal of these organic compounds can be upgraded by direct electron transfer to electrodes from exoelectrogens or the inclusion of dechlorinating species [24]. There is a proportional relationship among power production and the degradation of these organic compounds. This closed-circuit BMFC creates the optimum environment for the degradation of organic compounds. This system could have a negative impact on BMFC microbes if not used properly. During the remediation of the organic compound in BMFC, some common issue are encountered, such as cathodic pH becoming alkaline and anodic pH becoming acidic via water electrolysis [25]. Unequal nutrients distribution in the chamber, like nitrate and phosphate, accumulating in the cathode chamber and ammonium accumulating in the anode chamber are other issues encountered during the remediation. These issues not only effect the performance of BMFC but also the biological clogging. These issue can be resolved by reversal of electrodes polarity and with proper water circulation. The degradation of organic compounds is also influenced by the competitive reactions with nitrate and sulphate [26].

4. Electron Transfer Mechanism by Electrogens

The electrons transmission mechanism is essential in order to acquire a flawless knowledge for the application of BMFC at a large scale. In the anodic chamber of BMFC, organic substrates are reduced by microbes and transfer electrons to anodes, from where the electrons move to the cathode through external circuit to generate electricity [27]. Earlier, the microbes were exploited in the anodic chamber, but recently microbes are also exploited as biocathodes in the cathodic region to assist electrons transmission to the terminal electron acceptor (TEA) [28,29]. The power density, current density and coulombic efficiency can be measured by electron transfer rate. If the electrons transfer rate is higher than the electrons passing through the external circuit, more coulombic efficiency, power density and current density will be measured, leading to higher voltage production. The harvested bioenergy produced by the electron transfer towards electrodes from the respiration chain of electrogens is known as a new BMFC technology [30]. There are two means of electrons transfer in BMFC occupied by microbes: (i) direct electron transfer (direct contact between the microbes and the electrode surface) and (ii) indirect electron transfer (through the so-called electron mediators), as shown in Figure 3.

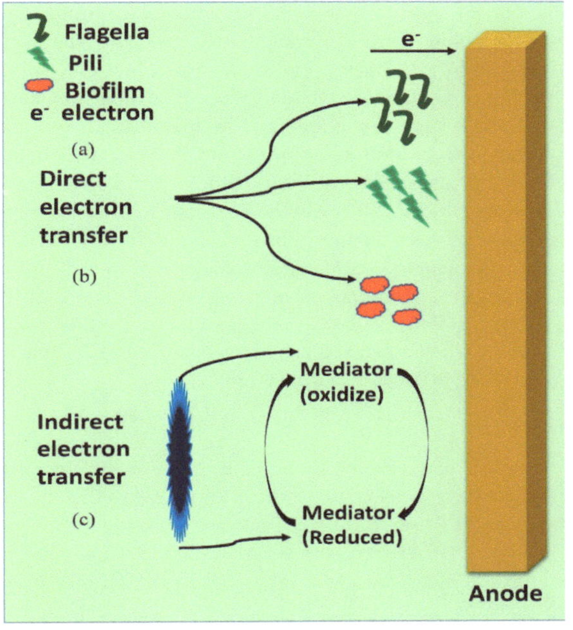

Figure 3. Proposed electron transfer mechanisms utilized direct electrons (**a**,**b**) and indirect electrons transfer (**c**) through electrogens using benthic microbial fuel cell.

Recently, the application of electro-autotrophs in the Bioelectrochemical Systems (BES) has attracted the attention of researchers. The exoelectrogens use the electrodes or extracellular insoluble mineral as terminal electron acceptor (TEA), while electro-autotrophs accept the electrons from electrodes or solid compounds for CO_2 reduction and produce multi-carbon compounds [31]. Gregory et al. first studied the electro-autotrophy in the *Geobacter*, which is a model exoelectrogen [32]. Most exoelectrogens are iron-oxidizing bacteria, which led to the hypothesis that dissimilatory iron-reducing bacteria can only accept the electrons from a cathode. Indeed, *Mariprofundus ferrooxydans* PV-1, *Acidithiobacillus ferrooxidans* and *Rhodopseudomonas palustris* have been selected as electro-autotrophs [33–35]. Furthermore, *Methanobacterium archaeon* strain IM1 and chemolithoautotrophic archea *Methanococcus maripaludis* were purified for electromethanogenesis with an electron donor (metallic iron) [36]. Many acetogenic bacteria like *Sporumosa acidovorans*, *Sporomusa silvacetica*, *Sporomusa sphaeroides*, *Sporumosa malonica*, *Moorella thermoacetica*, *Sporomusa ovate*, *Clostridium aceticum* and *Clostridium ljungdahlii* can also accept electrons from the cathode and reduce CO_2 to organic acids [37]. Some sulphate-reducing autotrophs are believed to accept the electrons from cathode and generate hydrogen (H_2) by reducing sulphate [38]. The cathodic electron consumption by bacteria causes anaerobic microbial-induced corrosion (MIC). The electro-autotrophs generate the corrosive hydrogen sulphide that results in chemically induced iron corrosion. The electro-autotrophs also stimulate the induced electrochemical corrosion by using cathodic hydrogen, which is generated by iron–water contact [39]. The benthic microbial fuel cells (BMFCs) were constructed for anaerobic exoelectrogenic enrichment, which separates the electrotrophic bacteria by opposing the anode to bio-cathode [40]. Recently, an MFC was developed initially with heterotrophic conditions that later alter with autotrophic conditions. After five batches of cultivation, the nonelectrochemical bacteria is dispersed into the liquid medium and only electro-autotrophs bacteria (*Geobacter*) were abundant in the MFC [41]. This electro-autotrophic process promotes the growth of exoelectrogens on the electrodes and reduces the number of nonelectrobiochemical bacteria, which finally increases the MFC's efficiency. The electro-autotrophic enrichment of the bio-cathode offers a simplified approach to purify the bio-chemical from various inoculum sources. Initially, bacteria are grown heterotrophically on fructose, glycerol and glucose, followed by acclimation to the medium, and CO_2 was provided as the sole electron acceptor [42]. The conventional cathode causes corrosion, denaturation and toxicity of material, but the bio-cathode is very cost-effective. The microbes must be chosen based on their capability to shift from heterotrophic to autotrophic metabolism. This pathway may help us to understand the metabolic pathways of different electron donors or acceptor microbes that have formed on bio-cathodes [43]. For the production of valuable organic and fuel commodities, pure culture was used because the diversified electro-autotrophs uptake the electrons from the negatively poised cathode for CO_2 reduction with heavier coulombic efficiencies. The mixed cultures primarily generate the complex products and acetates, which maintains the microbial metabolism. The surfeit of products was generated by employing a viable BES system with pure culture of *Clostridium ljungdahlii*. Overall, though, very little research has been focused on the electro-autotrophs, particularly the electrons transfer pathways from cathode to bacteria and their applications.

4.1. Direct Electron Transfer

Electrons should interact between the outer membrane of the microbes and the electrode. The biofilm or electrically conductive nanowires (pili and flagella) were found over the surface of the anode formed by electrogens [44]. The transmission of electrons takes place by direct interaction without any external mediator through an external membrane's cytochromes, nanowires and electron transport proteins in exchange with the microbial membranes. The external membrane's cytochromes are bonded with nanowires and allow electrogens to use an electrode as an electron acceptor. Furthermore, the direct electron transfer mechanism fully depends on the electron transport proteins, and they play a crucial role in electron transfer from cytoplasm to mitochondrial membrane. The drawback of this mechanism is the very poor electron transfer rate, because the active sites of electron transmission are deeply embedded within the proteins [45]. Recently, many electrochemical bacteria like Shewanella and Geobacter nanowires have been folded for better electrons transmission [46,47]. For effective and fast electron transfer (coulombic efficiency), the nanowires form an electroactive layer instead of a normal single layer. *Geobacter* species are diverse in their current production ability; *Geobacter hydrogenophilus* and *Geobacter metallireducens* produced higher current densities (0.2 $mAcm^{-2}$) than *Geobacter bremensis*, *Geobacter chapellei*, *Geobacter humireducens*, *Geobacter uraniireducen* and *Geobacter bemidjiensis*, which produced much lower current densities (0.05 $mAcm^{-2}$) [48]. Some electrogesns reported direct electron transfer to electrodes, such as *Geobacter sulfurreducens* [49], *Rhodopseudomonas palustris* [50], *Anaeromyxobacter dehalogenansc* [51], *Geobacter lovleyi* [52], *Pseudomonas aeruginosa* [53], *Thermincola potens* [54], *Shewanella oneidensis* [55], *Geothrix fermentans* [56], *Thermincola carboxydophila* [57], *Shewanella putrefaciens* [58], and *Escherichia coli* [59].

Much less is known about direct electron transfer pathways in the electro-autotrophic bacteria. From the experiments, it is confirmed that the Fe species uptake the electrons secreted by the cathodic biofilm. It is also ventured that *c-type cytochromes*, which are crucial constituents of Fe extracellular electron uptake, also play a vital part in the electron transmission from cathode to electro-autotrophs [60]. In the light of this hypothesis, the metaproteomics and metagenomics of the diversified microbial community inhibit the self-regenerating biocathode's effect whereby CO_2 is reduced via *c-type cytochromes* directly acquiring electrons from the *Chromatiaceae* family and other proteins related with Fe(II) oxidation [61]. The Fe(0)-corroding sulphate reducing microbes (SRM) could also uptake the electrons [62], and this discovery paved the way for scientists to use these microbes in biocathodic BES employments. So, this negative metabolic character can be turned into a sustainable positive biotechnological solution. Up to now, though, only some pure SRM cultures are used as electro-autotrophs. The cathodic biofilm of *Desulfovibrio desulfuricans* ATCC 27774 exhibited electro-autotrophic characteristics at an employed cathodic potential (E_{cath}) of −0.169 V vs. SHE. After 20 days, lactate was supplemented as the carbon source, not CO_2, and a stable negative current was measured [63]. Consequently, other species (*Desulfovibrio caledoniensis* and *Desulfovibrio paquesii*) of the genus *Desulfovibrio* were used for H_2 and cathodic current generation, employing lactate or bicarbonate as the carbon source and E_{cath} that enabled abiotic H_2 evolution [64].

The pure cultures of *Desulfovibrio piger* and *Desulfosporosinus orientis* displayed the electro-autotrophic properties at E_{cath} = −0.31 V vs. SHE, which has a higher positive potential than the neutral redox potential of H_2 evolution ($E^{0'}{}_{H+/H_2}$ = −0.41 V vs. SHE) and gaseous CO_2 supplemented as an inorganic source [65]. *Desulfovibrio piger* (SRM *Deltaproteobacterium*) is a H_2-oxidizing, Gram-negative, nonspore-forming electro-autotroph. It could oxidize organic matter, like lactate, pyruvate ethanol and, partly, acetate. Before this, its autotrophic metabolism effect on CO_2, was not reported for other *Desulfovibrio* species. *Desulfosporosinus orientis* (SRM class Clostridia) is an acetogenic, capable of executing anaerobic sulfate respiration, and is a spore-forming electro-autotroph. The broad range of energy sources, such as pyruvate, ethanol, formate, methanol, H_2, Fatty acids, lactate CO and CO_2, can be used by *D. orientis* [66]. It can use various TEAs, such as sulphite, sulphate, sulphur dioxide and thiosulfate [67].

In BES, for the first time *Desulfopila corrodens* strain IS4 was identified as an Fe(0)-corroding SRM [38]. By using an electron donor (metallic iron), this *Deltaproteobacterium* (Gram-negative) was quarantined from marine sediment. This strain performs very fast hydrogen generation and sulphate reduction by consuming iron as an energy source as compared to orthodox hydrogen-foraging *Desulfovibrio* species. In BES, by using CO_2 as the growth substrate at E_{cath} = −0.4 V vs. SHE, direct electron uptake was accomplished [38]. Currently, *Desulfobacterium autotrophicum* HRM2 (sulphate reducing bacteria) is being reported as an electro-autotroph at E_{cath} = −0.5 V vs. SHE. This *Deltaproteobacterium*, secluded from marine mud, is a fully SRM oxidizer having both directional pathways (Wood-Ljungdahl) and relating to the *c*-Cyt rich group [68]. *D. autotrophicum* HRM2 showed a high coulombic efficiency (83 ± 6%) and a capacity for acetate bio-electro synthesis [69].

4.2. Indirect Electron Transfer

Indirect electron transfer does not require direct physical interaction between the microbes and electrons acceptors. The small molecules and soluble mediator are involved in the inducement of this electron's transfer mechanism. In this mechanism, the electrons mediator enters into the microbes, where the electrons are extracted by a metabolic reaction of electrogens, and finally these electrons are transferred to an anode [70]. Initially, at the first BMFC operative phase, the presence of electron mediators was considered as important. The electron mediators auxiliary in the BMFC anodic chamber are produced by electrogens. Several types of species had been investigated, as the synthesis of self-mediators known as endo-electrogens mediators, such as phenazine and pyocyanin, could be secreted by *Shewanella* and *Pseudomonas* species [71]. The potential differences between several electron mediators and redox proteins were reported in many studies, which significantly affects the electron transfer efficiency of different species [72]. However, the tendency of electrons transfer is affected by different chemical compounds known as exoelectricigens mediators, such as anthracenedione, thionine, neutral red, humic acid, riboflavin and methylene blue [73–75]; both exo-electrogens and endo-electrogens are shown in Table 1. These electrogens are exploited to transfer the electrons from inside of the cell towards the electrode, and different microbes have a different capability to transfer electrons from cell to electrode.

Table 1. Performance of BMFC configuration through exoelectrogens and endoelectrogens with respect to power density.

Microorganisms	External Mediator	Power Density (mW m^{-2})	Configurations	Type of Electrons Transfer Mechanisms	References
Exoelectrogens microorganisms					
Shewanella oneidensis strain 14063	1–amino–2–Napthol	>40	Single chamber	Direct transfer	[76]
Shewanella oneidensis	Anthraquinone–2,6–disulfonate (AQDS)	24	Double chamber	Direct transfer	[77]
Klebsiella pneumoniae	HNQ as mediator biomineralized manganese as electron acceptor	–	–	Direct transfer	[78]
Pseudomonas species	phenazine–1–carboxamide	–	–	Indirect transfer	[79]
Pseudomonas aeruginosa	phenazine compounds	3322 ± 38	Single chamber	Direct transfer	[80]
Cellulomonas fimi	anthraquinone–2,6–disulfonate	38.7	Double chamber	Direct transfer	[81]
Lactococcus lactis	Riboflavin, flavins	–	Double chamber	Direct transfer	[82]
Geobacter sulfurreducens	*c–Cytochrome z, type IV pili*	3147	Double chamber	Direct transfer	[83]
Shewanella oneidensis DsP10	Anthraquinone–2,6–disulfonate (AQDS)	5000	Double chamber	Direct transfer	[77]
Rhodopseudomonas palustris DX-1	*c–Type cytochromes*	2720	Single chamber	Indirect transfer	[49]
Desulfovibrio desulfuricans ATTC	*c–Type cytochromes*	1580	Single chamber	Indirect transfer	[84]
Geobacter metallireducens	*c–Type cytochromes, OmcE and OmcB*	450	Single chamber	Indirect transfer	[85]
Desulfuromonas acetoxidans	*c–Type cytochromes*	2000	–	Indirect transfer	[13]
Klebsiella pneumonia	*2,6–Di–tert–butyl–p–benzoquinone*	199	–	–	[86]
Desulfovibrio alaskensis	*Transmembrane complexes, tetraheme cytochrome C3*	–	–	–	[87]
Pseudomonas aeruginosa	Phenazine–1–carboxamide, pyocyanin	4300	–	–	[88]
Thermincola ferriacetica	Anthraquinone–2,6–disulfonate	12,000	Single chamber	–	[89]
Shewanella putrefaciens	c–Type cytochromes including OmcA, MtrC, FAD transporter	492	Double chamber	Indirect transfer	[90]
Dechlorospirillum anomalous strain WD	*Anthraquinone–2,6–disulfonate hydrogen*	30	–	Direct transfer	[91]
Geobacter lovleyi	*Methyl viologen*	480	–	Indirect transfer	[92]
Chlorella vulgaris	*Methyl viologen, methylene blue*	30	Single chamber	Indirect transfer	[91]
Pseudomonas sp.	*Methylene blue*	979	Single chamber	Indirect transfer	[93]
Endoelectrogens microorganism					
Rhodoferax ferrireducens	–	158	Double chamber	Direct transfer	[94]
Klebsiela pneumoniae strain L17	–	34.77	Double chamber	Direct transfer	[95]
Nocardiopsis sp. KNU (strain), *Streptomyces enissocaesilis* KNU (K strains)	–	162 145	Double chamber	Direct transfer	[96]
Rhodoferax ferrireducens	–	–	Double chamber	Direct transfer	[97]
Escherichia coli strain K-12	–	215	Single chamber	–	[98]
Shewanella oneidensis	–	–	Single chamber	–	[99]
Pseudomonas aeruginosa	–	136 ± 87	Single chamber	–	[100]
Cellulomonas fimi	–	0.74 ± 0.07	Single chamber	Indirect transfer	[101]
Leptothrix discophora SP-6	–	70	–	Indirect transfer	[102]
Acinetobacter calcoaceticus	–	110	–	Indirect transfer	[50]
Escherichia coli	–	3390	–		[103]
Winogradskyella poriferorum	–	40	–	Indirect transfer	[104]
Pseudomonas fluorescens	–	210	Double chamber	Direct transfer	[105]
Citrobacter sp.	–	205	Double chamber	Indirect transfer	[106]
Lysinibacillus sphaericus	–	850	Double chamber	Direct transfer	[107]
Dechloromonas sp.	–	300	Double chamber	Indirect transfer	[108]
Arthrospira maxima	–	100	Double chamber	Direct transfer	[109]
Coriolus versicolor	–	3200	Single chamber	Indirect transfer	[110]

Very little is known about the electron uptake by acetogens from the cathode. Currently, by using a genetic system, it is being confirmed that *Clostridium ljungdahlii* (Gram-positive) exhibits protons pumps that cause proton motive force, which is necessary for its growth with CO_2 as a carbon source [111]. This gives clues about the energy conservation mechanism in the electro-autotrophic acetogens. In *Clostridium ljungdahlii*, the electron uptake mechanism is differently predicted, because it cannot synthesize quinones or *c*-type cytochromes [112]. By using genetic toolbox, the properties and electron uptake pathways of *Clostridium ljungdahlii* could be clearer, and also give information about the electron uptake pathways of many Gram-positive bacteria. The genomic sequence of acetogenic *Sporomusa ovata* (Gram-negative) is available now. Genes coding for type IV pili and *c*-type cytochromes are present in the genomic sequence, which are the two main parts of the extracellular electron transfer mechanism [113]. The *c*-type cytochromes are a precarious factor for the extracellular electron transfer mechanism in both electrotrophs and electrogenic types. In *Geobacter* spp., pili type IV are long strings that exhibit the metal-like conduction of long-range electron transfer. The gene coding for Ubiquinone also present in the genome of *Sporomusa ovate* is also crucial for the electron transfer pathway [114]. *Sporomusa ovate* has many extracellular electron transfer components, which proves that the electron uptake mechanisms of *Sporomusa ovata* are similar to those of other electrotrophic and electrigenic bacteria. *Sporomusa sphaeroides*-related acetogens showed direct electron transfer mechanisms. This showed that Gram-negative acetogens could use this strategy of electron transfer in different environments [115].

5. Performance of BMFC Affected by Organic Substrate

In BMFC, the chemical reaction is replaced by a microbial reaction where the organic substrates are utilized as fuel for feeding the microbes and generating renewable energy. All these microorganisms that grow are nourished by varieties of substrates, which include simple carbohydrates or polysaccharides, amino acids, organic acids, cellulose and lignocellulose [95]. Marine sediments and aqueous ones were also employed in BMFC as a substrate [116]. The substrate not only facilitates the microbes in producing the biofilm on the surface of the anode, but is also designed to increase the performance of the BMFC by producing higher coulombic efficiency and power density [117]. Moreover, the diverse substrate processes fully depend on the biodegradability factor. The power density of BMFC is directly proportional to the quantity of organic contents in the organic substrate and the biodegradation by electrogens of the microorganism [118]. The mechanism of organic substrate degradation through electrogens using BMFC is shown in Figure 4.

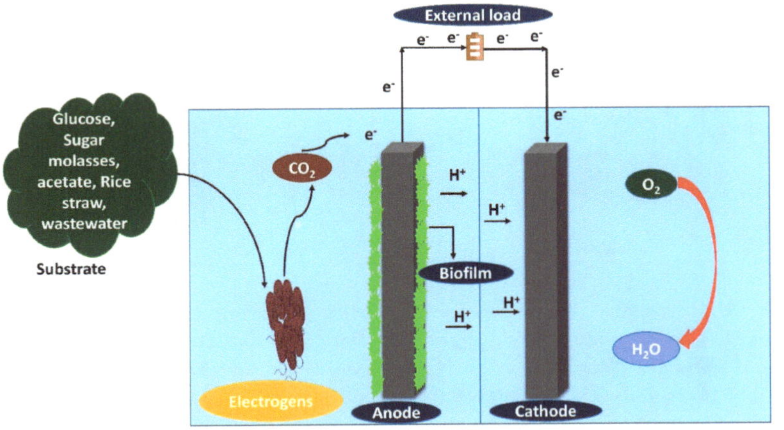

Figure 4. Representation of power generation by using organic contents as substrates by electrogens in a benthic microbial fuel cell.

Hassan et al. [119] studied the different organic substrates (glucose, fructose and sucrose) used in BMFC. Wang et.al. [120] developed a BMFC to generate a power density of about 12.7 mW/m^2 using an electron-mediating agent at pH 4, with the help of acidophilic bacterium, *Acidiphilium cryptum*, utilizing glucose as the organic substrate. The pure bacterial strain *Brevibacillus borstelensis* STRI1 produced a power density of about 188.5 mW/m^2 by using sugarcane molasses as the organic substrate [119]. The rice straw was also used as an organic substrate to generate a power density of about 293.33 ± 7.89 mW/m^2 [121]. The existing literature reveals the different kinds of organic waste being used as organic substrates, with their corresponding capacities for power density generation by electrogens, as shown in Table 2.

Table 2. Different substrates used in the BMFCs with corresponding power densities.

Waste Substrate	Electircigens	Power Density (mW/m^2)	Configurations	Type of Electrons Transfer Mechanisms	References
Glucose	*Acidiphilium cryptum*	12.7	Single chamber	Direct transfer	[120]
Cellulose	*Enterobacter cloacae*	5.4 ± 0.3	Double chamber	Direct transfer	[122]
Lactate	*Shewanella oneidensis* MR-1	0.3 × 10^{-2}	Single chamber	Indirect transfer	[123]
Lactate	*Geobacter sulfurreducens*	52 ± 4.7	-	Indirect transfer	[124]
Glucose	*Escherichia coli*	228	-	Indirect transfer	[125]
Malt extract	*Enterobacter cloacae*	9.3	-	Indirect transfer	[126]
Cellulose	*G. sulfurreducens* and *C. cellulolyticum*	83	Single chamber	Indirect transfer	[127]
Wheat straw	*Acidithiobacillus caldus*	123	Single chamber	-	[128]
Molasses	*B. borstelensis* STRI1	185.5	Single chamber	-	[119]
Sophorolipid with glucose and PBS	*Pseudomonas aeruginosa*	15.29	Single chamber	-	[129]
Glucose, fructose, and sucrose	*Saccharomyces cerevisiae*	72.77	Single chamber	-	[130]
Glucose in synthetic wastewater	–	1313	Double chamber	Direct transfer	[131]
xylose	*Geobacter sulfurreducens* *Escherichia coli,*	590	Double chamber	Direct transfer	[132]
Synthetic wastewater	*α–Proteobacteria, β–Proteobacteria, γ–Proteobacteria*	70	Double chamber	-	[133]
Sodium Fumarate	*Geobacter sulfurreducens*	–	Single chamber	-	[134]
Glucuronic acid	*Rhodococcus sp. and Paracoccus sp.*	2770	Double chamber	-	[135]
Xylose	*Clostridium spp. and Comamonas spp.*	1241	–	Direct transfer	[136]
Acetate	–	1430	–		[137]
Ethanol	Proteobacterium sp., Azoarcus sp. and Desulfuromonas sp.	40	–	Indirect transfer	[138]
Synthetic wastewater with molasses and urea	–	2.9	Single chamber		[139]
Cysteine	*Shewanella affinis*	39	–	–	[140]
Starch	*Clostridium butyricum or Clostridium beijerinckii*	–	–	–	[141]
Dye-containing wastewater in microbial desalination	*Bacillus subtilis, Aeromonas hydrophila subsp. hydrophila*	2.86	–	–	[142]
Rice straw	*Cellulose-degrading bacteria*	146	–	–	[121]
Coconut husk retting	*Ochrobactrum sp.*	362	Double chamber	Indirect transfer	[143]
Agriculture wastewater	*Shewanella oneidensis*	13	Double chamber	Indirect transfer	[144]
Rice paddy	*Geobacteraceae*	–	Double chamber	Indirect transfer	[145]
Chitin	*Bacillus circulans*	1.742	Double chamber	Indirect transfer	[146]

In the BMFC, various kinds of substrates could be employed; these substrates can be starch, petroleum-based compounds, cysteine, glucose, dairy-based, acetate, molasses, glutamic acid, food-based wastewater, river water and vegetable-based. The substrate selection is based on their biodegradability behaviors. The power production by BMFC depends upon the degradation rate by the bacteria and the quantity of organic contents in the substrates [147]. In BMFCs, there is a continuous generation of power which is impeded by access to nutrients in the anodic media. The nutrients in BMFCs are regularly supplied with fresh matter from the decay of microbes and animals, giving the BMFC an indefinite life span in theory [148]. In the BMFC, one biodegradable fuel was also the bio-battery, but with this the power generation ultimately drops with time. Some substrates only support a single form of organic material. Different types of chitin were also used in BMFC anode as substrates. Chitin 80 and chitin 20 produced optimum power of about 84 ± 10 and 76 ± 25 mW/m^2, respectively. The internal resistances of chitin 80 and chitin 20 were 650 ± 130 and 1300 ± 440, respectively. The electricity production could be enhanced by using substrates of precise size, and slowly degradable substrates. The substrates of precise size enhance the degradation surface area, and the slowly degradable substrates enhance the power production duration [146].

6. Conclusions

BMFC is a novel bio-technique that may be a potential solution to the two main problems, namely pollutants bioremediation and sustainable energy production. These BMFCs will open new possibilities for sustainable, cost-effective and controllable ways to generate power and bioremediate toxic pollutants. For power generation, there are two main routes of electron transfer: direct electron (physical contact between electrogens and anode) and indirect electron (conductive pili and flagella) transfer from the electrogens towards the anode of BMFC. The performance of BMFC depends on the use of different organic matters as the substrate. The novel BMFC technology will be encouraging for in situ pollutants bioremediation. The challenges of BMFCs will be addressed jointly by the efforts of scientists from many fields, such as environmental sciences, biotechnology, electrochemistry, electrical engineering, biology and material sciences.

Author Contributions: Conceptualization, M.F.U. and M.R.; writing—original draft preparation, M.F.U. and S.Z.A.; writing—review and editing, M.N.M.I., N.I. and M.R.; supervision, M.N.M.I., N.I., and M.R.; funding acquisition, M.R. All authors have read and agreed to the published version of the manuscript.

Funding: The authors are grateful to the Universiti Sains Malaysia for supporting this work through RUI grant (1001/PTEKIND/8011044).

Conflicts of Interest: The authors declare no conflict of interest.

References

1. Grey, D.; Garrick, D.; Blackmore, D.; Kelman, J.; Muller, M.; Sadoff, C. Water security in one blue planet: Twenty-first century policy challenges for science. *Philos. Trans. R. Soc. A Math. Phys. Eng. Sci.* **2013**, *371*, 20120406. [CrossRef] [PubMed]
2. Li, W.-W.; Yu, H.-Q. Stimulating sediment bioremediation with benthic microbial fuel cells. *Biotechnol. Adv.* **2015**, *33*, 1–12. [CrossRef] [PubMed]
3. Khan, M.E.; Khan, M.M.; Min, B.-K.; Cho, M.H. Microbial fuel cell assisted band gap narrowed TiO_2 for visible light-induced photocatalytic activities and power generation. *Sci. Rep.* **2018**, *8*, 1–12. [CrossRef]
4. Khan, M.E.; Han, T.H.; Khan, M.M.; Karim, M.R.; Cho, M.H. Environmentally sustainable fabrication of Ag@g–C_3N_4 nanostructures and their multifunctional efficacy as antibacterial agents and photocatalysts. *ACS Appl. Nano. Mater.* **2018**, *1*, 2912–2922. [CrossRef]
5. Khan, M.E.; Khan, M.M.; Cho, M.H. Environmentally sustainable biogenic fabrication of AuNP decorated-graphitic gC_3N_4 nanostructures towards improved photoelectrochemical performances. *RSC Adv.* **2018**, *8*, 13898–13909. [CrossRef]
6. Yaqoob, A.A.; Khatoon, A.; Mohd Setapar, S.H.; Umar, K.; Parveen, T.; Mohamad Ibrahim, M.N.; Ahmad, A.; Rafatullah, M. Outlook on the role of microbial fuel cells in remediation of environmental pollutants with electricity generation. *Catalysts* **2020**, *10*, 819. [CrossRef]
7. Yaqoob, A.A.; Mohamad Ibrahim, M.N.; Rafatullah, M.; Chua, Y.S.; Ahmad, A.; Umar, K. Recent advances in anodes for microbial fuel cells: An overview. *Materials* **2020**, *13*, 2078. [CrossRef]
8. Liew, K.B.; Daud, W.R.W.; Ghasemi, M.; Leong, J.X.; Lim, S.S.; Ismail, M. Non-Pt catalyst as oxygen reduction reaction in microbial fuel cells: A review. *Int. J. Hydrogen Energy* **2014**, *39*, 4870–4883. [CrossRef]
9. Abbas, S.Z.; Rafatullah, M.; Ismail, N.; Nastro, R.A. Enhanced bioremediation of toxic metals and harvesting electricity through sediment microbial fuel cell. *Int. J. Energy Res.* **2017**, *41*, 2345–2355. [CrossRef]

10. Martins, G.; Peixoto, L.; Brito, A.G.; Nogueira, R. Phosphorus–iron interaction in sediments: Can an electrode minimize phosphorus release from sediments? *Rev. Environ. Sci. Biotechnol.* **2014**, *13*, 265–275. [CrossRef]
11. Li, H.; Tian, Y.; Qu, Y.; Qiu, Y.; Liu, J.; Feng, Y. A pilot-scale benthic microbial electrochemical system (BMES) for enhanced organic removal in sediment restoration. *Sci. Rep.* **2017**, *7*, 1–9. [CrossRef] [PubMed]
12. Abbas, S.Z.; Rafatullah, M.; Ismail, N.; Syakir, M.I. The behaviour of membrane less sediment microbial fuel cell in the terms of bioremediation and power generation. *Malays. J. Microbiol.* **2018**, *14*, 108–112.
13. Reimers, C.E.; Girguis, P.; Stecher, H.A.; Tender, L.M.; Ryckelynck, N.; Whaling, P. Microbial fuel cell energy from an ocean cold seep. *Geobiology* **2006**, *4*, 123–136. [CrossRef]
14. Nielsen, M.E.; Reimers, C.E.; White, H.K.; Sharma, S.; Girguis, P.R. Sustainable energy from deep ocean cold seeps. *Energy Environ. Sci.* **2008**, *1*, 584–593. [CrossRef]
15. Hong, S.W.; Chang, I.S.; Choi, Y.S.; Kim, B.H.; Chung, T.H. Responses from freshwater sediment during electricity generation using microbial fuel cells. *Bioprocess Biosyst. Eng.* **2009**, *32*, 389–395. [CrossRef]
16. Rezaei, F.; Richard, T.L.; Brennan, R.A.; Logan, B.E. Substrate-enhanced microbial fuel cells for improved remote power generation from sediment-based systems. *Environ. Sci. Technol.* **2007**, *41*, 4053–4058. [CrossRef]
17. Yuan, Y.; Zhou, S.; Zhuang, L. A new approach to in situ sediment remediation based on air-cathode microbial fuel cells. *J. Soils. Sediments* **2010**, *10*, 1427–1433. [CrossRef]
18. Wu, M.; Xu, X.; Lu, K.; Li, X. Effects of the presence of nanoscale zero-valent iron on the degradation of polychlorinated biphenyls and total organic carbon by sediment microbial fuel cell. *Sci. Total. Environ.* **2019**, *656*, 39–44. [CrossRef]
19. Abbas, S.Z.; Rafatullah, M.; Ismail, N.; Syakir, M.I. A review on sediment microbial fuel cells as a new source of sustainable energy and heavy metal remediation: Mechanisms and future prospective. *Int. J. Energy Res.* **2017**, *41*, 1242–1264. [CrossRef]
20. Xu, X.; Zhao, Q.; Wu, M.; Ding, J.; Zhang, W. Biodegradation of organic matter and anodic microbial communities analysis in sediment microbial fuel cells with/without Fe (III) oxide addition. *Bioresour. Technol.* **2017**, *225*, 402–408. [CrossRef]
21. Khan, M.E.; Khan, M.M.; Cho, M.H. Recent progress of metal–graphene nanostructures in photocatalysis. *Nanoscale* **2018**, *10*, 9427–9440. [CrossRef] [PubMed]
22. Estevez-Canales, M.; Kuzume, A.; Borjas, Z.; Füeg, M.; Lovley, D.; Wandlowski, T.; Esteve-Núñez, A. A severe reduction in the cytochrome C content of *Geobacter sulfurreducens* eliminates its capacity for extracellular electron transfer. *Environ. Microbiol. Rep.* **2015**, *7*, 219–226. [CrossRef]
23. Khan, M.E.; Khan, M.M.; Cho, M.H. Green synthesis, photocatalytic and photoelectrochemical performance of an Au–Graphene nanocomposite. *RSC Adv.* **2015**, *5*, 26897–26904. [CrossRef]
24. Khan, M.E.; Khan, M.M.; Cho, M.H. Biogenic synthesis of a Ag–graphene nanocomposite with efficient photocatalytic degradation, electrical conductivity and photoelectrochemical performance. *New J. Chem.* **2015**, *39*, 8121–8129. [CrossRef]
25. Song, X.; Wang, W.; Cao, X.; Wang, Y.; Zou, L.; Ge, X.; Zhao, Y.; Si, Z.; Wang, Y. Chlorella vulgaris on the cathode promoted the performance of sediment microbial fuel cells for electrogenesis and pollutant removal. *Sci. Total. Environ.* **2020**, *728*, 138011. [CrossRef] [PubMed]
26. Li, W.; Quan, X.; Chen, L.; Zheng, Y. Application of slow-release carbon sources embedded in polymer for stable and extended power generation in microbial fuel cells. *Chemosphere* **2020**, *244*, 125515. [CrossRef]
27. Song, H.-L.; Zhu, Y.; Li, J. Electron transfer mechanisms, characteristics and applications of biological cathode microbial fuel cells—A mini review. *Arab. J. Chem.* **2019**, *12*, 2236–2243. [CrossRef]
28. Yang, Z.; Yang, A. Modelling the impact of operating mode and electron transfer mechanism in microbial fuel cells with two-species anodic biofilm. *Biochem. Eng. J.* **2020**, 107560. [CrossRef]
29. Abbas, S.Z.; Rafatullah, M.; Ismail, N.; Shakoori, F.R. Electrochemistry and microbiology of microbial fuel cells treating marine sediments polluted with heavy metals. *RSC Adv.* **2018**, *8*, 18800–18813. [CrossRef]
30. Jothinathan, D.; Mylsamy, P.; Bruno, L.B. Electricigens: Role and prominence in microbial fuel cell performance. In *Microbial Fuel Cell Technology for Bioelectricity*; Springer: Berlin/Heidelberg, Germany, 2018; pp. 169–185.
31. Tremblay, P.-L.; Angenent, L.T.; Zhang, T. Extracellular electron uptake: Among autotrophs and mediated by surfaces. *Trends Biotechnol.* **2017**, *35*, 360–371. [CrossRef]
32. Lee, C.S.; Robinson, J.; Chong, M.F. A review on application of flocculants in wastewater treatment. *Process Saf. Environ. Prot.* **2014**, *92*, 489–508. [CrossRef]

33. Carbajosa, S.; Malki, M.; Caillard, R.; Lopez, M.F.; Palomares, F.J.; Martín-Gago, J.A.; Rodríguez, N.; Amils, R.; Fernández, V.M.; De Lacey, A.L. Electrochemical growth of Acidithiobacillus ferrooxidans on a graphite electrode for obtaining a biocathode for direct electrocatalytic reduction of oxygen. *Biosens. Bioelectron.* **2010**, *26*, 877–880. [CrossRef] [PubMed]
34. Summers, Z.M.; Gralnick, J.A.; Bond, D.R. Cultivation of an obligate Fe (II)-oxidizing lithoautotrophic bacterium using electrodes. *mBio* **2013**, *4*, e00420-12. [CrossRef]
35. Bose, A.; Gardel, E.J.; Vidoudez, C.; Parra, E.; Girguis, P.R. Electron uptake by iron-oxidizing phototrophic bacteria. *Nat. Commun.* **2014**, *5*, 1–7. [CrossRef] [PubMed]
36. Beese-Vasbender, P.F.; Grote, J.-P.; Garrelfs, J.; Stratmann, M.; Mayrhofer, K.J. Selective microbial electrosynthesis of methane by a pure culture of a marine lithoautotrophic archaeon. *Bioelectrochemistry* **2015**, *102*, 50–55. [CrossRef]
37. Aryal, N.; Tremblay, P.-L.; Lizak, D.M.; Zhang, T. Performance of different *Sporomusa* species for the microbial electrosynthesis of acetate from carbon dioxide. *Bioresour. Technol.* **2017**, *233*, 184–190. [CrossRef]
38. Beese-Vasbender, P.F.; Nayak, S.; Erbe, A.; Stratmann, M.; Mayrhofer, K.J. Electrochemical characterization of direct electron uptake in electrical microbially influenced corrosion of iron by the lithoautotrophic SRB *Desulfopila corrodens* strain IS4. *Electrochim. Acta* **2015**, *167*, 321–329. [CrossRef]
39. Li, Y.; Xu, D.; Chen, C.; Li, X.; Jia, R.; Zhang, D.; Sand, W.; Wang, F.; Gu, T. Anaerobic microbiologically influenced corrosion mechanisms interpreted using bioenergetics and bioelectrochemistry: A review. *J. Mater. Sci. Technol.* **2018**, *34*, 1713–1718. [CrossRef]
40. Pisciotta, J.M.; Zaybak, Z.; Call, D.F.; Nam, J.-Y.; Logan, B.E. Enrichment of microbial electrolysis cell biocathodes from sediment microbial fuel cell bioanodes. *Appl. Environ. Microbiol.* **2012**, *78*, 5212–5219. [CrossRef]
41. Chen, K.-T.; Bai, M.-D.; Wu, S.-I.; Chen, C.-C.; Lu, W.-J.; Wan, H.-P.; Huang, C. Electro-autotrophs induced the growth of exoelectrogens on the anode in a microbial fuel cell. *Biochem. Eng. J.* **2019**, *141*, 29–34. [CrossRef]
42. Peng, T.; Feng, C.; Hu, W.; Chen, N.; He, Q.; Dong, S.; Xu, Y.; Gao, Y.; Li, M. Treatment of nitrate-contaminated groundwater by heterotrophic denitrification coupled with electro-autotrophic denitrifying packed bed reactor. *Biochem. Eng. J.* **2018**, *134*, 12–21. [CrossRef]
43. Jabeen, G.; Farooq, R. Bio-electrochemical synthesis of commodity chemicals by autotrophic acetogens utilizing CO2 for environmental remediation. *J. Biosci.* **2016**, *41*, 367–380. [CrossRef] [PubMed]
44. Reguera, G.; McCarthy, K.D.; Mehta, T.; Nicoll, J.S.; Tuominen, M.T.; Lovley, D.R. Extracellular electron transfer via microbial nanowires. *Nature* **2005**, *435*, 1098–1101. [CrossRef]
45. Zhao, F.; Slade, R.C.T.; Varcoe, J.R. Techniques for the study and development of microbial fuel cells: An electrochemical perspective. *Chem. Soc. Rev.* **2009**, *38*, 1926–1939. [CrossRef]
46. Malvankar, N.S.; Lovley, D.R. Microbial nanowires for bioenergy applications. *Curr. Opin. Biotechnol.* **2014**, *27*, 88–95. [CrossRef] [PubMed]
47. Abbas, S.Z.; Rafatullah, M.; Khan, M.A.; Siddiqui, M.R. Bioremediation and Electricity Generation by Using Open and Closed Sediment Microbial Fuel Cells. *Front. Microbiol.* **2019**, *9*, 3348. [CrossRef]
48. Rotaru, A.-E.; Woodard, T.L.; Nevin, K.P.; Lovley, D.R. Link between capacity for current production and syntrophic growth in *Geobacter* species. *Front. Microbiol.* **2015**, *6*, 744. [CrossRef]
49. Xing, D.; Zuo, Y.; Cheng, S.; Regan, J.M.; Logan, B.E. Electricity generation by *Rhodopseudomonas palustris* DX-1. *Environ. Sci. Technol.* **2008**, *42*, 4146–4151. [CrossRef]
50. Bond, D.R.; Lovley, D.R. Evidence for involvement of an electron shuttle in electricity generation by *Geothrix fermentans*. *Appl. Environ. Microbiol.* **2005**, *71*, 2186–2189. [CrossRef]
51. Strycharz, S.M.; Woodard, T.L.; Johnson, J.P.; Nevin, K.P.; Sanford, R.A.; Löffler, F.E.; Lovley, D.R. Graphite electrode as a sole electron donor for reductive dechlorination of tetrachlorethene by *Geobacter lovleyi*. *Appl. Environ. Microbiol.* **2008**, *74*, 5943–5947. [CrossRef]
52. Virdis, B.; Rabaey, K.; Yuan, Z.; Keller, J. Microbial fuel cells for simultaneous carbon and nitrogen removal. *Water Res.* **2008**, *42*, 3013–3024. [CrossRef]
53. Rabaey, K.; Boon, N.; Höfte, M.; Verstraete, W. Microbial phenazine production enhances electron transfer in biofuel cells. *Environ. Sci. Technol.* **2005**, *39*, 3401–3408. [CrossRef]
54. Wrighton, K.C.; Thrash, J.C.; Melnyk, R.A.; Bigi, J.P.; Byrne-Bailey, K.G.; Remis, J.P.; Schichnes, D.; Auer, M.; Chang, C.J.; Coates, J.D. Evidence for direct electron transfer by a Gram-positive bacterium isolated from a microbial fuel cell. *Appl. Environ. Microbiol.* **2011**, *77*, 7633–7639. [CrossRef] [PubMed]

55. Baron, D.; LaBelle, E.; Coursolle, D.; Gralnick, J.A.; Bond, D.R. Electrochemical measurement of electron transfer kinetics by *Shewanella oneidensis* MR-1. *J. Biol. Chem.* **2009**, *284*, 28865–28873. [CrossRef]
56. Titov, D.V.; Cracan, V.; Goodman, R.P.; Peng, J.; Grabarek, Z.; Mootha, V.K. Complementation of mitochondrial electron transport chain by manipulation of the NAD^+/NADH ratio. *Science* **2016**, *352*, 231–235. [CrossRef]
57. Mathis, B.J.; Marshall, C.W.; Milliken, C.E.; Makkar, R.S.; Creager, S.E.; May, H.D. Electricity generation by thermophilic microorganisms from marine sediment. *Appl. Microbiol. Biotecnol.* **2008**, *78*, 147–155. [CrossRef] [PubMed]
58. Lovley, D.R. The microbe electric: Conversion of organic matter to electricity. *Curr. Opin. Biotechnol.* **2008**, *19*, 564–571. [CrossRef]
59. Zhang, T.; Cui, C.; Chen, S.; Yang, H.; Shen, P. The direct electrocatalysis of *Escherichia coli* through electroactivated excretion in microbial fuel cell. *Electrochem. Commun.* **2008**, *10*, 293–297. [CrossRef]
60. Sydow, A.; Krieg, T.; Mayer, F.; Schrader, J.; Holtmann, D. Electroactive bacteria—Molecular mechanisms and genetic tools. *Appl. Microbiol. Biotechnol.* **2014**, *98*, 8481–8495. [CrossRef]
61. Wang, Z.; Leary, D.H.; Malanoski, A.P.; Li, R.W.; Hervey, W.J.; Eddie, B.J.; Tender, G.S.; Yanosky, S.G.; Vora, G.J.; Tender, L.M. A previously uncharacterized, nonphotosynthetic member of the Chromatiaceae is the primary CO2-fixing constituent in a self-regenerating biocathode. *Appl. Environ. Microbiol.* **2015**, *81*, 699–712. [CrossRef]
62. Xu, D.; Gu, T. Bioenergetics explains when and why more severe MIC pitting by SRB can occur. In Proceedings of the NACE International Corrosion 2011 Conference and Expo, Houston, TX, USA, 13–17 March 2011.
63. Cordas, C.M.; Moura, J.J.G. Sulphate reducing bacteria-electroactive biofilm formation. *Int. J. Med. Biol. Front.* **2011**, *17*, 295–312.
64. Arends, J.B.; Patil, S.A.; Roume, H.; Rabaey, K. Continuous long-term electricity-driven bioproduction of carboxylates and isopropanol from CO2 with a mixed microbial community. *J. CO2 Util.* **2017**, *20*, 141–149. [CrossRef]
65. de Campos Rodrigues, T.; Rosenbaum, M.A. Microbial electroreduction: Screening for new cathodic biocatalysts. *ChemElectroChem* **2014**, *1*, 1916–1922. [CrossRef]
66. Qian, Z.; Tianwei, H.; Mackey, H.R.; van Loosdrecht, M.C.; Guanghao, C. Recent advances in dissimilatory sulfate reduction: From metabolic study to application. *Water Res.* **2019**, *150*, 162–181. [CrossRef] [PubMed]
67. Berlendis, S.; Ranchou-Peyruse, M.; Fardeau, M.-L.; Lascourreges, J.-F.; Joseph, M.; Ollivier, B.; Aüllo, T.; Dequidt, D.; Magot, M.; Ranchou-Peyruse, A. *Desulfotomaculum aquiferis* sp. nov. and *Desulfotomaculum profundi* sp. nov., isolated from a deep natural gas storage aquifer. *Int. J. Syst. Evol. Microbiol.* **2016**, *66*, 4329–4338. [CrossRef] [PubMed]
68. Jiang, Y.; May, H.D.; Lu, L.; Liang, P.; Huang, X.; Ren, Z.J. Carbon dioxide and organic waste valorization by microbial electrosynthesis and electro-fermentation. *Water Res.* **2019**, *149*, 42–55. [CrossRef]
69. Jiang, Y.; Zeng, R.J. Bidirectional extracellular electron transfers of electrode-biofilm: Mechanism and application. *Bioresour. Technol.* **2019**, *271*, 439–448. [CrossRef] [PubMed]
70. He, L.; Du, P.; Chen, Y.; Lu, H.; Cheng, X.; Chang, B.; Wang, Z. Advances in microbial fuel cells for wastewater treatment. *Renew. Sustain. Energy Rev.* **2017**, *71*, 388–403. [CrossRef]
71. Bilal, M.; Wang, S.; Iqbal, H.M.N.; Zhao, Y.; Hu, H.; Wang, W.; Zhang, X. Metabolic engineering strategies for enhanced shikimate biosynthesis: Current scenario and future developments. *Appl. Microbiol. Biotecnol.* **2018**, *102*, 7759–7773. [CrossRef]
72. Li, Y.; Marshall, A.; Gostomski, P.A. Gaseous pollutant treatment and electricity generation in microbial fuel cells (MFCs) utilising redox mediators. *Rev. Environ. Sci. Bio/Technol.* **2014**, *13*, 35–51. [CrossRef]
73. Rahimnejad, M.; Najafpour, G.D.; Ghoreyshi, A.A.; Talebnia, F.; Premier, G.C.; Bakeri, G.; Kim, J.R.; Oh, S.-E. Thionine increases electricity generation from microbial fuel cell using *Saccharomyces cerevisiae* and exoelectrogenic mixed culture. *J. Microbiol.* **2012**, *50*, 575–580. [CrossRef] [PubMed]
74. Thygesen, A.; Poulsen, F.W.; Min, B.; Angelidaki, I.; Thomsen, A.B. The effect of different substrates and humic acid on power generation in microbial fuel cell operation. *Bioresour. Technol.* **2009**, *100*, 1186–1191. [CrossRef] [PubMed]
75. Rahimnejad, M.; Najafpour, G.D.; Ghoreyshi, A.A.; Shakeri, M.; Zare, H. Methylene blue as electron promoters in microbial fuel cell. *Int. J. Hydrogen Energy* **2011**, *36*, 13335–13341. [CrossRef]
76. Fernando, E.; Keshavarz, T.; Kyazze, G. Enhanced bio-decolourisation of acid orange 7 by *Shewanella oneidensis* through co-metabolism in a microbial fuel cell. *Int. Biodeterior. Biodegrad.* **2012**, *72*, 1–9. [CrossRef]

77. Ringeisen, B.R.; Henderson, E.; Wu, P.K.; Pietron, J.; Ray, R.; Little, B.; Biffinger, J.C.; Jones-Meehan, J.M. High Power Density from a Miniature Microbial Fuel Cell Using *Shewanella oneidensis* DSP10. *Environ. Sci. Technol.* **2006**, *40*, 2629–2634. [CrossRef]
78. Menicucci, J.; Beyenal, H.; Marsili, E.; Veluchamy, R.A.; Demir, G.; Lewandowski, Z. Procedure for Determining Maximum Sustainable Power Generated by Microbial Fuel Cells. *Environ. Sci. Technol.* **2006**, *40*, 1062–1068. [CrossRef]
79. Boon, N.; De Maeyer, K.; Höfte, M.; Rabaey, K.; Verstraete, W. Use of *Pseudomonas* species producing phenazine-based metabolites in the anodes of microbial fuel cells to improve electricity generation. *Appl. Microbiol. Biotechnol.* **2008**, *80*, 985–993.
80. Zhang, M.; Ma, Z.; Zhao, N.; Zhang, K.; Song, H. Increased power generation from cylindrical microbial fuel cell inoculated with *P. aeruginosa*. *Biosens. Bioelectron.* **2019**, *141*, 111394. [CrossRef]
81. Takeuchi, Y.; Khawdas, W.; Aso, Y.; Ohara, H. Microbial fuel cells using Cellulomonas spp. with cellulose as fuel. *J. Biosci. Bioeng.* **2017**, *123*, 358–363. [CrossRef]
82. Masuda, M.; Freguia, S.; Wang, Y.-F.; Tsujimura, S.; Kano, K. Flavins contained in yeast extract are exploited for anodic electron transfer by *Lactococcus lactis*. *Bioelectrochemistry* **2010**, *78*, 173–175. [CrossRef]
83. Lovley, D.R. Powering microbes with electricity: Direct electron transfer from electrodes to microbes. *Environ. Microbiol. Rep.* **2011**, *3*, 27–35. [CrossRef] [PubMed]
84. Ringeisen, B.R.; Ray, R.; Little, B. A miniature microbial fuel cell operating with an aerobic anode chamber. *J. Power Sources* **2007**, *165*, 591–597. [CrossRef]
85. Cordas, C.M.; Guerra, L.T.; Xavier, C.; Moura, J.J.G. Electroactive biofilms of sulphate reducing bacteria. *Electrochim. Acta* **2008**, *54*, 29–34. [CrossRef]
86. Leung, K.M.; Wanger, G.; El-Naggar, M.Y.; Gorby, Y.; Southam, G.; Lau, W.M.; Yang, J. *Shewanella oneidensis* MR-1 bacterial nanowires exhibit p-type, tunable electronic behavior. *Nano Lett.* **2013**, *13*, 2407–2411. [CrossRef]
87. Keller, K.L.; Rapp-Giles, B.J.; Semkiw, E.S.; Porat, I.; Brown, S.D.; Wall, J.D. New model for electron flow for sulfate reduction in *Desulfovibrio alaskensis* G20. *Appl. Environ. Microbiol.* **2014**, *80*, 855–868. [CrossRef]
88. Qiao, Y.; Li, C.M.; Bao, S.-J.; Lu, Z.; Hong, Y. Direct electrochemistry and electrocatalytic mechanism of evolved *Escherichia coli* cells in microbial fuel cells. *Chem. Comm.* **2008**, *11*, 1290–1292. [CrossRef]
89. Marshall, C.W.; May, H.D. Electrochemical evidence of direct electrode reduction by a thermophilic Gram-positive bacterium, *Thermincola ferriacetica*. *Energy. Environ. Sci.* **2009**, *2*, 699–705. [CrossRef]
90. Carmona-Martínez, A.A.; Harnisch, F.; Kuhlicke, U.; Neu, T.R.; Schröder, U. Electron transfer and biofilm formation of *Shewanella putrefaciens* as function of anode potential. *Bioelectrochemistry* **2013**, *93*, 23–29. [CrossRef]
91. Thrash, J.C.; Van Trump, J.I.; Weber, K.A.; Miller, E.; Achenbach, L.A.; Coates, J.D. Electrochemical stimulation of microbial perchlorate reduction. *Environ. Sci. Technol.* **2007**, *41*, 1740–1746. [CrossRef]
92. Kalathil, S.; Pant, D. Nanotechnology to rescue bacterial bidirectional extracellular electron transfer in bioelectrochemical systems. *RSC Adv.* **2016**, *6*, 30582–30597. [CrossRef]
93. Daniel, D.K.; Mankidy, B.D.; Ambarish, K.; Manogari, R. Construction and operation of a microbial fuel cell for electricity generation from wastewater. *Int. J. Hydrogen Energy.* **2009**, *34*, 7555–7560. [CrossRef]
94. Zhi-Dan, L.I.U.; Jing, L.; Zhu-Wei, D.U.; Li, H.-R. Construction of sugar-based microbial fuel cells by dissimilatory metal reduction bacteria. *Chin. J. Biotechnol.* **2006**, *22*, 131–137.
95. Liu, L.; Li, F.-b.; Feng, C.-h.; Li, X.-Z. Microbial fuel cell with an azo-dye-feeding cathode. *Appl. Microbiol. Biotechnol.* **2009**, *85*, 175–183. [CrossRef] [PubMed]
96. Hassan, S.H.A.; Kim, Y.S.; Oh, S.-E. Power generation from cellulose using mixed and pure cultures of cellulose-degrading bacteria in a microbial fuel cell. *Enzyme Microb. Technol.* **2012**, *51*, 269–273. [CrossRef]
97. De Vet, S.J.; Rutgers, R. From waste to energy: First experimental bacterial fuel cells onboard the international space station. *Microgravity Sci. Technol.* **2007**, *19*, 225–229. [CrossRef]
98. Zheng, X.; Nirmalakhandan, N. Cattle wastes as substrates for bioelectricity production via microbial fuel cells. *Biotechnol. Lett.* **2010**, *32*, 1809–1814. [CrossRef]
99. Marsili, E.; Baron, D.B.; Shikhare, I.D.; Coursolle, D.; Gralnick, J.A.; Bond, D.R. *Shewanella* secretes flavins that mediate extracellular electron transfer. *Proc. Natl. Acad. Sci. USA* **2008**, *105*, 3968–3973. [CrossRef]

100. Ali, N.; Anam, M.; Yousaf, S.; Maleeha, S.; Bangash, Z. Characterization of the electric current generation potential of the pseudomonas aeruginosa using glucose, fructose, and sucrose in double chamber microbial fuel cell. *Iran. J. Biotechnol.* **2017**, *15*, 216. [CrossRef]
101. Khawdas, W.; Watanabe, K.; Karatani, H.; Aso, Y.; Tanaka, T.; Ohara, H. Direct electron transfer of *Cellulomonas fimi* and microbial fuel cells fueled by cellulose. *J. Biosci. Bioeng.* **2019**, *128*, 593–598. [CrossRef]
102. Powell, E.E.; Mapiour, M.L.; Evitts, R.W.; Hill, G.A. Growth kinetics of *Chlorella vulgaris* and its use as a cathodic half cell. *Bioresour. Technol.* **2009**, *100*, 269–274. [CrossRef]
103. Cournet, A.; Délia, M.-L.; Bergel, A.; Roques, C.; Bergé, M. Electrochemical reduction of oxygen catalyzed by a wide range of bacteria including Gram-positive. *Electrochem. Commun.* **2010**, *12*, 505–508. [CrossRef]
104. Raghavulu, S.V.; Goud, R.K.; Sarma, P.N.; Mohan, S.V. Saccharomyces cerevisiae as anodic biocatalyst for power generation in biofuel cell: Influence of redox condition and substrate load. *Bioresour. Technol.* **2011**, *102*, 2751–2757. [CrossRef] [PubMed]
105. Erable, B.; Vandecandelaere, I.; Faimali, M.; Delia, M.-L.; Etcheverry, L.; Vandamme, P.; Bergel, A. Marine aerobic biofilm as biocathode catalyst. *Bioelectrochemistry* **2010**, *78*, 51–56. [CrossRef]
106. Nguyen, T.A.; Lu, Y.; Yang, X.; Shi, X. Carbon and Steel Surfaces Modified by *Leptothrix discophora* SP-6: Characterization and Implications. *Environ. Sci. Technol.* **2007**, *41*, 7987–7996. [CrossRef] [PubMed]
107. Wu, S.J.; Hu, Z.H.; Zhang, L.L.; Yu, X.; Chen, J.M. A novel dichloromethane-degrading *Lysinibacillus sphaericus* strain wh22 and its degradative plasmid. *Appl. Mcrobiol. Biotechnol.* **2009**, *82*, 731–740. [CrossRef]
108. Xu, S.; Liu, H. New exoelectrogen *Citrobacter* sp. SX-1 isolated from a microbial fuel cell. *J. App. Microbiol.* **2011**, *111*, 1108–1115. [CrossRef] [PubMed]
109. Inglesby, A.E.; Beatty, D.A.; Fisher, A.C. *Rhodopseudomonas palustris* purple bacteria fed *Arthrospira maxima* cyanobacteria: Demonstration of application in microbial fuel cells. *RSC. Adv.* **2012**, *2*, 4829–4838. [CrossRef]
110. Wu, C.; Liu, X.-W.; Li, W.-W.; Sheng, G.-P.; Zang, G.-L.; Cheng, Y.-Y.; Shen, N.; Yang, Y.-P.; Yu, H.-Q. A white-rot fungus is used as a biocathode to improve electricity production of a microbial fuel cell. *Appl. Energy* **2012**, *98*, 594–596. [CrossRef]
111. Huang, H.; Chai, C.; Yang, S.; Jiang, W.; Gu, Y. Phage serine integrase-mediated genome engineering for efficient expression of chemical biosynthetic pathway in gas-fermenting *Clostridium ljungdahlii. Metab. Eng.* **2019**, *52*, 293–302. [CrossRef]
112. Klask, C.-M.; Kliem-Kuster, N.; Molitor, B.; Angenent, L.T. Nitrate Feed Improves Growth and Ethanol Production of *Clostridium ljungdahlii* with CO_2 and H_2, but Results in Stochastic Inhibition Events. *Front. Microbiol.* **2019**. [CrossRef]
113. Mateos, R.; Escapa, A.; Vanbroekhoven, K.; Patil, S.A.; Moran, A.; Pant, D. Microbial electrochemical technologies for CO_2 and its derived products valorization. In *Microbial Electrochemical Technology*; Elsevier: Amsterdam, The Netherlands, 2019; pp. 777–796.
114. Kato, S.; Yumoto, I.; Kamagata, Y. Isolation of acetogenic bacteria that induce biocorrosion by utilizing metallic iron as the sole electron donor. *Appl. Environ. Microbiol.* **2015**, *81*, 67–73. [CrossRef] [PubMed]
115. Krishna, K.V.; Swathi, K.; Hemalatha, M.; Mohan, S.V. Bioelectrocatalyst in Microbial Electrochemical Systems and Extracellular Electron Transport. In *Microbial Electrochemical Technology*; Elsevier: Amsterdam, The Netherlands, 2019; pp. 117–141.
116. Reimers, C.E.; Tender, L.M.; Fertig, S.; Wang, W. Harvesting Energy from the Marine Sediment−Water Interface. *Environ. Sci. Technol.* **2001**, *35*, 192–195. [CrossRef]
117. Chae, K.-J.; Choi, M.-J.; Lee, J.-W.; Kim, K.-Y.; Kim, I.S. Effect of different substrates on the performance, bacterial diversity, and bacterial viability in microbial fuel cells. *Bioresour. Technol.* **2009**, *100*, 3518–3525. [CrossRef] [PubMed]
118. Borole, A.P.; O'Neill, H.; Tsouris, C.; Cesar, S. A microbial fuel cell operating at low pH using the acidophile *Acidiphilium cryptum. Biotechnol. Lett.* **2008**, *30*, 1367–1372. [CrossRef] [PubMed]
119. Hassan, S.H.A.; Zohri, A.A.; Kassim, R.M.F. Electricity generation from sugarcane molasses using microbial fuel cell technologies. *Energy* **2019**, *178*, 538–543. [CrossRef]
120. Wang, Z.; Lee, T.; Lim, B.; Choi, C.; Park, J. Microbial community structures differentiated in a single-chamber air-cathode microbial fuel cell fueled with rice straw hydrolysate. *Biotechnol. Biofuels* **2014**, *7*, 9. [CrossRef]
121. Hassan, S.H.A.; Gad El-Rab, S.M.F.; Rahimnejad, M.; Ghasemi, M.; Joo, J.-H.; Sik-Ok, Y.; Kim, I.S.; Oh, S.-E. Electricity generation from rice straw using a microbial fuel cell. *Int. J. Hydrogen Energy* **2014**, *39*, 9490–9496. [CrossRef]

122. Rezaei, F.; Xing, D.; Wagner, R.; Regan, J.M.; Richard, T.L.; Logan, B. E Simultaneous Cellulose Degradation and Electricity Production by Enterobacter cloacae in a Microbial Fuel Cell. *Appl. Environ. Microbiol.* **2009**, *75*, 3673. [CrossRef]
123. Manohar, A.K.; Mansfeld, F. The internal resistance of a microbial fuel cell and its dependence on cell design and operating conditions. *Electrochim. Acta* **2009**, *54*, 1664–1670. [CrossRef]
124. Jung, S.; Regan, J.M. Comparison of anode bacterial communities and performance in microbial fuel cells with different electron donors. *Appl. Microbiol. Biotechnol.* **2007**, *77*, 393–402. [CrossRef]
125. Zou, Y.; Xiang, C.; Yang, L.; Sun, L.-X.; Xu, F.; Cao, Z. A mediatorless microbial fuel cell using polypyrrole coated carbon nanotubes composite as anode material. *Int. J. Hydrogen Energy* **2008**, *33*, 4856–4862. [CrossRef]
126. Mohan, Y.; Manoj Muthu Kumar, S.; Das, D. Electricity generation using microbial fuel cells. *Int. J. Hydrogen Energy* **2008**, *33*, 423–426. [CrossRef]
127. Ren, Z.; Steinberg, L.M.; Regan, J.M. Electricity production and microbial biofilm characterization in cellulose-fed microbial fuel cells. *Water. Sci. Technol.* **2008**, *58*, 617–622. [CrossRef] [PubMed]
128. Hassan, S.H.A.; Van Ginkel, S.W.; Kim, S.-M.; Yoon, S.-H.; Joo, J.-H.; Shin, B.-S.; Jeon, B.-H.; Bae, W.; Oh, S.-E. Isolation and characterization of Acidithiobacillus caldus from a sulfur-oxidizing bacterial biosensor and its role in detection of toxic chemicals. *J. Microbiol. Methods* **2010**, *82*, 151–155. [CrossRef] [PubMed]
129. Shen, H.-B.; Yong, X.-Y.; Chen, Y.-L.; Liao, Z.-H.; Si, R.-W.; Zhou, J.; Wang, S.-Y.; Yong, Y.-C.; OuYang, P.-K.; Zheng, T. Enhanced bioelectricity generation by improving pyocyanin production and membrane permeability through sophorolipid addition in *Pseudomonas aeruginosa*-inoculated microbial fuel cells. *Bioresour. Technol.* **2014**, *167*, 490–494. [CrossRef] [PubMed]
130. Jafary, T.; Rahimnejad, M.; Ghoreyshi, A.A.; Najafpour, G.; Hghparast, F.; Daud, W.R.W. Assessment of bioelectricity production in microbial fuel cells through series and parallel connections. *Energy Convers. Manag.* **2013**, *75*, 256–262. [CrossRef]
131. Akman, D.; Cirik, K.; Ozdemir, S.; Ozkaya, B.; Cinar, O. Bioelectricity generation in continuously-fed microbial fuel cell: Effects of anode electrode material and hydraulic retention time. *Bioresour. Technol.* **2013**, *149*, 459–464. [CrossRef]
132. Mäkinen, A.E.; Lay, C.-H.; Nissilä, M.E.; Puhakka, J.A. Bioelectricity production on xylose with a compost enrichment culture. *Int. J. Hydrogen Energy.* **2013**, *38*, 15606–15612. [CrossRef]
133. Aldrovandi, A.; Marsili, E.; Stante, L.; Paganin, P.; Tabacchioni, S.; Giordano, A. Sustainable power production in a membrane-less and mediator-less synthetic wastewater microbial fuel cell. *Bioresour. Technol.* **2009**, *100*, 3252–3260. [CrossRef]
134. Dumas, C.; Basseguy, R.; Bergel, A. Microbial electrocatalysis with *Geobacter sulfurreducens* biofilm on stainless steel cathodes. *Electrochim. Acta* **2008**, *53*, 2494–2500. [CrossRef]
135. Catal, T.; Li, K.; Bermek, H.; Liu, H. Electricity production from twelve monosaccharides using microbial fuel cells. *J. Power Sources* **2008**, *175*, 196–200. [CrossRef]
136. Rismani-Yazdi, H.; Christy, A.D.; Dehority, B.A.; Morrison, M.; Yu, Z.; Tuovinen, O.H. Electricity generation from cellulose by rumen microorganisms in microbial fuel cells. *Biotechnol. Bioeng.* **2007**, *97*, 1398–1407. [CrossRef] [PubMed]
137. Logan, B.; Cheng, S.; Watson, V.; Estadt, G. Graphite fiber brush anodes for increased power production in air-cathode microbial fuel cells. *Environ. Sci. Technol.* **2007**, *41*, 3341–3346. [CrossRef] [PubMed]
138. Kim, J.R.; Jung, S.H.; Regan, J.M.; Logan, B.E. Electricity generation and microbial community analysis of alcohol powered microbial fuel cells. *Bioresour. Technol.* **2007**, *98*, 2568–2577. [CrossRef] [PubMed]
139. Kargi, F.; Eker, S. Electricity generation with simultaneous wastewater treatment by a microbial fuel cell (MFC) with Cu and Cu–Au electrodes. *J. Chem. Technol. Biotechnol.* **2007**, *82*, 658–662. [CrossRef]
140. Logan, B.E.; Murano, C.; Scott, K.; Gray, N.D.; Head, I.M. Electricity generation from cysteine in a microbial fuel cell. *Water Res.* **2005**, *39*, 942–952. [CrossRef]
141. Niessen, J.; Schröder, U.; Scholz, F. Exploiting complex carbohydrates for microbial electricity generation–a bacterial fuel cell operating on starch. *Electrochem. Commun.* **2004**, *6*, 955–958. [CrossRef]
142. Kalleary, S.; Abbas, F.M.; Ganesan, A.; Meenatchisundaram, S.; Srinivasan, B.; Packirisamy, A.S.B.; krishnan Kesavan, R.; Muthusamy, S. Biodegradation and bioelectricity generation by microbial desalination cell. *Int. Biodeterior. Biodegrad.* **2014**, *92*, 20–25. [CrossRef]

143. Jayashree, C.; Arulazhagan, P.; Kumar, S.A.; Kaliappan, S.; Yeom, I.T.; Banu, J.R. Bioelectricity generation from coconut husk retting wastewater in fed batch operating microbial fuel cell by phenol degrading microorganism. *Biomass. Bioenerg.* **2014**, *69*, 249–254. [CrossRef]
144. Nimje, V.R.; Chen, C.-Y.; Chen, H.-R.; Chen, C.-C.; Huang, Y.M.; Tseng, M.-J.; Cheng, K.-C.; Chang, Y.-F. Comparative bioelectricity production from various wastewaters in microbial fuel cells using mixed cultures and a pure strain of *Shewanella oneidensis*. *Bioresour. Technol.* **2012**, *104*, 315–323. [CrossRef]
145. Kouzuma, A.; Kasai, T.; Nakagawa, G.; Yamamuro, A.; Abe, T.; Watanabe, K. Comparative metagenomics of anode-associated microbiomes developed in rice paddy-field microbial fuel cells. *PLoS ONE* **2013**, *8*, e77443. [CrossRef] [PubMed]
146. Gurav, R.; Bhatia, S.K.; Choi, T.-R.; Jung, H.-R.; Yang, S.-Y.; Song, H.-S.; Park, Y.-L.; Han, Y.-H.; Park, J.-Y.; Kim, Y.-G. Chitin biomass powered microbial fuel cell for electricity production using halophilic *Bacillus circulans* BBL03 isolated from sea salt harvesting area. *Bioelectrochemistry* **2019**, *130*, 107329. [CrossRef] [PubMed]
147. Pushkar, P.; Mungray, A.K. Exploring the use of 3 dimensional low-cost sugar-urea carbon foam electrode in the benthic microbial fuel cell. *Renew. Energy* **2020**, *147*, 2032–2042. [CrossRef]
148. Juan, J.G.; Keegan, G.C.; Marcus, O.G.; Sage, E.R.; Peter, R.G.; Michael, A.C. Benthic microbial fuel cells: Long-term power sources for wireless marine sensor networks. In Proceedings of the SPIE 7666, Sensors, and Command, Control, Communications, and Intelligence (C3I) Technologies for Homeland Security and Homeland Defense IX, Orlando, FL, USA, 5 May 2010. [CrossRef]

MDPI
St. Alban-Anlage 66
4052 Basel
Switzerland
Tel. +41 61 683 77 34
Fax +41 61 302 89 18
www.mdpi.com

Membranes Editorial Office
E-mail: membranes@mdpi.com
www.mdpi.com/journal/membranes

www.ingramcontent.com/pod-product-compliance
Lightning Source LLC
Chambersburg PA
CBHW041359210926
51850CB00013B/3123
9783036527697